T0299766

Object Oriented
Data Analysis

MONOGRAPHS ON STATISTICS AND APPLIED PROBABILITY

Editors: F. Bunea, R. Henderson, N. Keiding, L. Levina, R. Smith, W. Wong

Recently Published Titles

For more information about this series please visit: https://www.crcpress.com/Chapman–HallCRC-Monographs-on-Statistics–Applied-Probability/book-series/CHMONSTAAPP

Object Oriented Data Analysis

J.S. Marron and Ian L. Dryden

CRC Press
Taylor & Francis Group
Boca Raton London New York

CRC Press is an imprint of the
Taylor & Francis Group, an **informa** business

A CHAPMAN & HALL BOOK

First edition published 2022
by CRC Press
6000 Broken Sound Parkway NW, Suite 300, Boca Raton, FL 33487-2742

and by CRC Press
2 Park Square, Milton Park, Abingdon, Oxon, OX14 4RN

Library of Congress Cataloging-in-Publication Data

Names: Marron, James Stephen, 1954- author. | Dryden, I. L. (Ian L.), author.
Title: Object oriented data analysis / J.S. Marron and Ian L. Dryden.
Description: [Boca Raton] : Taylor & Francis Group, LLC, [2021] | Includes bibliographical references and index. | Summary: "Object Oriented Data Analysis (OODA) provides a useful general framework for the consideration of many types of Complex Data. It is deliberately intended to be particularly useful in the analysis of data in complicated situations which are typically not easily represented as an unconstrained matrix of numbers"-- Provided by publisher.
Identifiers: LCCN 2021023347 (print) | LCCN 2021023348 (ebook) | ISBN 9780815392828 (hardback) | ISBN 9781032114804 (paperback) | ISBN 9781351189675 (ebook)
Subjects: LCSH: Object-oriented methods (Computer science) | Quantitative research. | Statistics--Methodology.
Classification: LCC QA76.9.O35 M369 2021 (print) | LCC QA76.9.O35 (ebook) | DDC 005.1/17--dc23
LC record available at https://lccn.loc.gov/2021023347
LC ebook record available at https://lccn.loc.gov/2021023348

ISBN: 978-0-8153-9282-8 (hbk)
ISBN: 978-1-032-11480-4 (pbk)
ISBN: 978-1-351-18967-5 (ebk)

DOI: 10.1201/9781351189675

Typeset in Nimbus
by KnowledgeWorks Global Ltd.

Dedication

To our families for their ongoing strong support over the many years it took to fully develop these ideas, and to the many colleagues who have played a vital role in shaping this approach to data analysis.

Contents

Preface

This book is intended as a resource for researchers in the development of novel statistical and data science methodology. At the time of this writing, *Big Data* is a very popular area of study. While Big Data does indeed present major statistical challenges, an even greater challenge is dealing effectively with *Complex Data* which is the main motivation for *Object Oriented Data Analysis*. The latter is a framework that facilitates inter-disciplinary research through new terminology for discussing the often many possible approaches to the analysis of complex data. Such data are naturally arising in a wide variety of areas. This book aims to provide ways of thinking that enable the making of sensible choices. The main points are illustrated with many real data examples, based on the authors' personal experiences, which have motivated the invention of a wide array of analytic methods.

A generally relevant comment is that most statistical problems can be solved in many sensible ways. Simon Sheather elegantly summarized that state of affairs (applicable to the many methods discussed in this book) as: "every dog has its day". The point is that any method that has been seriously advocated by someone has situations where it gives excellent performance, but also situations where it can be quite poor. The challenge is to understand the properties of each well enough to guide good choices. The material in this book will provide the reader with useful insights and a general framework to assist in this process.

A fundamental theme throughout the book, that has not been deeply explored elsewhere is *modes of variation*. That provides a novel terminology and framework for understanding many aspects of Object Oriented Data Analysis.

While the mathematics goes far beyond the usual in statistics (including differential geometry and even topology), the book is aimed at accessibility by graduate students. There is deliberate focus on ideas over mathematical formulas. An exception is the detailed linear algebra development in Chapter 17. While many references to various aspects of OODA are given, it should be noted we have deliberately not attempted to be comprehensive in those. Our aim instead is to simply provide useful starting points that interested researchers can use for their own bibliographic searches.

The historical background of the Object Oriented Data Analysis terminology is discussed in Section 18.1.

Much of the material that went into this book, including data sets, and the code to generate most of the graphics can be found in the the web companion to this book at Marron (2020). Many of those require Marron's Matlab Software, available at: Marron (2017b). The companion website also contains further references and links to other software packages.

Acknowledgments

Many of the ideas and presentation style have been developed during the teaching of a graduate course entitled Object Oriented Data Analysis and have been taught roughly every other year at the University of North Carolina since 2005. There were some important precursors, including a related course taught at Cornell University in 2002. Two such courses were offered at the Statistical and Mathematical Sciences Institute in 2010 and 2011 with lecturers (beyond the authors of this book) including Hans-Georg Müller, James O. Ramsay, and Jane-Ling Wang. The course was also taught at the National University of Singapore in 2015.

Several events have played a pivotal role in the development of Object Oriented Data Analysis. One was the Statistical and Mathematical Sciences Institute program on "Analysis of Object Data" during 2010–2011, with co-organizers Hans-Georg Müller, James O. Ramsay and Jane-Ling Wang. Another pivotal workshop was the November 2012 "Statistics of Time Warpings and Phase Variations" at the Mathematical Biosciences Institute, with co-organizers James O. Ramsay, Laura Sangalli and Anuj Srivastava.

General research in this area by the authors has been supported over the years by a number of grants from the National Science Foundation, including DMS-9971649, DMS-0308331, DMS-0606577, DMS-0854908 and IIS-1633074, and the Engineering and Physical Sciences Research Council grants EP/K022547/1 and EP/T003928/1.

The material on sounds as data objects was kindly provided by Davide Pigoli. The authors are especially grateful to Stephan F. Huckemann, John T. Kent, James O. Ramsay, and Anuj Srivastava for providing formal reviews of early drafts that fundamentally impacted the final version. Additional useful comments on various drafts have been provided by Iain Carmichael, Benjamin Elztner, Thomas Keefe, Carson Mosso, Vic Patrangenaru, Stephen M. Pizer, Davide Pigoli, and Piercesare Secchi.

The authors are grateful to John Kimmel, for helpful advice at many points, and for his patience over the long time it took for this book to come together.

CHAPTER 1

What Is OODA?

The fields of human endeavor currently known as *statistics*, *data science*, and *data analytics* have been radically transformed over the recent past. These transformations have been driven simultaneously by a massive increase in computational capabilities coupled with a rapidly growing scientific appetite for ever deeper understanding and insights. The notion of forming a *data matrix* provides a useful paradigm for understanding important aspects of how these fields are evolving. In particular, the currently popular context of *Big Data* has several quite different facets, ranging from *low dimension high sample size* areas (the basis of classical mathematical statistical thought, which is perhaps typified by sample survey and census data), through both high dimension and sample sizes (common for internet scale data sets of many types), and on to *high dimension low sample size* contexts (frequently encountered in areas such as genetics, medical imaging and other types of extremely rich but relatively expensive measurements). The pressing need to analyze data in this wide array of contexts has generated many exciting new ideas and approaches.

Yet a deeper look into these developments suggests that the organization of data into a matrix may itself be imposing limitations. In particular, there is a growing realization that the challenges presented by Big Data are being eclipsed by the perhaps far greater challenges of *Complex Data*, which are typically not easily represented as an unconstrained matrix of numbers. *Object Oriented Data Analysis* (OODA) provides a useful general framework for the consideration of many types of Complex Data. It is deliberately intended to be particularly useful in the analysis of data in complicated situations, diverse examples of which are given in the first two chapters. The phrase OODA in this context was coined by Wang and Marron (2007). An overview of the area was given in Marron and Alonso (2014). For more discussion of Big Data and its relation to statistics, see Carmichael and Marron (2018) and many interesting viewpoints in the special issue edited by Sangalli (2018).

The OODA viewpoint is easily understood through taking *data objects* to be the *atoms* of a statistical analysis, where atom is meant in the sense of elementary particle, studied in several contexts of increasing complexity:

- In a first course in statistics atoms are numbers, and the goal is to develop methods for understanding of variation in populations of numbers.

- A more advanced course, termed *multivariate analysis* in the statistical culture, generalizes the atoms, i.e. the data objects from numbers to vectors and involves a host of methods for managing uncertainty in that context. For example

2 WHAT IS OODA?

see Mardia et al. (1979), Muirhead (1982), and Koch (2014) (for a more up to date treatment).

- At the time of this writing a fashionable area in statistics is *Functional Data Analysis* (FDA), where the goal is to analyze the variation in a population of curves. A good introduction to this vibrant research area, where functions are the data objects, can be found in Ramsay and Silverman (2002, 2005). A case study, illustrating many of the basic concepts of FDA which are useful for understanding OODA is given in Section 1.1.

- OODA provides the next step in terms of complexity of atoms of a statistical analysis to a wide array of more complicated objects. The important example of shapes as data objects is considered in the case study of Section 1.2. A wide variety of other examples, which highlight the breadth of OODA, appears in Chapter 2.

Note that each of the above areas can be thought of as containing the preceding ones as special cases. For example, multivariate analysis is the case of FDA where the functions are discretely supported. Similarly multivariate analysis and FDA are special cases of OODA. In later chapters it is useful to recall that OODA includes these predecessors as special cases. This is because often simple multivariate examples are used for maximal clarity in the illustration of concepts and methods, but the ideas are useful more generally for OODA.

A good question is: What is the value added to applied statistics and data science from the concept of OODA and its attendant terminology? The terminology is based on very substantial real-world experience with a wide variety of complex data sets. A fact that rapidly becomes clear in the course of interdisciplinary research is that there frequently are substantial hurdles in terminology. Especially at the beginning of such endeavors, it can feel like collaborators are even speaking different languages, so often serious effort needs to be devoted to the development of a common set of definitions just to carry on a useful discussion. An added complication is that for complex data contexts, it is frequently not obvious how to even "get a handle on the data". Usually there are many options available, which are most effectively decided upon through careful discussion between domain scientists and statisticians. In such discussions, the issue of *what should be the data objects?* has proven to frequently lead to useful choices, thus resulting in an effective and insightful data analysis.

Real data examples demonstrating data object choices in a variety of contexts are given in the following and Chapter 2. In particular, Section 1.1 introduces curves as data objects. A more complex variant involves curves with interesting variation in phase in place of, or in addition to, the usual FDA amplitude variation discussed in Section 2.1. A mathematically deeper case is considered in Section 1.2 where *shapes* are the data objects which require special treatment as shapes are most naturally viewed as points on a curved manifold. Section 2.2 considers a perhaps even more challenging data set of tree-structured data objects. The data objects in Section 2.3 are recordings of sounds, in particular human spoken words, which bring special challenges in the choice of data objects. Finally, in Section 2.4, a fun example with images of faces as data objects is considered.

It is seen that the notion of data objects provides a particularly useful format for discussing *modes of variation* that give insights about population structure. This term is formally defined in Section 3.1, but until then the meaning should be intuitively clear from the context.

One more general aspect of OODA is that there are frequently three major phases of this type of data analysis:

1. *Object Definition*. This is the phase where the fundamental issue of what should be the data objects is addressed. A number of examples of this phase are provided in the rest of this chapter and also in further examples in other sections.

2. *Exploratory Analysis*. Here the goal is to find perhaps surprising population structure in data, often using some type of visualization method. A wide variety of examples and methods for exploratory analysis are given in the rest of this Chapter and in Chapters 2, 4, 5–10. Exploratory analysis frequently only appears sparingly in most classical statistics courses, but is usually more prominent in machine learning. However it has a strong statistical tradition, going back well before the ideas nicely summarized in Tukey (1977).

3. *Confirmatory Analysis*. While many great discoveries have been made using exploratory methods, it is also very easy to make discoveries that are not real, in the sense of being non-replicable sampling artifacts. For this reason it is very important to validate such discoveries. This critical topic and many variants of approaches to it are discussed in detail in the very large classical statistical literature. Some less well known aspects, that are particularly relevant to OODA are discussed in Chapter 13.

A companion website to this book, containing links to available software, the Matlab or R programs used to generate most of the figures in this book, and additional graphics can be found at Marron (2020).

Further discussion on other ideas and nomenclature related to OODA can be found in Chapter 18. Additional big picture discussion of data science and statistics can be found in Marron (2017a) and Carmichael and Marron (2018).

1.1 Case Study: Curves as Data Objects

An interesting example of functional data analysis (viewed here as an important special case of OODA) is the *Spanish Mortality* data, first studied from an FDA viewpoint in Section 2 of Marron and Alonso (2014). Such data sets are available at the Human Mortality Database of Wilmoth and Shkolnikov (2008). For a given population (e.g. citizens of one country) mortality data are generally a matrix with rows and columns indexed by years and ages. The matrix entries are the chance of a person of each age dying in the given year, calculated as the number of deaths divided by the number of people for that year–age pair. Here we study mortality of males in Spain, mostly because there are interesting features in the data, due to recent Spanish history.

This data set provides good illustration of the issue of Object Definition, because there are several data object choices to be made in the analysis of this data. First, since these probabilities range over several orders of magnitude, logarithms are useful to provide good visual separation across a wide range of scales. Particularly strong interpretability comes from the choice of $\log_{10}$ of the probability (e.g. -2 corresponds to a probability of 0.01 as opposed to about 0.135 for the natural log). The utility of this data object choice is demonstrated in Figure 1.1, where the raw probabilities are shown in the left panel (with much interesting structure missed since this is very nearly 0 for the important younger age groups) with $\log_{10}$ mortality in the right (highlighting important contrasts among the younger ages). Second there are two different ways to turn the matrix of data into functional data. One is to consider data objects to be curves of mortality as a function of age, with curves indexed by year. The other is (the matrix transpose) where the mortality is viewed as a function of year, with data object curves indexed by age. In this analysis, the former choice is used, because it gives the best illustration of the usefulness of OODA concepts and also gives an interesting narrative. The latter choice is considered in Figure 17.7. An analysis that also integrates female mortality in an interesting way can be found in Feng et al. (2018). The choice here results in $n = 95$ curves corresponding to the years 1908–2002. Ages considered here are 0 through 98, since larger ages are problematic due to occasional small population sizes. The raw data are shown as overlaid curves in Figure 1.1. There the curves are distinguished using the standard graphical technique of a rotating color palate (in this case the default 7 colors in Matlab).

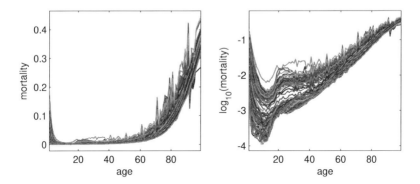

Figure 1.1 *Spanish Mortality curves as a function of age. Raw male mortality is in the left panel, with $\log_{10}$ mortality on the right. Years are distinguished using a rotating color palette. Shows age effects and large variation (factors of more than 10 for some age groups) across years, as well as the data object choice of $\log_{10}$ mortality being the more useful scaling of the data.*

This view already shows interesting aspects of the data. For example, being born is a risky activity, with a high mortality rate. However, the chance of dying falls off rapidly, up until the teen years when risky behavior tends to begin. Then through adulthood the death rate slowly increases, becoming quite high in old age. Also note the bundle of curves is quite thick, with the axes indicating approximately a 10 fold change over the years, begging an investigation into how things have changed over time. This is easily provided in Figure 1.2, by applying a different color scheme to the curves in the right panel of Figure 1.1. Here time ordering of the curves is highlighted through coloring with a rainbow scheme to indicate years, starting with magenta ([1 0 1] in RGB coordinates) for 1908 and ranging through violet, blue, cyan, green, yellow, and orange to red ([1 0 0] in RGB) for 2002.

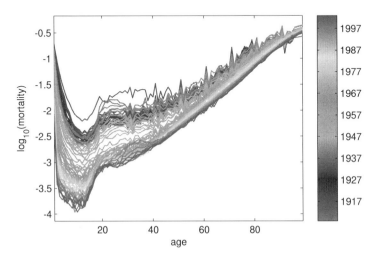

Figure 1.2 *Spanish Mortality curves using a rainbow color scheme to indicate progression in time (over years 1908–2002). Shows major improvements in mortality over this time range.*

This shows a very clear overall improvement over the years in mortality, due mostly to improvements in medicine and public health. Note also that these improvements have benefited younger people more than the old, as there is not yet much treatment available for aging. As happens frequently with OODA data, additional visual insights come from careful decomposition of the variation present in these curves, through a *Principal Component Analysis* (PCA). See Chapters 4 and 17 and Jolliffe (2002) for background information concerning the many ways this method is used. One important use of PCA is to gain insight into how data objects relate to each other. Insight comes from considering the data as lying in an abstract *point cloud* in $d - 99$ dimensional space, where low-dimensional projections frequently visually illustrate key relationships (e.g. clustering of data objects). An often useful first step of a PCA is *mean centering*, which essentially moves the point cloud so that it is centered at the origin. As seen in Figure 1.3, this centering operation itself can provide an informative decomposition of the data into the mean and residuals about the mean.

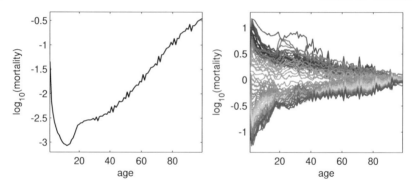

Figure 1.3 *Left panel is the mean mortality curve. Right panel contains the mean residuals, where the mean is subtracted from each curve, using the same color scheme. Shows that age effects are essentially common for all (i.e. over time), in the sense of appearing in the mean. Improvements over time appear in the residuals, with overall most improvement for the young.*

   The left panel of Figure 1.3 shows the mean curve, computed as the point-wise mean of the curves in Figure 1.2. The right panel contains the mean residuals, which are computed by subtracting the mean from each of the data curves, while retaining the original year coloring. Note that the mean curve contains many of the important features of the raw data, especially those related to age. In particular, the danger of being born together with low mortality for the young with increasingly higher mortality for the old are all properties of the mean. These essentially do not appear in the mean residuals, indicating that these are population properties which have not changed much over time. A perhaps surprising aspect of the mean is the occasional blips that appear. One might think these are random noise, but note that they are quite periodic and in fact appear at decades. This is a function of historically poor record keeping. The early lack of birth certificates for the full population led to some uncertainty of age at the time of death for some, with subsequent rounding to decades which is clearly visible. The mean residuals also reflect an important aspect of the population structure, being driven by the changes over time. Most important are the dramatic improvements in mortality that have been made over the course of this study. This view also makes it clear that the young have benefited the most with that benefit decreasing as a function of age.
   PCA is usefully understood as decomposing the mean centered data in the right panel of Figure 1.3 into insightful modes of variation (this concept is formally defined in Section 3.1.4). One such mode is the variation revealed by the first principal component as shown in Figure 1.4. Insight comes from thinking of the above mentioned point cloud, where each data object (curve in this case) is a point. The PCA modes of variation are developed by seeking orthogonal directions of maximal variation within the point cloud. The first PC *direction* is the unit (i.e. norm 1) vector, based at the sample mean, which maximizes the variance of the data projected onto that vector. This direction is easily computed as the first eigenvector of the sample covariance matrix (defined at (3.5)). The entries of that vector (which

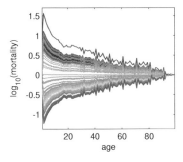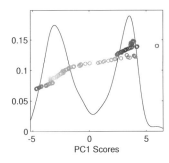

Figure 1.4 *PC1 mode of variation plot (left) and scores distribution plot (right). The former shows that this dominant mode of variation reflects most of the overall improvement in mortality. Scores plot shows most of the improvements happened relatively rapidly, plus highlights the 1918 Flu Pandemic (violet outlier on the right) and the Spanish Civil War (light blue sharp trend to the right).*

indicate how it relates to the variables, i.e. features, of the data set) are called the *loadings*. Visual insight into these loadings comes from the *mode of variation plot* in the left panel of Figure 1.4. The horizontal axis indexes the variables, which are ages in this case, and the curves are all multiples of the eigenvector. In particular the curves are projections of the data curves onto the direction vector. These are the columns of the rank 1 matrix that is the product of the column vector of loadings times the row vector of *scores*, which are the projection coefficients of each data object onto the eigenvector. In classical multivariate analysis the scores are also called the principal components. This matrix is the (least squares) best rank 1 approximation of the mean residual matrix shown in the right panel of Figure 1.3. This PC1 view highlights the *dominant* mode of variation, which nicely reflects the major overall improvement in mortality. In addition, as life and death record keeping has improved over time, the decline in age rounding effects is reflected in the decadal spikes pointing upwards (early) and downwards (later). In particular, the rounding was present earlier, not later, so it shows up partially in the mean in Figure 1.3, and then as this contrast in Figure 1.4 (left panel).

The right panel of Figure 1.4 is the PC1 *scores distribution plot*, which will be used frequently in the following to display detailed information as to how the data objects relate to each other. Each circle represents one score using the same color scheme. Horizontal coordinates indicate the score and vertical coordinates indicate order in the data set, in this case the year. The magenta color of the top circle is the year 1908 and the red color of the lowest circle is for the year 2002. The overall leftward trend again shows the overall improvement in mortality over these years. The black curve shows a *kernel density estimate*, which can be thought of as a smooth histogram. The vertical axis records the heights of this curve. Detailed discussion of kernel density estimation is in Chapter 15. See Wand and Jones (1995) for a more in-depth overview. This type of display of one-dimensional distributions, which includes both the actual data points and the smooth histogram, is used many other times in the following. In this case it shows much higher density of scores in the higher and lower regions, which is another way of seeing that most of the overall transition from higher to lower mortality was relatively rapid. A couple of smaller-scale aspects are also clear in this scores plot. The violet year, farthest to the right was the year 1918, when many people around the world died during a flu pandemic, which until recently was the largest ever well-documented epidemiological event worldwide. Also notable is the shift toward higher mortality (i.e. to the right) shown as light blue, which was the time of the Spanish Civil War, just before World War II (in which Spain was not a combatant).

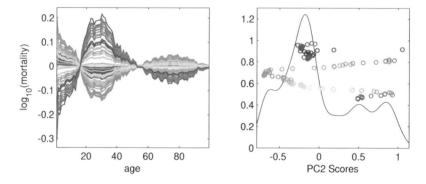

Figure 1.5 *PC2 mode of variation (left) and scores distribution (right), using the same format as Figure 1.4. The loadings plot shows this second mode of variation provides a contrast between the 20–45-year-olds with the rest. The scores plot shows the deep effects of the flu pandemic, the Spanish civil war and automotive death rate.*

Figure 1.4 showed the first mode of variation in the mortality data called PC1. An interesting complementary mode of variation is the second PC, as shown in Figure 1.5. This represents the direction of second strongest variation (in the sense of being orthogonal to the first direction) measured again in terms of variance of projections. It is computed as the second eigen direction of the sample covariance matrix. The PC2 mode of variation plot (left panel) shows that this direction highlights differences between the 20–45-year-old cohort, with the union of the young and the old. The color pattern is harder to interpret in this mode, but is very clear in the scores distribution plot (right panel). Note that the 20–45-year-olds suffered even stronger effects from both the pandemic and also the war, as they died at a substantially higher rate than usual in those times. Another interesting feature is the growing mortality for this cohort in the 1960s to 1980s (green to orange). This period corresponds to growing access to automobiles, and apparently the idea that young males are the group most prone to risky automobile behavior. Note that in the final years, the direction of this trend has fortunately reversed, which has been ascribed to much improved car safety (such as seat belts) and also to major improvements in roads.

The concept of modes of variation as determined by PCA loadings and scores is explored more deeply in Section 3.1.

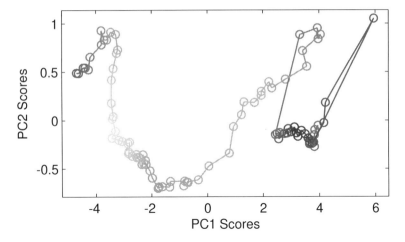

Figure 1.6 *Scatterplot of PC1 vs. PC2 scores for the Spanish Mortality data. This shows many of the above historical trends in a single plot.*

Figure 1.6 shows a scatterplot of the bivariate distribution of the PC1 and PC2 scores, which provides a useful and concise summary of both modes of variation, i.e. of much of the structure in this data set. The one-dimensional PC1 scores distribution in the right panel of Figure 1.4 is on the horizontal axis, while the vertical axis has the corresponding PC2 scores distribution from the right of Figure 1.5. This is the two-dimensional projection of the data onto the plane with maximal variation. Note that the circles representing the data objects (i.e. the mortality curves) are connected with line segments in time order, which facilitates keeping the progression of years in mind when interpreting the plot. The overall improvement in mortality, with the exceptions of flu and war, are clear from the main leftwards progression. Variation over time of the contrast between the 20–45-year-olds and the rest are also clear on the vertical axis, nicely highlighting the flu, war and automobile effects.

For this data set, the most interesting views are in the first two PC components. For others, more components can also be quite insightful. A useful summary of several PC components is a matrix of such scatterplots, with the axes carefully coordinated over both rows and columns. The diagonal of such a display is most useful when it shows some sort of 1-d distributional summary, e.g. the combination of jitter plots (the colored circles) and kernel density estimates used to show the distribution of scores in the right panels of Figures 1.4 and 1.5. Jitter plots are discussed in more detail in Section 4.1. Further examples of such matrices of scatterplots can be found in Figures 4.4, 4.12, 4.13, and many other places in later chapters.

Mortality rates for other countries can be explored in a similar way. For example mortality data from Switzerland (also available in Wilmoth and Shkolnikov (2008)) show similar flu pandemic and automobile effects as observed here, but neither the data rounding (due to a longer period of good record keeping) nor the war caused mortality effects are visible as expected.

1.2 Case Study: Shapes as Data Objects

A particularly deep and important example of shapes as data objects is the *Bladder-Prostate-Rectum* data, motivated by the challenge of planning radiation treatment of prostate cancer described in Chaney et al. (2004).

1.2.1 The Segmentation Challenge

Radiation treatment of cancer is quite effective, and administered over the course of a number of days. The goal is to provide a maximal radiation dose to the prostate while minimizing the impact on nearby sensitive organs such as the attached bladder and the rectum, which is adjacent. A major radiation treatment planning challenge is that (even within the same person) the locations of all 3 organs vary widely on the critical time scale of days. Computed Tomography (CT) images are useful for visually locating these organs on a given day, with CT preferred over Magnetic Resonance images due to its superior accuracy of location. However *segmentation*, i.e. finding the set of *voxels* (three-dimensional analogs of pixels) inside each organ, was a challenging problem because of poor contrast and noise, as shown in Figure 1.7. That is one slice of a 3-d stack of images, showing a side view of the hip region for one patient. The color scheme of CT is the same as for x-rays, so dense objects such as bones show up as white. Thus the upper right of Figure 1.7 shows the tailbone, and a hipbone passes through this slice in the lower center. Black indicates the least dense regions which are gas bubbles in the rectum, which is the curved lighter region containing the darkest spots starting near the top center and curving down below and to the left of the tail bone. The lighter gray region between the top of the rectum and the small hip bone is the bladder. The prostate, which is the target of the treatment, is a light gray region between the hip bone, the bladder and the lowest visible section of the rectum.

Segmentation of the prostate is quite challenging because of very poor contrast with surrounding objects (it is essentially the same shade of gray and has both lighter and darker regions nearby) and because of the relatively high noise level. For these reasons, incorporation of anatomical knowledge is essential to the segmentation process. *Manual segmentation* achieves this through an anatomically trained technician drawing the boundary of an object on each slice of the 3-d image. The union of the interior voxels aggregated over slices then gives a segmentation of the object. An example of that process is in Figure 1.8, which shows two views of a manual segmentation of the bladder in Figure 1.7. The left panel shows how voxels are aggregated across slices, using a view orthogonal to that where the drawing was done. The right panel is a rotated view of the highlighted collection of blue colored voxels without the CT image, giving a clear impression of the 3-d object.

While manual segmentation is quite effective at locating these organs for

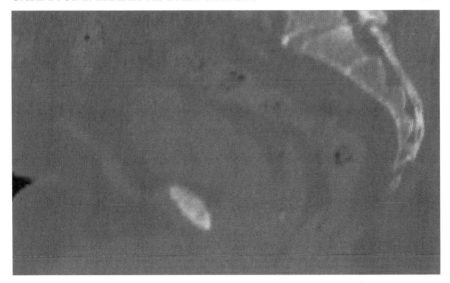

Figure 1.7 *One slice of 3-d CT image in Bladder-Prostate-Rectum data. Bones are white, black gas bubbles indicate the rectum. Bladder and prostate are light gray near the center and lower center. This image shows that automatic segmentation is very challenging.*

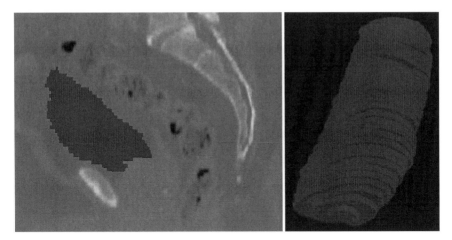

Figure 1.8 *Left panel shows the results of a manual segmentation of the bladder, performed sequentially on orthogonal slices. Right panel shows a rotated view of the same bladder, to highlight the 3-d aspect of the segmentation.*

planning radiation treatment, it is time-consuming and hence it is not practical to repeat this manual operation many times over the course of radiation treatment (i.e. in a clinical setting). This has motivated a lot of research on automatic segmentation of these organs, much of which was developed in the references cited at the end of this section. The key idea is to incorporate anatomical information into the training process, using a Bayesian statistical model. The starting point for this is a *shape representation*, i.e. a parametric model for each organ.

1.2.2 General Shape Representations

In some contexts shape is conveniently represented by *landmark configurations*, i.e. a set of points that correspond across members of the data set, which can be readily found on each. The statistical analysis of landmark configuration shape data objects was pioneered by Kendall (1984) and Bookstein (1986). For introduction to the large literature on that, see Dryden and Mardia (2016). The fundamental idea is illustrated by a toy data set of triangles in $\mathbb{R}^2$ as data objects in Figure 1.9. An intuitive representation of each triangle is the configuration of the $\mathbb{R}^2$ coordinates of the vertices (a 6-tuple), which are natural landmarks. However many triangles with different configurations have the same shape. In particular, the triangles to the left of the dashed line are all translations, rotations, and scalings of each other, i.e. all have the same shape. Two other sets of common shapes appear between the vertical lines, and to the right of the dot-dashed line. The mathematical device of *equivalence relation* provides a convenient formulation of the notion of shape. Calling two triangular configurations *equivalent* when they are translations, rotations, and scalings of each other results in *equivalence classes*. These are the sets of all triangles which can be translated, rotated, and scaled into each other, i.e. triangles of the same shape. These equivalence classes of identified triangles then become the shape data objects. Spaces of equivalence classes are widely studied in differential geometry, where they are called *quotient spaces*. Common synonyms for the equivalence classes are *fibers* (frequently used here in Chapter 8) and *orbits* (appearing often here in Chapter 9). As discussed in Section 8.4 and in Section 4.3.4 of Dryden and Mardia (2016), the natural geometry

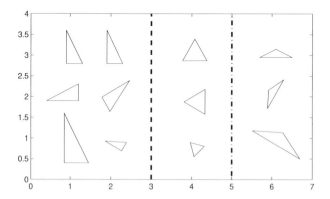

Figure 1.9 *Toy data set of triangles in $\mathbb{R}^2$, to illustrate shapes as data objects. Lines separate three equivalence classes (i.e. fibers or orbits) with respect to translation, rotation, and scaling.*

of the quotient space of triangle shapes is the sphere S^2 (see (3.2) for a formal definition). Sections 8.2 and 8.4 contain a broader discussion of shape quotient spaces, where it is seen that many of those are also curved. This provides strong motivation for studying data objects lying on curved manifolds, as done below and in more depth in Section 8.3.

While landmark approaches are useful for many tasks, they are typically less useful in many medical imaging situations, such as soft tissues, where landmarks that correspond across cases can be hard to find, with often very few obvious choices apparent. Hence, there has been much research devoted to *boundary representations*. In the computer graphics world a very common boundary representation is a triangular mesh, see e.g. Owen (1998). A major challenge to the use of mesh representations in shape statistics is *correspondence*, i.e. relating the mesh parameters (e.g. triangle vertices) across instances of shape data objects. Two important approaches to this are Active Shape Models, see Cootes et al. (1994) for a good introduction, and the entropy-based ideas of Cates et al. (2007). Another major formulation of boundary representations is through Fourier methods, e.g. as in Kelemen et al. (1999). For sufficiently smooth shapes, Kurtek et al. (2013) have shown that superior representation comes from enhancing boundary representations by also including surface normal vectors in the data objects.

1.2.3 Skeletal Shape Representations

As discussed in Siddiqi and Pizer (2008), a *medial representation* can provide improvements for a number of imaging tasks. The key idea is to base the representation on the more robust concept of 3-d solids instead of on 2-d boundary surfaces. For the reasons discussed in Chapter 3 of Siddiqi and Pizer (2008), the concept of medial locus has been generalized to give *skeletal representations*. As noted in Pizer et al. (2013) the enhanced flexibility of skeletal representations allows for superior fits to data. A skeletal representation of one bladder, prostate and rectum instance is illustrated in Figure 1.10.

The left panel of Figure 1.10 shows the interior components of three skeletal representations, one for each organ. Each has a set of yellow dots, called *skeletal atoms*, connected by green line segments, which are a discretization of the *skeletal sheet*, the 2-d surface which is approximately medial in the sense of being equidistant from both boundaries. Each skeletal atom has *spokes*, shown as cyan and magenta line segments, extending from the skeletal sheet to the boundary of the organ. Skeletal atoms at the edge of the sheet each have one additional spoke shown in red, extending to the edge of the organ. The central panel of Figure 1.10 adds three colored meshes (yellow for the bladder, green for the prostate, red for the rectum) which indicate the boundary of each that is implied by the interior components as a quadrilateral mesh that connects the ends of the spokes. The right panel shows the boundary more explicitly by coloring the panels of the quad meshes and using a light source shading in the same colors. The skeletal model is a parametric model of shape, whose parameters are the 3-d locations of the yellow

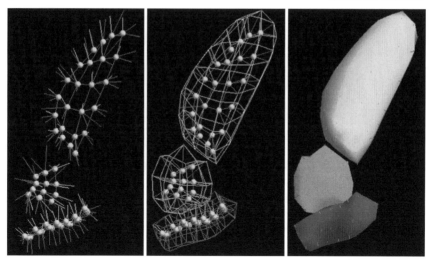

Figure 1.10 *Skeletal representation of a single bladder-prostate-rectum. Left panel shows the central skeletal sheets, atoms and spokes for each shape object. Center panel adds the implied boundaries as quad meshes, using yellow for the bladder, green for the prostate, red for the rectum. Right panel represents the implied boundaries using a light source rendering.*

atoms, the lengths of the spokes, and the angles of the spokes, each of which is represented as a point on the sphere S^2.

The data objects in this OODA case study are chosen to be the skeletal models represented by the locations of k atoms in $\mathbb{R}^3$, l positive spoke lengths in $\mathbb{R}_+$ and m directions on S^2. For CT images where a manual segmentation has been performed, the skeletal shape model can be fit to the binary image shown in blue in Figure 1.8 (i.e. the various parameters estimated), using direct methods such as least squares. However as discussed above, for clinical applications such as radiation treatment planning, with a need for a technician to perform this operation several tens of times for one course of treatment, manual segmentation is prohibitively expensive. This motivated the work cited at the end of this section, on automating fitting of skeletal models (as shown in Figure 1.10) directly to raw CT images (as shown in Figure 1.7). As discussed above, this requires incorporation of something akin to anatomical information. That is done using a Bayesian statistical approach. Essentially some manual segmentations are used to train a prior distribution using OODA, which is combined with a likelihood that is based on a new CT image, to generate a posterior distribution which is maximized over the parameters of the skeletal shape representation, to give an automatic segmentation.

1.2.4 Bayes Segmentation via Principal Geodesic Analysis

The Bayes implementation employed in this type of application differs somewhat from most modern Bayes applications. On one hand, the underlying probability distributions are very basic, since only conjugate Gaussian priors, likelihood, and hence posteriors are used. This is a strong contrast with the complicated models involving Markov chain Monte Carlo methods that are currently very prevalent in applications of Bayes methods. On the other hand, this Bayes application is relatively deep in two ways. First the number of parameters to fit is typically much higher then the number of training instances, i.e. it lies in the high dimensional OODA domain discussed in a general way in Chapter 14. The second complication is the non-Euclidean nature of the reparameterizations, caused mostly by each spoke naturally lying on the surface of the sphere S^2. As this research has progressed, the high dimensionality has been handled by a variety of methods related to PCA. More challenging is that skeletal parameterized data objects are naturally elements of a space of the form $\mathbb{R}^{3k} \times \mathbb{R}^l_+ \times (S^2)^m$ (i.e. tuples of k real numbers, l positive reals, and m points on the sphere). Such spaces are called *manifolds* in differential geometry (see Section 8.2 for an introduction to aspects of this topic needed for OODA) and are usefully thought of as curved surfaces (e.g. the surface of a sphere).

The need to address the first complication (the high dimension) in the bladder-prostate-rectum segmentation challenge described above has led to a series of developments in terms of analogs of PCA for data lying on the manifolds of skeletal representations. The Principal Geodesic Analysis (PGA) of Fletcher et al. (2004) represents an important early advance in this work. The main idea of PGA is to consider the Euclidean PCA basis as a set of orthogonal lines that (sequentially) best fit the data. In PGA these best fitting lines are replaced by best fitting *geodesics* (e.g. great circles on S^2) which are a natural analog of lines. The results of a PGA, based upon $n = 17$ skeletal representations (collected over a sequence of days) from a single patient are shown in Figure 1.11.

Figure 1.11 reveals clinically interesting modes of variation of these organs within this person. The left column (first mode of variation) seems to reflect vertical shift variation driven by the rectum. The second mode (middle column) shows twisting, while the third (right column) is about emptying and filling of the bladder. This input led to the Bayes segmentation method giving very effective automatic segmentation. That was the basis for the successful start-up company Morphormics, which was subsequently purchased by the radiation treatment equipment manufacturer Accuray.

More recently there has been a series of improvements to PGA, motivated by a succession of deeper and deeper integrations of statistical ideas with differential geometry. Detailed discussion of this progression appears in Section 8.3. While this discussion has focused mostly on segmentation using skeletal shape representations, much important related work has been done on classification as discussed in Chapter 11 and on confirmatory analysis which appears here in Chapter 13. A good overview of the usefulness of skeletal representations, especially in

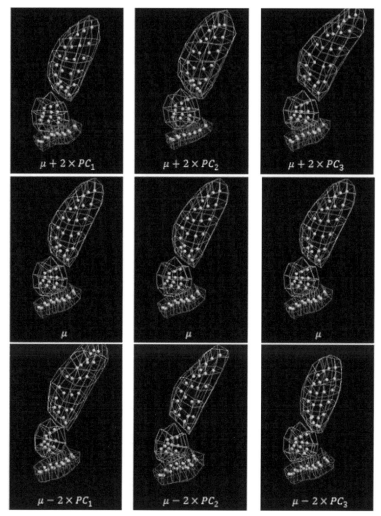

Figure 1.11 *Principal Geodesic Analysis of Bladder-Prostate-Rectum variation within one person. Columns give visual impression of first 3 PGA modes of variation. All three plots in the second row are the Fréchet mean (notion of center defined in (7.5)). Top row shows the three +2 standard deviation departures from the mean, and bottom row shows the corresponding -2 standard deviation departures. This gives three interpretable and sensible modes of variation.*

comparison to other types of representations can be found in Pizer et al. (2013, 2014), Schulz et al. (2016), and Hong et al. (2016).

The bladder-prostate-rectum research that lies at the core of the discussion of this section was developed in a series of papers. That includes Chaney et al. (2004), Broadhurst et al. (2005), Davis et al. (2005), Pizer et al. (2005a,b, 2006,

2007), Lu et al. (2007), Stough et al. (2007), Jeong et al. (2008), Merck et al. (2008), and Feng et al. (2010).

CHAPTER 2

Breadth of OODA

This chapter illustrates the breadth of OODA through relatively brief overviews of quite diverse applications.

2.1 Amplitude and Phase Data Objects

A challenging situation in FDA is when the curve data objects are misaligned, as shown in the top panel of Figure 2.1 for a Proteomics data set called the *TIC Curves* here. Many statistical methods can be strongly impacted by misalignment. An example of the impact of misalignment on the sample mean in FDA is shown using the Shifted Betas toy data in Figure 5.17. A quite different type of impact on PCA appears in Figure 9.2. As noted in Marron et al. (2014b), FDA approaches to dealing with alignment issues are sometimes called *curve registration*, because it is very useful in situations where the curve data objects are clearly misaligned.

There are many approaches to the curve registration challenge, with an overview provided in the survey paper Marron et al. (2015). Most methods in the area involve tuning parameters that have proven to be tricky to choose in a fully automatic way, as illustrated in Figure 9.1. This problem has been solved using the OODA way of thinking, as discussed in Chapter 9. In particular, that approach is based on unusually deep mathematical ideas based on the Fisher-Rao metric, which resulted in a rigorous methodology that is hence fully automatically useful.

An interesting example of curve registration, from Koch et al. (2014) and Marron et al. (2014a), is shown in Figure 2.1. The data objects here are proteomics mass spectrometry profiles from Ho (2011), a larger study of bio-markers in Acute Myeloid Leukemia. A detailed description of this data set including a number of pre-processing steps (such as median smoothing and interpolation to an equally spaced grid) can be found in Koch et al. (2014). Essentially there are 5 patients, represented as colors, with 3 replicate curves for each patient, thus 15 curves in all, shown in the top part of the top panel. Each curve shows Total Ion Counts (TIC), for an equally spaced grid of 2001 mass to charge ratios (horizontal coordinate). The TIC curves have many peaks, which correspond to various peptides. A common goal of mass spectrometry analyses is curve registration, i.e. finding deformations, sometimes called *warpings* (intuitively thought of as stretchings and compressings of the horizontal axis), to properly align the peaks so that they chemically correspond. In most contexts it is hard to quantitatively assess the performance of a given registration, but this data set is special because the locations of several of the actual peptide peaks have been (laboriously) found for each curve

DOI: 10.1201/9781351189675-2

using additional information as detailed in Koch et al. (2014). These peak locations, for each of the 15 curves, are indicated by peak numbers (1–14), with colors
corresponding to the curves. The peak numbers are sorted vertically by height of
the corresponding peak and connected with gray line segments to give some visual
correspondence. It is hard to see much pattern, showing this to be a challenging
curve registration problem.

As noted above, there are a number of approaches to this type of data challenge,
with several such analyses of this data set discussed in Marron et al. (2014a).
The bottom panel of Figure 2.1 shows the results of registration of these same

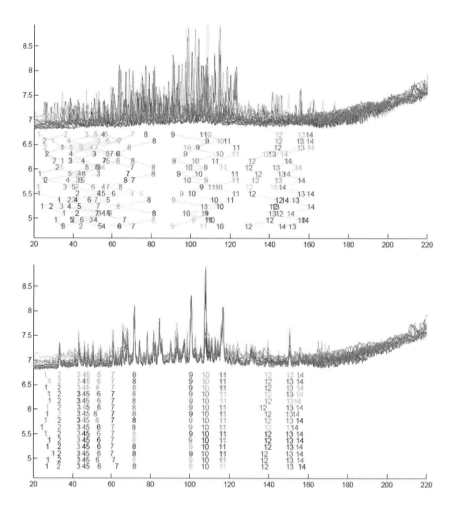

Figure 2.1 *Top panel contains raw TIC Curves data, with a labeling of certain important
peaks in the lower part of the panel. Bottom panel shows a Fisher-Rao registration of the
TIC curves. Numbers under the curves indicate peak locations, showing that the registration has been mostly quite effective.*

TIC curves using the Fisher-Rao method proposed in Srivastava et al. (2011) and Kurtek et al. (2012) (discussed in more detail in Section 9.1), using only the curves themselves and not the peak location information. The colored numbers reveal that this is a particularly challenging problem, because the peaks have quite different heights across patients. Peak 10 is particularly challenging as it is quite low for the red patient (especially compared to nearby very tall peaks), yet is the highest peak for other patients. Note the alignment is not perfect for every numbered peak, but it is still of impressively high quality. Roughly comparable quality has been obtained using a linear registration approach that is integrated with clustering in Bernardi et al. (2014b), and by a Bayesian approach in Cheng et al. (2014).

An important point made in the overview of Marron et al. (2015) is that curve registration methods are useful more generally than simply to align curves. While in some contexts, such as that of Figure 2.1, the phase component is merely *nuisance variation* to be dealt with but of no intrinsic interest, there are many other situations where the warps themselves represent useful modes of variation. In such contexts it is insightful to consider different types of data objects for OODA. In particular, *amplitude data objects*, whose variation is contained in the aligned curves, and *phase data objects* which are the warps used to achieve the alignment. Depending on the context either or both choices of data object can be of primary interest or either could represent just nuisance variation.

The notions of amplitude and phase data objects are illustrated using the *Bimodal Phase Shift* example in Figure 2.2. The upper left panel shows a simulated functional data set, where every data object (curve) has two peaks and is a multiple of a beta mixture probability density. A rainbow color scheme is used to distinguish the curves, in order of how separated the peaks are. The peaks have both different heights showing substantial amplitude variation, and also quite different locations reflecting strong phase variation. These two types of mode of variation are decomposed in a useful way by the warping functions shown in the bottom right panel, computed using the Fisher-Rao method, described in Section 9.1. The vertical axis is the same as in the upper left panel. Rescaling that axis using the magenta warp functions moves the magenta peaks inwards, and using the red warp functions moves the red peaks outwards. The top right panel shows the amplitude data objects, i.e. aligned curves. A careful look shows that the random peak heights are linearly related with the left peak being high when the right peak is low. This set of amplitude data objects consists of just a single one-dimensional mode of variation. The warps in the lower right panel can be thought of as the phase data objects, although they are not easy to interpret. Enhanced interpretation of the variation in the phase data objects comes from the view in the lower left panel. That is an application of each of the warps to the Fréchet mean (discussed in Section 7.7) template from the Fisher-Rao calculation, which nicely reflects the one-dimensional phase variation.

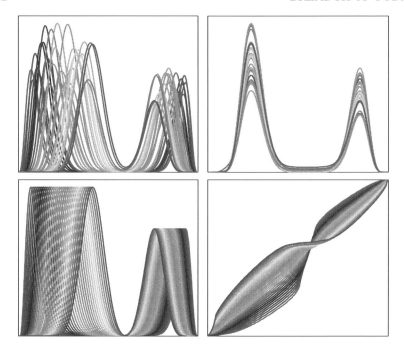

Figure 2.2 *Bimodal Phase Shift data (top left panel) showing decomposition into amplitude (top right panel) and phase (bottom left panel) modes of variation. Decomposition is based on the warping functions (bottom right panel). Rainbow color scheme highlights the phase mode, with red for closest peaks through magenta for farthest peaks.*

As clearly demonstrated in Figures 9.2, 9.3, 9.9, and 9.11, decompositions of the type shown in Figure 2.2 can be much more useful than a standard PCA in FDA, which tends to both mix the amplitude and phase components, and also to spread the phase variation over a large number of components, because it is a nonlinear mode of variation from that viewpoint. As discussed in Marron et al. (2014b, 2015), amplitude-phase decomposition is useful in many FDA applications. As noted above, for some of these, such as the TIC data shown in Figure 2.1, the amplitude data objects are the focus of the analysis, and the phase data objects can be viewed as nuisance parameters. However in other situations, for example when analyzing neural spike train data (as discussed in Wu et al. (2014)) the phase data objects are of primary interest, and the amplitude data objects are the nuisance component. In still other situations, both amplitude and phase data objects are important, as is their joint variation. These include the variation in the AneuRisk65 artery shape data in Sangalli et al. (2014a), and in the juggling data discussed in Ramsay et al. (2014).

Figure 2.3 shows some of the analysis of the *Juggling* data from Lu and Marron (2014a). The starting point was positional recordings of location over time of the hand of a juggler, which were reduced to time series of acceleration curves, as discussed in Ramsay et al. (2014). These traces were cut into cycles and time registered, to obtain the 113 curves shown in the far left panel of Figure 2.3. Thus the data objects in this OODA are time registered 1-d acceleration curves. Figure 3 of Lu and Marron (2014a) shows a variety of PCA type scores plots. Most of these seem to indicate a homogeneous population. The middle left panel of Figure 2.3 shows the version based on the method of Principal Nested Spheres (PNS), from Jung et al. (2012a). As further described in Sections 8.5 and 9.2, PNS makes special use of the fact that Fisher-Rao warp data objects naturally lie on a high dimensional sphere. The value added of using this method which takes the curvature of the sphere properly into account, is that it shows two clear clusters, which are highlighted using the graphical technique of *brushing*, i.e. visually separating the cluster through the use of colors and symbols. See Section 9.2 for more discussion (based on Yu et al. (2017a)) of how and why PNS provides enhanced statistical analysis of Fisher-Rao phase data objects. The clusters shown in the center-left panel of Figure 2.3 represent important underlying structure in the data. This is clear from the two right-hand panels of Figure 2.3, which show actual vertical and horizontal locations of the paths (orthographic projections) corresponding to these clusters, using the same colors. These are clearly two quite different types of motions present in the data, which correspond to "better controlled" and "less well-controlled" cycles.

Figure 2.3 *Analysis of the Juggling data. Far left panel shows the input acceleration curves. Center left is the Principal Nested Spheres scatterplot, revealing two distinct clusters highlighted by brushing. Right panels verify these clusters represent two different types of cycles.*

Figure 2.3 uses parts of Figures 2, 3, and 4 from Lu and Marron (2014a).

2.2 Tree-Structured Data Objects

A very different example of OODA is *trees*, in the sense of graph theory, as data objects. An interesting data set, where each data object is a representation of the set of arteries in one person's brain, was collected by Bullitt and Aylward (2002); Aylward and Bullitt (2002). While a long term goal is to study pathologies, including stroke tendency or brain cancer, such cases were deliberately screened out of this data set to focus on normal variation within the population. Interesting quantities that are useful for various comparisons below are age and gender. Section 10.1 gives an overview of various analytic approaches that feature improving abilities to distinguish age and gender.

These data objects were acquired using a modality of Magnetic Resonance

Imaging called Magnetic Resonance Angiography (MRA). This modality flags
motion as white, so the flow of blood through the arteries shows up very well.
This is seen in Figure 2.4 as the white spots, where the different panels show
adjacent horizontal slices of the 3-d image.

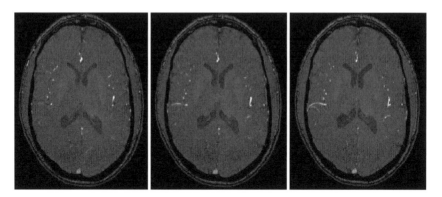

Figure 2.4 *Three adjacent slices of an MRA image for a single subject. Arteries show up
as white dots and curves.*

A major contribution of Aylward and Bullitt (2002) was the development of a
3-d tube tracking algorithm which was used to generate reconstruction of a given
artery tree. At this point the data object is the union of many small spheres, whose
centers follow the central curve of each arterial branch, and whose radii are the
branch radius at that point. This tree representation, from the MRA shown in Figure 2.4 can be seen in Figure 2.5. The three panels show different rotations of the
same set of arteries. The left panel is a fairly large rotation and the right panel is a
small one, with the closest vessels moved to the left and right, respectively.

Figure 2.5 *Three views of the arterial tree for the subject in Figure 2.4, showing the 3-d
structure through somewhat different rotations.*

Such data object representations have been computed for approximately 100
people, in the Brain Artery data set. For example, three more of these for three
different subjects are shown in Figure 2.6. The original study was a little larger,
but some were deleted due to MRA acquisition problems.

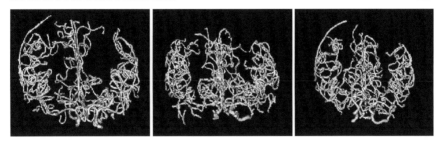

Figure 2.6 *Artery tree data objects for three additional subjects.*

Data objects of this type present major challenges to doing statistical analysis. For example, it is really not clear how to define even the sample mean of such a set of objects. Understanding variation about the mean, e.g. as done by PCA in Section 1.1, is a further challenge. A number of approaches to this are discussed in Chapter 10, which studies these trees in the more general context of graphs as data objects.

2.3 Sounds as Data Objects

Another example of OODA is *sounds* as data objects, which have been studied in a particularly deep way in a series of papers analyzing human speech based on digital recordings. Hadjipantelis et al. (2012, 2015) investigated Mandarin Chinese using a mixed effect model to develop relations between dialects which were consistent with linguistic ideas. Coleman et al. (2015) used these methods to extrapolate back in time to estimate how archaic languages may have sounded. Pigoli et al. (2018) analyze the relationships between modern romance languages, yielding insights well beyond those available from classical textual linguistic analysis (such as studied in Section 10.2). In addition a transformation is proposed that provides an estimated reconstruction of how a given speaker would sound speaking a different language. Tavakoli et al. (2019) combined these analyses with spatial smoothing to produce a dialectic map of the United Kingdom. Shiers et al. (2017) developed a sound-based evolutionary tree for romance languages and dialects.

A typical first step in those analyses is to decompose the raw digital recording of the sound into a spectrogram, which is a moving window version of the Fourier transform, giving a frequency representation in time, as shown in Figure 2.7, from the study of Pigoli et al. (2018), kindly provided by Davide Pigoli. The top panel is the raw recording of one person saying the word "deux" (two) in French.

Frequently, the focus is on human speech from the viewpoint that aspects such as pitch and timing are nuisances to be removed from the analysis. For that choice of data objects, those effects are removed by reducing the spectrogram to appropriately defined time and frequency covariance matrices (which are finite representations of covariance functions). Mean vectors also sometimes play an important role. Color heat-map representation summaries (as discussed in Section 6.1) of five covariance matrices (with entries colored according to the bars on the right,

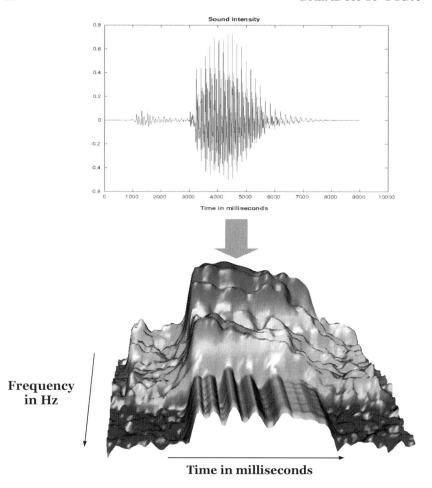

Figure 2.7 *Summarization of raw recording of a human speech sound of "deux" in French, top panel, into a corresponding spectrogram showing time and frequency information with color coding height, shown in the bottom panel.*

all using the same scale to facilitate comparison) from Pigoli et al. (2018) are shown in Figure 2.8, also from Davide Pigoli. For each language these summaries are based on aggregating sounds for the spoken digits (1–10). An exploratory visual comparison of these suggest some similarities (e.g. American and Castilian Spanish) and also some stark contrasts such as Portuguese from the others. Confirmatory analysis of these points and a number of others using permutation testing methods can be found in Pigoli et al. (2018).

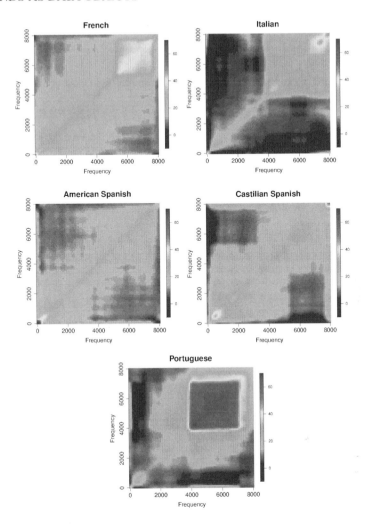

Figure 2.8 *Covariance representation summaries of speech sound from five different languages/dialects. Note strong differences between them, with potentially interesting historical and geographical connections.*

In the overall area of sounds as data objects, there is another interesting parallel to the phenomenon noted in Section 2.1, that depending on the context either phase or amplitude data objects could be the major focus of the analysis with the other considered to be nuisance variation. In particular, the above work focused on a particular type of analysis of sounds as data objects, where the goal was to study human speech by a variety of speakers. As the human brain does when parsing speech, they deliberately chose data objects which focused on aspects of the sound that are about meaning of the words, which means generally treating issues such as pitch, volume, and timing as nuisances to be mathematically ignored. This a strong contrast with the area of Music Data Analysis, which has been deeply studied in Weihs et al. (2016) where timing, volume and pitch are actually of keen interest as the data objects.

Statistical analysis of covariance matrices as data objects is particularly challenging, and discussed further in Section 8.7.

2.4 Images as Data Objects

The field of image analysis is very large. Statistics has traditionally appeared there in several ways. Early work, with famous papers including Geman and Geman (1984) and Besag (1986), tended to focus on aspects of mostly a single image, with tasks such as denoising, segmentation, and registration being predominant. However, those fields are now relatively mature, so a currently more important role for statistical ideas comes at the population level which yields a very rich source of potential data objects. For example, the shapes studied in Section 1.2 and the trees considered in Section 2.2 are two types of data objects derived from images.

But in other situations the images themselves can be treated as data objects. An example of this is the *Faces* data that appears in Benito et al. (2017), which studies a data set of $n = 108$ images (actually 248×186 gray level photographs) of students from the University of Carlos III in Madrid. There is quite a lot of variation among the faces, yet the human perceptual system is good at distinguishing gender. In that paper, male vs. female classification of these data is carefully studied. As discussed in Section 5.4 of Benito et al. (2017), manual affine registration was used to put each face into a common location in its image. Then the gray level pixels of the images were *vectorized* (by stacking columns, an operation sometimes denoted as *vec*) into a single long vector, and various classification methods were used to try to understand the difference between males and females. *Classification*, also sometimes called *discrimination*, is an important OODA topic discussed in Section 11. The classification methods used on this face data set were linear methods, as those yield the best interpretation of the results.

Particularly good results came from *Distance Weighted Discrimination* (DWD) (proposed by Marron et al. (2007) and studied here in Section 11.4) as shown in Figure 2.9. DWD is discussed in more detail and compared with other classification methods in Chapter 11. The lower panel of Figure 2.9 shows the DWD scores, i.e. the projections of the data onto the DWD separation direction (the normal vector to the DWD separating hyperplane) using a format similar to that of the right panel of Figure 1.4. The red plus signs correspond to the females and the blue circles are the males, which are completely separable using DWD. Also shown are three kernel density estimates. That for the full population appears in black (mostly underneath the others). Female and Male sub-densities (i.e. rescaled according to sub-sample size) are shown as red and blue, respectively. The top panel gives insight into what DWD is doing with the images, by showing a representative set of 8 reconstructions (i.e. the vectors are converted back into an image) from 8 equally spaced points (locations shown as the 8 equally spaced black bars in the bottom panel) along the DWD separating vector.

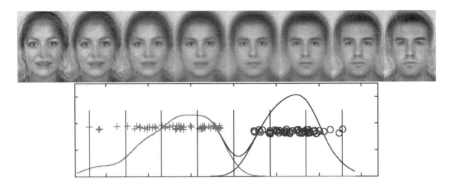

Figure 2.9 *Results of DWD discrimination between males and females in the Faces data. Bottom panel shows distribution of DWD scores. Top panel contains 8 reconstructions of faces, corresponding to the 8 points along the DWD separating vector shown as vertical bars in the bottom. Shows clear insight as to how DWD separates males from females.*

The array of faces in the top panel is quite compelling. They look clearly very female on the left side, quite androgynous in the middle, and clearly male on the right. Also apparent in perhaps the second and third panels from either end is the idea from Langlois and Roggman (1990) that average faces tend to be more beautiful. In addition, note that farther to the right corresponds to stronger masculinity. Note that this collection of faces represents yet another type of mode of variation, which focuses on the female male differences, instead of on maximal variation as in the PCA analysis of the Spanish Mortality data discussed in Section 1.1.

Another example of images as data objects is the Cornea Curvature data studied in Section 16.2.1. That analysis uses a much different data object representation.

CHAPTER 3

Data Object Definition

This chapter and the next together discuss basic aspects of OODA. There are two main themes. The first theme is the three phases of OODA (object definition, exploratory analysis, and confirmatory analysis) that were introduced at the beginning of Chapter 1. Object definition is discussed here, while exploratory and confirmatory analysis are detailed in Chapter 4. The second theme is *modes of variation*, which have been informally discussed through most of Chapters 1 and 2, and will be formally defined here in Section 3.1.4. This chapter also provides an overview of methods discussed in more detail in later chapters.

3.1 OODA Foundations

3.1.1 OODA Terminology

Any OODA starts with data object selection. This typically has two main components, *determination* of data objects and their numerical *representation*. Determination involves choice of focus of the analysis. One example is the Spanish Mortality data of Section 1.1 and the choice between age indexed curves over years and year indexed curves over age, as well as the choice of $\log_{10}$ mortality. Another example is choosing whether to focus on amplitude and / or phase variation in Section 2.1. It is useful to consider the notion of *object space* as the conceptual space containing all potential data objects, e.g. the space of curves for these two examples. Fundamental OODA insights come from simultaneously considering the parallel notion of *feature space*, which contains the practical numerical representations, such as *feature vectors*, as illustrated in Section 3.1.2.

Feature vectors are often aggregated into a *data matrix* which is a useful framework for organizing data analytic thoughts. One of the matrix dimensions typically represents the *cases*, i.e. the *elements* of a statistical sample, which are also sometimes called *observations* or *individuals*. Some potentially confusing cross-cultural terminology has arisen in bioinformatics, where a complex biological experiment is used to collect each measurement, i.e. feature vector, which itself is frequently called a *sample* (by biologists, in stark contrast to the statistical use of the term applying to the entire data set). The other matrix dimension is used to index *features* or numerical descriptors of each data object, with *variables* being a common synonym, as in Table 3.1. Another synonym is *traits* which is common in biological applications and is particularly useful in Chapters 6 and 17.

DOI: 10.1201/9781351189675-3

	Number	Synonyms
Cases	n	elements of a statistical sample, observations, individuals, biological samples, experimental units
Features	d	variables, descriptors, traits

Table 3.1 *Commonly used synonyms for cases and features.*

An important issue is that there is a distinct dichotomy in personal preference as to which data matrix dimension is which. From the classical linear algebraic point of view, where vectors are columns, it makes the most sense to treat each data object as a column vector, and then to horizontally concatenate these (i.e. bind the columns), resulting in columns as data objects, with rows representing numerical features. However, from the equally classical statistical tabulation viewpoint, it is perhaps more natural to put variables (i.e. features or traits) in the columns and to hence use row vectors as the data objects.

Keeping this distinction in mind is critical to having meaningful technical conversations, especially when linear algebra is involved. OODA terminology makes this straightforward, by first agreeing whether it will be rows or columns that are the data objects. This choice is often closely connected with software preference. Much mainstream statistical analysis is done using R and SAS, where rows as data objects are the convention. More mathematically oriented work is often done in Matlab where columns as data objects is the more natural choice. Columns as data objects is typical in bioinformatics as well, although this convention appears to be largely driven by the fact that typical data sets tend to have many more variables (features or traits) than cases, which were easiest to store in early versions of Excel in that format. The convention in this book is columns as data objects.

Another point of varying conventions is the letters used to denote the dimensions of the data matrix. Again this is context dependent, with choices like m and n appearing in some areas. Statisticians generally agree that n should be used for sample size, i.e. for the number of data objects. Quite common also is p for the number of variables (features, traits). Less clear is what p might stand for. Some say it stands for *predictors*, but this seems limited to mostly regression contexts. Others suggest *parameters*, which makes sense in contexts such as regression where the mean is the focus, but not for consideration of covariance matrices (which typically involve many more than p parameters). The convention here is d standing for *dimension* of the data object vectors, and so with columns as data objects the data matrix is a $d \times n$ matrix throughout this book. Further matrix notation appears at (3.4).

3.1.2 Object and Feature Space Example

As noted in Marron and Alonso (2014) and mentioned above, a useful framework for understanding relationships between data objects is through the twin concepts of *object* and *feature* spaces. The object space contains the raw curves, images, shapes or trees, while the feature space contains some sort of numerical representation often in vector form. As pointed out in Telschow et al. (2014), the feature space can also be called *descriptor space*, because that is where statistical description, such as the mean and variation about the mean as in the following example, are naturally computed.

The interplay between these spaces is illustrated using the simple *2-d Toy* FDA example shown in Figure 3.1. The data objects are the $n = 24$ very simple functions shown as black piece-wise lines in the left panel of Figure 3.1, the conceptual object space. This piece-wise linear functional form is used here because it is *two dimensional*, in the sense that each data curve is entirely determined by heights of the two x signs plotted on the vertical lines. Each curve has the constant values of height x_1 on [0,1] and x_2 on [2,3], and is piece-wise linear between the x signs (on [1,2]).

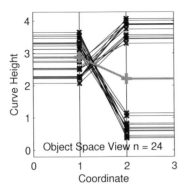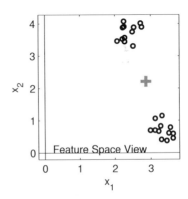

Figure 3.1 *Simple 2-d Toy Example illustrating object space (left panel) showing data objects as (simple piece-wise linear) curves, and feature space representing the same data objects as points (circles in the right panel). The sample mean is shown in green in both panels.*

The relationships between these data objects is clearly illustrated in the feature space view shown in the right panel, where each black circle represents one piece-wise linear function, using a conventional (x_1, x_2) scatterplot. Note that every point (not just the data points) in the feature space has a representation as a piece-wise curve and vice versa, i.e. there is a one to one correspondence between these spaces. This view clearly shows two very distinct clusters, which are also apparent in the left panel, at least after seeing the right panel. In general higher dimensional cases, the feature space is usually of at least somewhat higher dimension, and thus is more challenging to visualize. However, it is often very useful to still think in terms of such a point cloud in feature space as a device for considering relationships between data objects (e.g. the clusters apparent here). This concept was used to explain the PCA modes of variation in Figures 1.4 and 1.5 for the Spanish Mortality data. Graphical devices for visualizing such point clouds

include matrices of scatterplots as introduced in Figure 1.6 and discussed further
in Section 4.1.

Figure 3.1 also shows the sample mean in green. This is computed as the conventional vector mean in the feature space, shown as the green plus sign. The
corresponding piece-wise curve is shown in green in the object space in the left
panel. Note that the green curve is also the point-wise mean of the data curves.

It was seen in Section 1.1 that PCA can provide an insightful decomposition
into modes of variation. This is explored in the context of this same 2-d toy example in Figure 3.2. The first principal mode of variation is usefully understood in
the feature space as based on the direction from the mean (i.e. the red line through
the green plus sign), that maximizes projected variation. The projection of each
black data point onto the red line is a magenta plus sign, which is the point on the
line closest to the data point, connected by a cyan line segment. The coefficients
of this projection (relative to the unit vector pointing in the direction of the red
line) are the scores (e.g. these quantities were highlighted for the Spanish Mortality data in the right panels of Figures 1.4 and 1.5). The entries of the direction
vector are the loadings that determine the shape of the mode of variation shown
in the left panel of Figure 3.1. The red line has been carefully chosen to maximize
the sample variance of the scores, thus giving the direction of maximal variation
in the data. By the Pythagorean theorem, it is easily seen that this solution is the
same as minimizing the sum of the squared lengths of the cyan line segments.

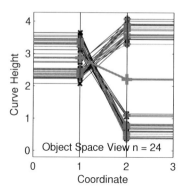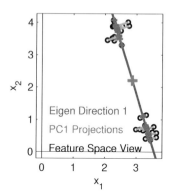

Figure 3.2 *First principal component mode of variation for the 2-d Toy Example. Red line
in the right panel is the direction of maximal variation, cyan line segments show projections to magenta plus signs. Left panel shows corresponding projected piece-wise lines in
magenta. This plot shows how PC1 is the one-dimensional approximation which captures
most of the variation in the data.*

The left panel of Figure 3.2 shows these projected points as curves. As the
magenta plus signs in the right panel are close to the black circles, the magenta
piece-wise lines in the left panel are close to their corresponding black curves.
These magenta curves are the best one-dimensional approximation of the data,
in that they are (signed) multiples of the same curve (with respect to the mean)

that has been determined by the loadings. Thus in the left panel the Coordinate 1 (x_1) heights are negatively correlated with the Coordinate 2 (x_2) heights. This negative correlation is of course also reflected by the circles in the right panel. The magenta curves are also usefully interpreted as a mode of variation, with relatively small variation in the first coordinate (x_1) and much larger negatively correlated variation in the second coordinate (x_2).

Also insightful is the second mode of variation, defined in terms of the second principal component, shown in Figure 3.3. The yellow line in the right panel shows the second principal component direction, which is orthogonal to the red line in Figure 3.2 (generally the orthogonal line with maximal variation, but in this 2-d example it is the only choice). Magenta line segments indicate the operation of projection of each black data point onto the yellow line, which results in a cyan plus sign. Note that lengths of these magenta line segments are the same as the distances between the green center point and the magenta plus signs in the right panel of Figure 3.2. Furthermore the distances between the cyan plus signs and the green center point on the right of Figure 3.3 are the lengths of the short cyan line segments on the right in Figure 3.2.

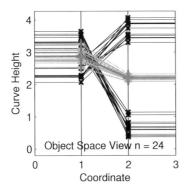 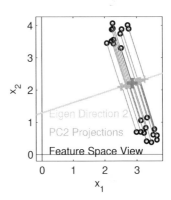

Figure 3.3 *Second principal component mode of variation for the 2-d Toy Example. Yellow line in the right panel is the PC direction, with magenta line segments showing cyan projections. Second mode of variation is shown as cyan piece-wise lines in the left panel.*

The left panel of Figure 3.3 shows the corresponding mode of variation as the cyan piece-wise curves. Note that there is far less visual variation present which is not surprising because this is the direction of minimal variation. Also note that the x_1 heights (Coordinate 1) are positively correlated with the x_2 heights (Coordinate 2). This is consistent with the fact that the cyan pluses in the right panel lie on an upward sloping line.

A useful summary of the decomposition into modes of variation shown in the above plots appears in Figure 3.4. The raw data object piece-wise lines shown in black in the upper left are the sum of the components shown in the other panels: the sample mean in green in the upper right, magenta PC1 projections (first mode of variation) in the lower left, cyan PC2 projections (second mode of variation) in the lower right. Note that the magenta first mode focuses on the clustering aspect of the data, in addition to the negative correlation of x_1 and x_2 and greater variability in x_2 compared to x_1. The cyan second mode of variation is much smaller in magnitude, and contains the smaller scale positive correlation between x_1 and x_2 as well as being driven more by x_1 than by x_2. Note that several effects have been picked up by each mode of variation, as is typical in many applications.

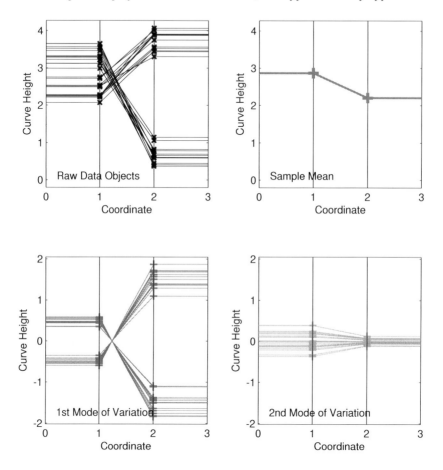

Figure 3.4 *PC decomposition of the 2-d Toy Example. Raw data objects in the upper left, mean in the upper right (green), 1st mode in the lower left (magenta), 2nd in the lower right (cyan). Raw curves are the sum of others, showing insightful decomposition of variation.*

3.1.3 Scree Plots

Another aspect of this data set that is clearly visible in Figure 3.4 is that the first mode of variation (magenta curves in the lower left) contains much more of the overall variation in the data than the second mode (cyan curves, lower right). This apportionment of variation is often worth numerically quantifying using sums of squares and then graphically displaying the results as a *scree plot*, several of which are shown in Figure 3.5. A scree plot summarizes the variation (measured as sums of squares) of each PC mode, as a function of the component index plotted as a solid red piece-wise line. This is frequently (as here) scaled so the total is one, thus displaying the proportion of variation for each mode. It is also frequently useful to study the proportion of variation explained by the approximation of the data made by each mode together with its predecessors. This is the cumulative sum of the red values, shown here using a dashed blue piece-wise line. A few of the values are also highlighted using circles (for the red single mode points) and plus signs (for the blue cumulatives). These represent modes that are explicitly displayed in the respective analyses elsewhere in this book.

The upper left panel of Figure 3.5 shows the scree plot for the 2-d Toy data set used in Figures 3.1–3.4. This shows that the first PC mode of variation has 98.3% of the variation, which is consistent with the visual impression in the lower panels of Figure 3.4. Note that there are only two components for this data set, so the dashed blue cumulative curve is actually at height one starting at mode index 2.

The scree plot for the Spanish Mortality data from Figures 1.1–1.6 is in the upper right panel of Figure 3.5. Once again the first mode contains a very large share of the variation (95.8%), which is consistent with the fact that the vertical axis for displaying the first mode of variation in the left panel of Figure 1.4 is an order of magnitude larger than for the second mode in Figure 1.5. A careful look (together with the overall wiggliness of the curves in Figure 1.2) suggests there may be additional variation in other modes of variation although it is visually hard to distinguish from 0. When it is important to understand such modes of variation that are orders of magnitude smaller, a log transformation of the scree plot is useful (the same graphical principle illustrated in Figure 1.1).

The bottom left panel of Figure 3.5 is the scree plot for the Tilted Parabolas 10-d toy data which will be studied in Figures 4.1 and 4.2. This shows a couple of large modes, followed by a number of smaller ones. The large modes are classically thought of as "signal", and the smaller ones as "noise". That conceptual ideal is the basis of the use of the term "scree" here (apparently coined by Cattell (1966)). In the natural world, as steep cliffs erode, rocky rubble called scree tends to collect at their base. The idea of the graphic is that the signal appears as a steep cliff, while the noise is the scree collected next to it. It has been proposed to try to find "elbows" (some say "knees") in the scree plot to select the "right" number of principal components, but this can be a very slippery operation. For example, the visual impression of knees in graphics can change completely when transforming the scree plot, e.g. by looking at the log proportions. The lower left panel of Figure 3.5 also serves to illustrate problems with this approach, as most would say that it suggests one signal mode of variation, yet the analysis in Figure 4.1 shows *two* modes of variation representing clear underlying population structure.

To make the point that not all scree plots have a large first mode, the scree plot for the Pan Cancer gene expression data set which will be studied in Figures 4.9–4.11 is shown in the bottom right panel. This time the first mode only explains less than 20% of the variation, and many components are substantial players. Here the

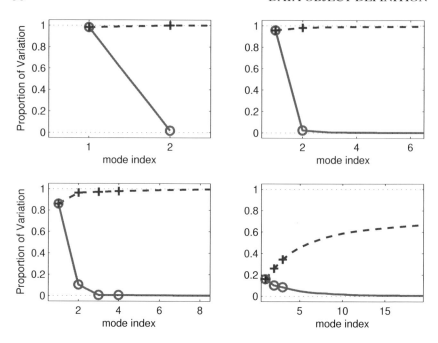

Figure 3.5 *Scree plots, showing how variation is apportioned among PC modes of varia-tion for the 2-d Toy Example data from Figures 3.1–3.4 (upper left), the Spanish Mortality data from Figures 1.1–1.6 (upper right), the 10-d Tilted Parabolas data from Figures 4.1 and 4.2 (lower left), and the Pan-Cancer gene expression data set from Figures 4.9–4.11. Solid red curve shows individual proportions of variance and blue dotted curve is the cor-responding cumulative.*

cumulative dashed blue curve shows that the first 20 components together explain only around 60% of the variation. This is because gene expression is associated with very many diverse biological processes in living cells, so its overall variation is intrinsically high-dimensional.

3.1.4 Formalization of Modes of Variation

Figure 3.4 uses the 2-d Toy data to highlight the usefulness of the concept of *modes of variation* in OODA, which has been informally used in the above discussion and also in Chapters 1 and 2. One candidate for a more formal definition is a direction in feature space, such as that shown as a red line in the right panel of Figure 3.2 and a yellow line in the right panel of Figure 3.3. These directions are the loadings vectors whose entries determine the shape of the colored curves in the left panels of those figures. However, the variation shown in these examples involves more than just the direction in which it occurs. In particular, another important aspect of that variation is the two cluster population structure, which shows up in the PC1 scores (red plus signs) in the right panel of Figure 3.2. That also determines the clear cluster structure in the curves in the left panel. This suggests that a more useful definition of mode of variation could be the union of the loadings and the scores. However, consideration of the amplitude and phase variation through the Bimodal Phase Shift data in Figure 2.2, as well as the shape variation illustrated in Figure 1.11, suggest that in general such linear summaries of loadings and scores are not sufficiently general. Instead the needed intuition comes more generally from consideration of some type of one-dimensional set of objects in the object space. This motivates:

Definition: A *mode of variation* of a sample of data objects is a set of potential members of the object space that provides a simple summary of one component of the variation. This summary is in some sense one dimensional, e.g. reasonably representable by a single real-valued parameter.

Such summarizing sets are bundles of curves representing two modes of variation of the Spanish Mortality data in the left panels of Figures 1.4 and 1.5. Summarizing sets, representing modes of variation using the center and two extreme shapes are shown for the Bladder-Prostate-Rectum data in Figure 1.11. Examples of nonlinear modes of variation again displayed using bundles of curves approximating the Bimodal Phase Shift data appear in Figure 2.2. A summarizing set illustrating a mode of variation connecting females to males in image space appears for the Faces data in the top part of Figure 2.9. The notion of mode of variation is revisited at several points in the following, perhaps most notably in Chapters 6, 9, and 17.

3.2 Mathematical Notation

Notation that is useful in the next section, as well as at many later points includes:

- The set of real numbers, $\mathbb{R}$.
- The standard d-dimensional Euclidean space of column vectors (denoted as lower case bold letters),

$$\mathbb{R}^d = \left\{ \boldsymbol{x} : \boldsymbol{x} = \left(\begin{array}{ccc} x_1 & \cdots & x_d \end{array} \right)^t, \ x_1, \cdots, x_d \in \mathbb{R} \right\}. \qquad (3.1)$$

- Matrix transpose indicated by a superscript t.
- The sphere whose surface dimension is d (essentially points on the surface of the solid unit ball in $\mathbb{R}^{d+1}$),

$$S^d = \left\{ \boldsymbol{u} \in \mathbb{R}^{d+1} : \|\boldsymbol{u}\|_2 = 1 \right\}. \qquad (3.2)$$

- The L^p norm on $\mathbb{R}^d$,

$$\|\boldsymbol{x}\|_p = \left(\sum_{j=1}^{d} |x_j|^p \right)^{1/p}. \qquad (3.3)$$

- The Cartesian product, $\times$.

- The set of $d \times n$ matrices (using bold capital letter notation),

$$\mathbb{R}^{d \times n} = \left\{ \boldsymbol{X} : \boldsymbol{X} = \begin{bmatrix} x_{1,1} & \cdots & x_{1,n} \\ \vdots & \ddots & \vdots \\ x_{d,1} & \cdots & x_{d,n} \end{bmatrix}, \ x_{11}, \cdots, x_{dn} \in \mathbb{R} \right\}. \quad (3.4)$$

- Given a $d \times n$ data matrix $\boldsymbol{X}$, the sample covariance matrix (using the notation $\widehat{var}_i$ and $\widehat{cov}_{i,i'}$ to denote the sample variance and covariance of the i-th and i'-th rows of $\boldsymbol{X}$ formally defined at (17.18) and (17.19)),

$$\widehat{\boldsymbol{\Sigma}} = \begin{bmatrix} \widehat{var}_1 & \widehat{cov}_{1,2} & \cdots & & \widehat{cov}_{1,d} \\ \widehat{cov}_{2,1} & \widehat{var}_2 & \ddots & & \vdots \\ \vdots & \ddots & \ddots & & \widehat{cov}_{d-1,d} \\ \widehat{cov}_{d,1} & \cdots & & \widehat{cov}_{d,d-1} & \widehat{var}_d \end{bmatrix}. \quad (3.5)$$

3.3 Overview of Object and Feature Spaces

Figures 3.1–3.4 and many others focus on the display of curves as data objects which can be thought of as visual representations of vectors. The relevant plots are all piece-wise linear plots, where the heights of the vertices are the entries of the vectors. Such plots have been called *parallel coordinate plots* by Inselberg (1985, 2009), who advocated them as a general multivariate analysis visualization tool.

In FDA, other representations of curves besides digitization are also commonly used, often based on mathematical *basis* ideas. These include:

- *Fourier.* This orthogonal basis is very useful for curves which are smooth and periodic. Insightful discussion of Fourier methods in the context of time series analysis can be found in Bloomfield (2000) and Brillinger (1981).

- *Orthogonal Polynomials.* There are many such orthogonal bases for curve space. Many useful facts can be found in the classical book by Szegő (1975). A very useful and easily accessible summary of many important aspects can be found in Gradshteyn and Ryzhik (2015). A specific example of orthogonal basis data object representation, using a tensor product of polynomials and a Fourier basis, to analyze the Cornea Curvature data set is in Section 16.2.1.

- *B-splines.* There are many variants of these typically smooth curves, which provide flexible and effective representations of smooth data objects. See Eilers and Marx (1996), Stone et al. (1997), and Ruppert et al. (2003) for good overviews of statistical aspects of this area. An important classical B-spline reference is de Boor (2001).

- *Wavelets.* This orthonormal basis can give efficient data object representation for curves with varying amounts of smoothness in different locations. See the book Frazier (2006) for introduction to this area. Other important references include Mallat (1989), Daubechies (1992), Donoho and Johnstone (1994), and Donoho et al. (1995). As discussed in Mallat (2009), a combination of sparsity ideas and wavelets has been particularly useful in image denoising. Different types of useful insights into wavelet curve estimation come from exact risk calculation in Marron et al. (1998), and using spectral ideas in Marron (1999).

The object feature space concept is also useful for these curve representations, where again the object space consists of curves, but now the feature space is the space of basis coefficients. Data analysis methods such as PCA still tend to work quite well when performed on the vectors of basis coefficients in that feature space, together with insightful visualization of modes of variation seen in the object space, in the spirit illustrated in Figures 3.1–3.4. As noted above, a particularly deep example of this type, where cornea curvature images on a disk are represented by a carefully chosen orthonormal basis is discussed in Section 16.2.1.

Another very important aspect of data object representation is *transformation*. The utility of this was illustrated in Figure 1.1, where it was seen that $\log_{10}$ mortality gave much clearer insights than were available from the raw mortality. Data transformation is further studied in Section 5.3.

Sangalli et al. (2014b) gave an interesting discussion of the importance of *sufficiency* in data object choice. A related issue, very important to mathematical statistical analysis of OODA is the choice of data object space, which includes an appropriate metric. For example, in FDA, there are many ways to measure distance between curves, e.g. there is the whole family of L^p norms. Much of the literature has been dominated by the choice $p = 2$ because of its close relationship with classical least squares and its tractability. However, when robustness issues (discussed in Chapters 7 and 16) are important $p = 1$ can be very useful. Furthermore Devroye and Györfi (1985) offer good reasons why L^1 is more natural in the case of probability densities as data objects. In some cases, such as the occasional need to strongly penalize thin spike departures, the choice $p = \infty$ can be more useful. In other situations performance of derivatives are critical, so Sobolev type norms are the most sensible choice. Marron and Tsybakov (1995) explored error criteria that correspond to human visual impression. Panaretos and Zemel (2019); Zemel and Panaretos (2019); Panaretos and Zemel (2020) show that the Wasserstein metric provides those good visual properties in a natural way, using optimal transport ideas. More discussion of these error criteria can be found in Sections 7.3 and 15.2. However, these OODA object space issues run deeper than just the mathematical statistics. In particular, as seen in Chapter 7, even simple data analytic notions such as population center can depend critically on such choices. Piercesare Secchi nicely summarized this set of ideas as: "Experimental units only become data objects after embedding in an appropriate space", which is the role of the object space.

Table 3.2 revisits the examples in Sections 1.1–2.4 and above, with the goal of clarifying the "data object" terminology and the respective roles of the conceptual object and feature spaces.

This table emphasizes the respective roles of the object and feature space. For example

- *Mortality Example* (Section 1.1). Each object curve is represented by a 99

Example	Choice of data objects	Appropriate?	Object space	Feature space	Methods used	Results
Mortality	Raw Mortality vs Age curve	× Obscures detail	Curve space	$\mathbb{R}^{99}_+$	Mean & PCA	Figs. 1.1–1.6
	$\log_{10}$ Mortality vs Age curve	✓ Reveals detail	Curve space	$\mathbb{R}^{99}$		
Prostate	Skeletal shapes	✓	3-d S-rep models	$\mathbb{R}^k \times \mathbb{R}^l_+ \times (S^2)^m$	PGA	Fig. 1.11
Proteomics	Raw curves	× Obscures detail	Curve space	$\mathbb{R}^{2001}$		
	Aligned curves	✓	Curve space	$\mathbb{R}^{2001}$	Align true markers	Fig. 2.1
	Warp functions	✓	1-d Diffeomorphisms	S^{2000}		
	Aligned & warps	✓	Curve × Diffeomorphisms	$\mathbb{R}^{2001} \times S^{2000}$		
Juggling	Warping of acceleration	✓	1-d Diffeomorphisms	S^{d-1}	PNS	Fig. 2.3
Speech	Sound recording	× Obscures detail				
	Spectrogram	× Obscures detail				
	Covariance of spectrogram	✓	Covariance functions	PSD_m	Group means & test	Fig. 2.8
Faces	Unregistered faces	× Too blurred	Set of all 2-d images	$\mathbb{R}^{248 \times 186}$		
	Registered faces	✓	Set of all 2-d images	$\mathbb{R}^{248 \times 186}$	DWD	Fig. 2.9
2-d toy example	Curves	✓	Curves	$\mathbb{R}^2$	Mean & PCA	Fig. 3.1–3.4

Table 3.2 *Specifics illustrating choices of conceptual object and feature spaces.*

dimensional feature vector (in $\mathbb{R}^{99}$) for the 99 ages 0–98. As observed in Figure 1.1, the $\log_{10}$ variable choice reveals more useful data structure.

- *Proteomics Example* (Section 2.1). Here there are multiple reasonable choices of data objects, which are curves or warps (i.e. diffeomorphisms, see Section 9.1.2 for further discussion). The curves are represented as 2001 dimensional feature vectors, while the Fisher-Rao warping functions are usefully represented in the feature space as points on the high-dimensional sphere S^{2000}.

- *Speech Example* (Section 2.3). Choices of conceptual data objects here include digital sound intensities as shown in the top panel of Figure 2.7, spectrograms in the bottom panel, and covariance functions of frequencies. As noted in Section 2.3 the latter is the best choice for revealing important aspects of human speech. Their discretization results in the set of $m \times m$ symmetric *positive semidefinite* (also called *nonnegative definite*) matrices, PSD_m, being the natural feature space.

- *Faces Example* (Section 2.4). The conceptual data objects are the 248×186 two-dimensional images. The feature representation came from reorganizing each image matrix into a $(248 \times 186) = 46,128$ dimensional feature vector.

3.3.1 Example: Probability Distributions as Data Objects

Another data object representation issue arises in the case of probability distributions as data objects. Some commonly used representations of distributions in probability theory are shown in Figure 3.6, for the bimodal mixture distribution $\frac{1}{3}N(120, 1200) + \frac{2}{3}N(280, 1200)$. The left panel shows the *probability density function* representation. This form is most useful for highlighting important distributional aspects such as modality, as pointed out for example by Silverman (1986) and Scott (2015) and also as is clear from this figure. The center panel shows the *cumulative distribution function* of the same distribution, which is computed by

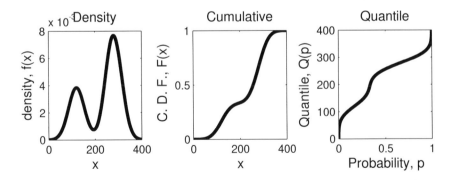

Figure 3.6 *Different representations of the same bimodal normal mixture distribution. Left panel is the probability density, center is the corresponding cumulative, right is the quantile i.e. inverse cumulative.*

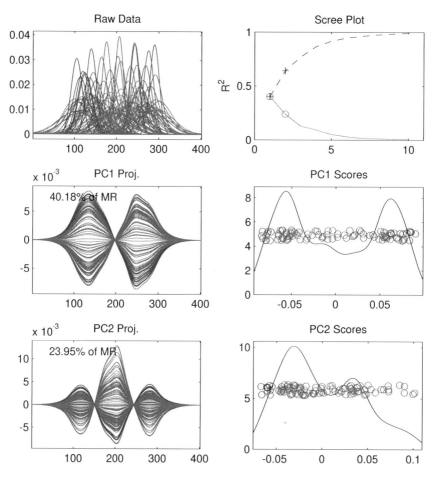

Figure 3.7 *Raw data (upper left panel) and PCA of a family of Gaussian densities. Colors more blue for larger mean, and more red for larger variance. Scree plot in the upper left panel, and the modes of variation (2 lower right panels), show very poor PCA summarization of this data set.*

integrating the density, and has strong utility for the task of calculating probabilities such as p-values. The right panel shows the corresponding *quantile function*, which is just the inverse function of the cumulative distribution function. See Parzen (2004) for discussion of many appealing properties of the quantile function.

Which representation is preferable strongly depends on the context. A perhaps not well understood point, illustrated in Figures 3.7 and 3.8, is that for sets of distributions where variation in the mean and/or variance are important modes, the quantile function representation can easily give the most efficient and insightful PCA decomposition. The upper left panel of Figure 3.7 shows a data set of

$n = 100$ curve data objects, each of which is a Gaussian density, with a wide range of means and variances. The data objects are colored, with more blue colors coding larger means, and more red representing larger variances (thus lower curves). Hence the small mean and variance densities (upper left) are black, with the large mean and variance curves colored magenta (lower right). The top right panel of Figure 3.7 shows the scree plot for the PCA of these curves, which indicates quite poor signal compression, in the sense that the first few modes of variation contain a fairly small fraction of the variation, i.e. the signal power of this data set is spread widely across the PCA spectrum. This is disappointing as there are in some sense only two modes of variation (only differing means and variances). The reason for this appears in the other left panels, which show the PCA loadings plots as introduced in the left panels of Figures 1.4 and 1.5. Unlike the analysis of the mortality data shown in Section 1.1 this decomposition does not provide a very intuitive view of the variation in this data set. In particular, neither of the first two modes reflect the two intuitively obvious modes of variation (means and variances).

As shown in Figure 3.8, the quantile representation of this set of probability distributions is far more amenable to PCA. The upper left plot in Figure 3.8 shows the quantile representations of the densities in the top left panel of Figure 3.7, using the same colors. The PCA scree plot of these curves in the upper right panel shows that the full data set is described completely by only the first two PC components, which is a stark contrast to the poor PCA signal compression seen for the density representation of these data in Figure 3.7. Furthermore the first component contains well over 95% of the variation.

Note that the PCA loading plots shown in the remaining left panels of Figure 3.8 are quite interpretable. PC1 is clearly a vertical shift mode of variation, which nicely captures the shifting mean component of variation. That follows from the fact that the vertical axis in the right (quantile) panel of Figure 3.6 is the same as the horizontal axes in the center (cumulative) and left (density) panels. PC2 in Figure 3.8 captures the variation in variances by reflecting differing slopes which correspond to differing widths of the densities. The scores distribution plots in the middle and lower right panels of Figure 3.8 cast additional insights into these modes of variation. Note that the PC1 scores show increasing blue toward the right, with black and red on the left, and blue and magenta on the right, corresponding to the left to right colors in the upper right panel of Figure 3.7. Also, the PC2 scores show increasing red toward the right, going from black and blue to red and magenta, similar to the downward coloring in the density representations for increasing variance.

Another approach to the dominant mean shift variation in this example is the use of variation in warping functions to explicitly target phase variation, as seen in Section 2.1, and discussed in detail in Chapter 9. Hron et al. (2016) and Menafoglio et al. (2018) propose other promising approaches to the analysis of probability distributions as data objects, which similarly result in modes of variation that are more insightful than those captured by PCA.

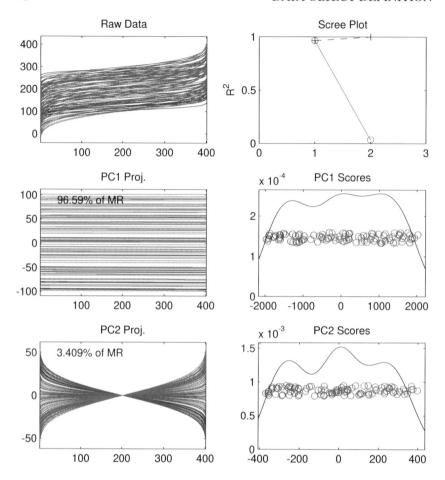

Figure 3.8 *PCA of the quantile representations of the same set of distributions as in Figure 3.7. Shows much more efficient PCA summarization of variation in this data set, as well as more interpretable modes of variation, using this improved choice of data objects.*

There are situations where explicit feature representation of data objects can be side-stepped. An example is when only distances between data objects are measured. There are many methods for handling such situations, discussed in Chapter 7.

CHAPTER 4

Exploratory and Confirmatory Analyses

This chapter discusses the second and third phases of OODA (exploratory analysis and confirmatory analysis) from the beginning of Chapter 3.

4.1 Exploratory Analysis–Discover Structure in Data

Data visualization, as illustrated for example using the Spanish Mortality data in Figures 1.1–1.6, is a very important part of exploratory data analysis. A personal opinion is that it should represent a larger part of statistical training, and of funded research, than it currently does. The present state seems to be driven by statistical models and goals (for example analyzing causality) becoming increasingly complex, which led to a tendency to co-opt a large share of attention in the field. However visualization is not only important for exploratory analysis and understanding how data objects relate to each other as demonstrated in Figures 1.1–1.6, it is frequently also important for the effective choice of data object (e.g. whether or not to transform), and further provides important reality checks.

Important references on data visualization include Tufte (1983), Cleveland et al. (1985), Cleveland (1993), and Tukey (1990). These works contain many useful ideas and discussion of what comprises good graphics, although they can sometimes be overly prescriptive. The rest of this section considers two specific types of data visualization that are critical to OODA.

A perhaps too often ignored, but frequently critical, step in OODA is the study of marginal distributions. Visualizations of marginal distributions, e.g. by histograms or QQ plots, are common when there is time for careful analysis of classical small scale data sets. This often proves very useful in handling variables with strong natural skewness, indicating a potential benefit from transformation (see Section 5.3 for much more on this), and also in the case of strong outliers, which depending on the context can either be deleted or handled through the use of robust methods (see Chapter 16).

A reason that this step seems challenging in high-dimensional contexts is that there are generally just too many variables (i.e. features or traits) to humanly comprehend the population structure of all of them. A careful analyst will try to look at some representatives, but it may not be obvious how to choose those. This problem is addressed using the graphical device of *marginal distribution plots* in Section 5.1, where case studies on the Spanish Mortality data (from Section 1.1) and the Drug Discovery data set are provided.

As illustrated in Section 1.1, PCA is an effective and commonly used tool for exploring modes of variation. These give insights into how data objects relate to each other, such as exploring potential clusters. Ramsay and Silverman (2002, 2005) made it clear that PCA is a powerful tool for understanding variation in FDA, i.e. curves as data object contexts. Less well known is that this insightful idea was first published in Rao (1958), in the context of analysis of growth curves. Basics of PCA are described in Chapter 17. Important ideas discussed there include the fact that the good idea of PCA has been rediscovered (and generally given different names) a number of times, and that the misconception that PCA is only useful for Gaussian data sets (because one motivation of it is via Gaussian likelihood ideas) is seriously misleading. The latter point is also clear from several of the examples given in this section.

4.1.1 Example: Tilted Parabolas FDA

Figure 4.1 shows an FDA toy example to illustrate the concept of decomposition into modes of variation, in the spirit of Figure 3.4. The $n = 50$ input raw data curves are shown in the top left panel. These are simulated to have an approximately parabolic shape, but variation of several types is included as well, hence the name *Tilted Parabolas*. Each curve is really just a parallel coordinates plot (as discussed in Section 3.1) of a collection of 10 dimensional vectors, but conceptually it makes sense to think of a bundle of smooth curves.

The object space - feature space concept illustrated in Figures 3.1–3.3 is useful here, except that explicit visualization of the feature space is not done because that space is $\mathbb{R}^{10}$ for this data. Nonetheless, it is still useful to think of statistical analysis as being done in that space on the cloud of points that represents the bundle of curves, while looking at the corresponding object space (i.e. curves) view. Colors are based on the Matlab default rotating color palette, with the same colors used in the other panels for visual correspondence.

The top center panel of Figure 4.1 shows the first natural statistical summary: the sample mean. Again, it is useful to think of that curve as the object space representation of the mean of the cloud of points in the feature space ($\mathbb{R}^{10}$). The mean curve can also be considered as the point-wise mean of the curves in the top left panel. The top right panel shows the mean residuals, which are a visualization of the curves that correspond to shifting the point cloud in $\mathbb{R}^{10}$ so that it is mean centered at the origin. These residuals already highlight an interesting aspect of the data: the parabolic shape of the curves is driven entirely by the mean, and not the variability about the mean.

The next three rows show the results of a PCA decomposition into modes of variation, of the same type shown in Figures 1.4–1.5 and 3.2–3.3.

The first mode of variation is the left plot in the second row. As discussed in Section 1.1, this is based on finding the direction in the feature space ($\mathbb{R}^{10}$) that maximizes the projected variation (in the sense illustrated in Figure 3.2), projecting each mean residual curve onto that direction, and then showing the resulting set of curves (as the projection coefficient multiplied by the direction vector). Recall that this set of curves are columns of a rank-one matrix. In particular, they are all multiples of the same curve (which is the curve representation of the direction

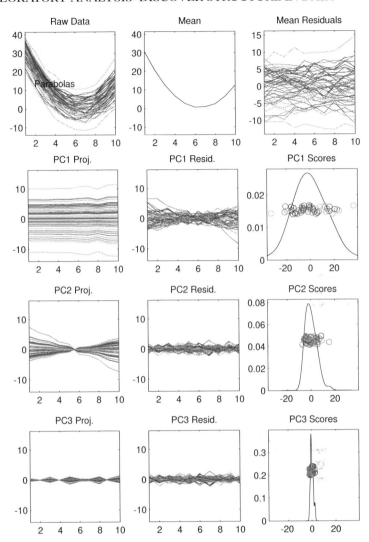

Figure 4.1 *Tilted Parabolas 10-d toy example to illustrate concept of modes of variation in FDA. Top row has the raw data curves on the left, mean curve in the center, and mean residuals on the right. Remaining rows show PC components, with mode of variation plots (projections) on the left, residuals center and distribution of scores (projection coefficients) on the right. This demonstrates insights available by decomposing data into modes of variation.*

vector in the feature space). This clearly shows that the first mode of variation is essentially a vertical shift. With this knowledge in hand, that mode can clearly be seen also in the mean residuals on the top right, as well as in the raw data on the left.

The right panel in the second row, shows the distribution of the projection co-efficients, i.e. the *scores*, again with corresponding colors (e.g. the gold followed by gold and red on the right correspond to the same colored curves on the bottom of the left-hand panel). The format of these scores distribution plots is the same as that used in Figures 1.4, 1.5, 5.1, and 5.2, where each score is represented with a symbol, and the black curve is a smooth histogram. Because there is no special ordering in this data set, the height of the points in the PC scores plots can be con-sidered to be random. Using such displays with simulated random heights is the *jitter plot* idea proposed by Tukey and Tukey (1990) as a device for visualizing one-dimensional data sets. In many displays here (also called jitter plots) order in the data is used for heights as that can reveal aspects of the data which would disappear using random heights. Specific examples of this include Section 5.1.1 and the discussion of Figure 11.3. The center panel (second row of Figure 4.1) shows the corresponding PC1 residual curves, each of which is just the centered residual minus its PC1 projection. Note that these are also the projections of the mean residuals onto the hyperplane orthogonal to the PC1 direction.

The third row shows the second mode of variation. The left-hand panel is the object space representation of the projections of the PC1 residuals (2nd row mid-dle panel) onto the 2nd PC direction in $\mathbb{R}^{10}$. Note that this also shows a very interpretable mode of variation, a random tilt. This mode is much harder to see in either the raw data curves, or the mean residuals, demonstrating the ability of PCA to find interesting modes that are not visually apparent in the raw data. The PC2 scores (i.e. the projection coefficients) in the right panel show much less variation than for PC1, and the PC2 residuals in the center panel also show relatively less variation.

The fourth row shows PC3, i.e. third mode of variation. The loadings plot in the left panel looks rather random. This is because the data were simulated as

$$(x-6)^2 + 4Z_{1,j} + 0.5Z_{2,j}(x-5) + Z_{x,j}, \qquad (4.1)$$

for $j = 1, \cdots, 50$, where $x = 0.5, 1.5, \cdots, 9.5$, and where the $Z_{x,j}$ are indepen-dent standard Gaussian random variables. Note the coefficients are deliberately chosen to make these components correctly ordered in PCs 1,2,3. Since the noise terms $(Z_{0.5,j}, \cdots, Z_{9.5,j})^t \in \mathbb{R}^{10}$ follow an isotropic Gaussian distribution, the PC3 direction is random. The relatively small scale of the noise is also clear from the tightness of the PC3 scores shown in the right panel. Also the PC3 residuals in the center panel show that PC3 explains relatively little of the variation in the PC2 residuals above, again because the noise is isotropic, and thus evenly distributed among the remaining directions in $\mathbb{R}^{10}$. Again these PC3 residuals are the PC2 residuals above minus the PC3 projections to the left.

As discussed in Section 3.1.3, a useful viewpoint on these issues comes from various sums of squares (in the spirit of Analysis of Variance). The fact that the

PC1 projections explain most of the variance is quantified by the sum of squares of the PC1 projections (left, 2nd row) representing 86% of the sum of the mean residual sum of squares. The visual impression that the PC2 projections (left, 3rd row) contain less variation is clear from the sum of squares being only 10.4%. The remaining sum of squares (i.e. summed over all remaining PC components, which is also the sum of the residuals shown center, 4th row) is only 3.6%, confirming that all of the remaining variation is quite small. The spherical nature of the remaining variation is confirmed by the PC3 variation explained being only 0.7%. Graphical comparison of these numbers is provided by the scree plot in the lower left panel of Figure 3.5.

Figure 4.1 also provides an additive decomposition of variation as highlighted in Figure 3.4. In particular, the raw data in the top left panel is the sum of the mean in the top center, the modes of variation in the remaining left panels, plus the residuals in the bottom center panel.

As discussed in detail in Section 17.1.2, the PC direction vectors used in the above data decomposition are easily computed, using either an eigen analysis of the covariance matrix, or equivalently a Singular Value Decomposition of the mean residual matrix.

A graphical point worth discussion here is the axes used in Figure 4.1. In particular (except for the first row) the vertical axes in the first two columns, as well as the horizontal axes in the third column, are deliberately taken to be the same (even across the rows). Such a view is quite nonstandard for most graphics packages, which generally adhere to the goal of trying to use as much of the graphics space as efficiently as possible, in particular minimizing *white space*. While the minimization of white space is generally a sensible default, in this context it does have an intuitive cost as demonstrated in Figure 4.2, which is a replotting of the bottom 3 rows of Figure 4.1, but this time using axes that minimize white space. The difference between these two figures is perhaps most strong in the bottom row, which is easily understood as small scale noise artifacts in Figure 4.1, but visually appear to be equal players in Figure 4.2. In particular the important decrease in variation for the higher PC components become much harder to see (only discernible by carefully studying the axis labels). The relative *shapes* (horizontal shift in the first mode, tilt in the second, noise in the third) are now highlighted, at the cost of it being harder to interpret *relative variation*. However, this effect can be easily mitigated by also including a scree plot, as introduced in Figure 3.5, which includes this data set in its lower left panel. Understanding variation using a scree plot has already been illustrated in Figures 3.2 and 3.6.

One more issue about white space, is that when trying to put a number of plots on a single page, it can make sense to also eliminate the white space *between* plots. The *trellis graphic* ideas of Becker et al. (1996) provide appropriate ways to do this. Examples appear in Figure 8.10, in the context of the DNA Molecule data, and in Figures 16.8–16.10 for the Cystic Fibrosis GWAS data.

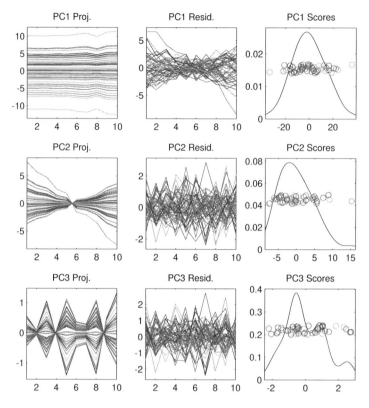

Figure 4.2 *Same FDA of Tilted Parabolas data as in Figure 4.1, showing more typical axis choice to minimize white space. Note this loses the intuitive illustration of relative amount of variation in the modes.*

4.1.2 Example: Twin Arches FDA

Another FDA toy example is the *Twin Arches* data shown in Figure 4.3, whose format is very similar to Figure 4.1. This time the $n = 50$ data curves shown in the upper left panel, are parallel coordinate plots of vectors in $\mathbb{R}^{50}$. Details of the construction are given below, but at this point consider the data in an exploratory spirit. Apparent is a somewhat higher background noise level than for the Twin Arches data, and also some strong structure in the data. The center panel shows the sample mean which this time is essentially constant, so the mean residuals in the left panel are very similar to the original data curves (again using the same axes makes this visually apparent, which would be much harder to see with white space minimizing axes, as in Figure 4.2).

Again the second row shows the PC1 mode of variation. The PC1 loadings plot on the left seems to capture the main twin arch structure, but note that some of these projection coefficients are essentially zero. The PC1 scores in the right panel make it clear that there are actually three strong clusters in the PC1 direction, i.e.

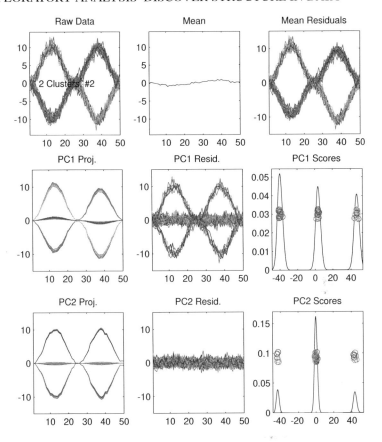

Figure 4.3 *Twin Arches toy FDA example in 50-d, illustrating ability of PCA to find insightful modes of variation. This time the mean is negligible and the strong arched structure are artifacts of clustering, i.e. non-Gaussian structure in the data.*

this is a very non-Gaussian mode of variation. Perhaps more surprising is the object space representation of the PC1 residuals in the center panel. While some of the residual curves seem to be 0 plus noise, others seem to retain the same arched structure, for reasons discussed below.

The PC2 mode of variation plot, on the left in the third row, may also at first be surprising. This is because it looks similar to the PC1 mode just above. But they cannot be similar since these direction vectors in $\mathbb{R}^{50}$ (recall each such plot consists of multiples of a single vector) must be orthogonal. A careful look at the colors reveals what is happening. Notice for PC1, the generally gold color goes up at the first arch and down at the second. Suggesting this function is roughly a sine wave. In PC2 the mostly blue color goes upwards for both arches, while the more red curves go downwards for each, which is a direction orthogonal to the PC1 eigenvector. By the way, these colors have not been deliberately assigned, but are just artifacts of the random generation of the curves, together with over-plotting effects, where the color tends to be dominated by the last plotted curves.

As suggested by the PC2 residual curves, the next components are pure Gaussian noise, with PC directions looking quite random, which thus are not shown here.

As noted in the survey paper by Febrero-Bande and Oviedo de la Fuente (2012), a number of FDA software packages aim to integrate PCA with noise reduction in a single step, as pioneered by Rice and Silverman (1991). These include perhaps most notably the FDA package accompanying Ramsay and Silverman (2002, 2005), and the PACE package started by Yao et al. (2005). While this process is critically important in many high noise cases, as well as in the case of uneven and sparse sampling (of the horizontal coordinates of the curves), in perhaps surprisingly many cases such as these two examples, it can be enough to simply do naive PCA on the data. The reason seems to be that when the true underlying direction is smooth, the noise is averaged out by PCA. In Gaydos et al. (2013) a variation of PCA, which maximizes smoothness instead of variation, is proposed and integrated with PCA in an interesting way.

An important principle of multivariate analysis is that joint distributions can contain much richer structure than is apparent from the marginals. This concept can be used for a more clear understanding of the structure of the Twin Arches toy data illustrated in Figure 4.3 by studying bivariate projections in addition to the univariate scores distributions shown in the right-hand column. Such a view is the *scatterplot matrix* shown in Figure 4.4, which is in the same format as shown in Figure 1.6 for the Spanish Mortality data. This view shows the distributions of the 1-d projection coefficients (PC scores) along the diagonal, with in particular the first two being the same as the lower two in the right column of Figure 4.3. The off-diagonal plots show corresponding two-dimensional plots. For example, the top center panel is the scatterplot of the PC1 versus PC2 scores. Note this is closely linked with the panel below (the horizontal axes are the same, so e.g. the left cluster in the PC2 scores is the same as the left cluster in the scatterplot), and with the plot to the left (where the PC1 score axis becomes the vertical axis so the left cluster in the PC1 scores is the bottom cluster in the scatterplot). This PC1 versus PC2 scatterplot gives a clear view of the underlying structure in this case, there are actually 4 clusters, which project down to 3 clusters in each of the PC1 and PC2 directions. Note that the center left plot is just the transpose of the top center plot. Since this does not convey much new information, the below diagonal plots are sometimes replaced by other graphics.

The remaining column and row all show the 3rd component. The univariate PC3 scores distributions appear to be Gaussian, which is consistent with how the data were generated. Note that a more conventional white space minimizing choice of axes is used this time, so a careful look at the axis labels is helpful to see that this mode contains much less variation than PC1 and PC2. Again the scatterplots in the right column share the same horizontal axis, so the actually spherical clusters have been strongly stretched by this axis choice. Similarly, for the scatterplots in the bottom row, which are just transposes.

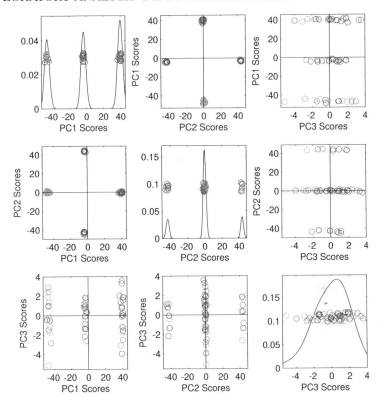

Figure 4.4 *Scatterplot matrix view of the Twin Arches data from Figure 4.3. Shows clusters apparent in 1-d marginal scores distribution plots come from marginalizing 4 actual clusters.*

Scatterplot matrix views tend to be very insightful, and are recommended for most situations where relationships between data objects are relevant. A natural question is: if 2-d projections show more than 1-d projections, why not also consider higher dimensional projections? While potential improvements appear to be obvious, and implementation is fairly straightforward for 3-d, it does come with substantial overhead, such as the need for dynamic graphics. These take substantial energy to both implement and to visually explore, which can be a large drawback for routine data analysis tasks. For projections of dimension higher than 3, visualization becomes much more challenging, and is thus not frequently done.

4.1.3 Case Study: Lung Cancer Data

While toy examples, such as those in Figures 4.1–4.4 can give many insights, it is also important to consider real data sets. An interesting example is the set of curves (called the *Lung Cancer RNAseq* data) shown in Figure 4.5. These come from a study of lung cancer, and in particular was an early data set collected as

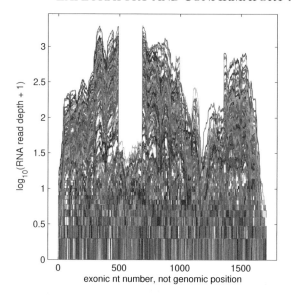

Figure 4.5 *Raw* log_{10} *count curves of Lung Cancer RNAseq data. Colors are standard rotating palette used for good contrast of curves. It is not easy to discern population structure.*

part of The Cancer Genome Atlas (TCGA), Weinstein et al. (2013). The focus here is on the gene CDKN2A, which has long been known to be involved in many types of cancer. The horizontal axis represents the region on the chromosome that is used to produce the RNA measured here (using the RNAseq technology described in Wang et al. (2009)). For each of the $d = 1709$ locations, the vertical axis shows the counts (on the $\log_{10}(\cdot + 1)$ scale) of pieces of amplified RNA molecules that match the chromosome at this location. There are $n = 180$ such curves. The $\log_{10}$ scale is useful, and is the data object representation choice used in the rest of this section, since these counts range over 3 orders of magnitude. An artifact of this $\log_{10}$ scaling for this data set is that very small counts, such as 1 and 2 occupy a large chunk of the bottom of the plot, since $\log_{10}(1+1) \approx 0.301$ and $\log_{10}(2+1) \approx 0.477$. The same rotating palette of seven colors used in Figures 4.1–4.4 is used here as well. The curves seem somewhat chunky in nature, in particular being substantially lower over some intervals, because these coding regions do not appear in a contiguous region on the chromosome, but instead are separated into intervals called *exons*. The union of these exonic regions are used as the horizontal axis in Figure 4.5.

While there is a lot of variation in these curves, it is hard to discern much structure, although they vary over several orders of magnitude. As illustrated above a PCA scores scatterplot, shown for the Twin Arches data in Figure 4.6, is useful for understanding relationships between data objects. Even the 1-d scores distributions on the diagonal already show multi-modal structure, which is quite

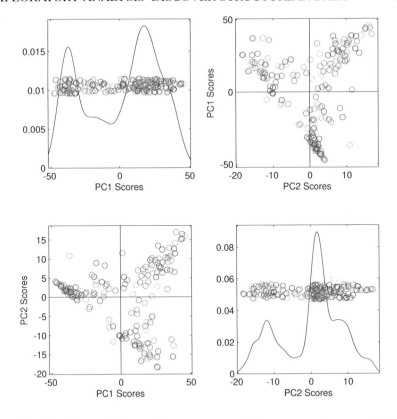

Figure 4.6 *Lung Cancer RNAseq PCA scatterplot matrix. Colored circles represent feature space view of corresponding curves in Figure 4.5. Shows three clear clusters in the data.*

apparent in the black smooth histograms. But, as illustrated in Figure 4.4, the 2-d off-diagonal scatterplots do a much better job of highlighting the multi-modal structure of the data, where three clusters are immediately apparent. Statistical significance of these clusters is formally explored in Section 13.2.1. Note that in the spirit of the phenomenon illustrated in Figure 4.4 the top two clusters are combined in the PC1 scores, and the right two clusters combine in the PC2 scores. Only the first 2 components are shown in Figure 4.6, because the 3rd and 4th components are driven by a few outlying cases, which are not further studied here. In particular, the first two components explain 82% and 8%, respectively, while the 3rd and 4th explain 3% and 2%. Another interesting pattern is the points in the second through fourth quadrants appear to be bounded below by a line. An open question is what generates this apparent structure in the data.

Insight into the drivers of these clusters comes from a technique called *brushing* in Becker and Cleveland (1987). The idea is to use colors to keep track of subsets of the data in multiple graphics. This is illustrated in Figure 4.7, which shows the same distribution of points as in Figure 4.6, with colors that have now been

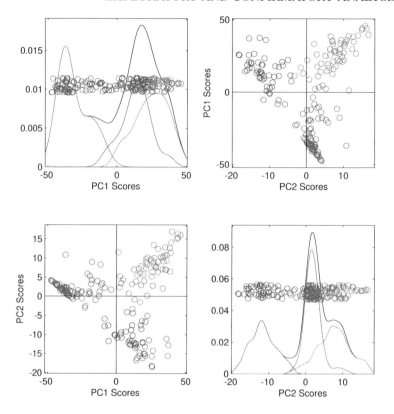

Figure 4.7 *Brushed PCA scatterplot matrix of Lung Cancer RNAseq data, using colors to highlight the three clusters. Also shows sub-density estimates, for visual separation, in the 1-d distributions on the diagonal.*

manually chosen to highlight the three clusters. Automatic versions of clustering can also be done using a variety of methods as discussed in Chapter 12.

Note that the visual impression of these clusters in the 1-d distributions shown on the diagonal is now enhanced with appropriately colored versions of the smooth histogram which focus on each cluster. These are *sub-densities* in the sense that the area under the main black curve is 1, and the areas under each colored curve, which are proportional to the cluster sizes, sum to 1. Note that in regions where one cluster is dominant, the colored sub-density is the same as the black overall density, and thus overplots it (e.g. red on the left side of the PC1 scores and blue on the left side of the PC2 scores). In other regions, the relative curve heights give a clear visual impression of the corresponding cluster proportions.

This brushing technique is often especially insightful when used across a variety of graphical displays. An example of this appears in Figure 4.8. These are the same curves shown in Figure 4.5, but now the color scheme developed in Figure 4.7 is used. Note that the red curves are all substantially lower than the others.

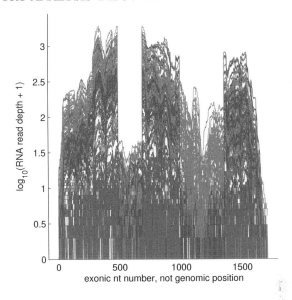

Figure 4.8 *Same Lung Cancer RNAseq curves as in Figure 4.5, now using brushed colors from Figure 4.7. Shows clusters are very important, in particular representing alternate splicing.*

In the early days of gene expression these cases would have been labeled as *unexpressed* (which actually means expression at a much lower level, recall the log scale on the vertical axis). Note that for most exons the blue and brown curves both have high expression values. There is an exon to the left of center where all cases seem unexpressed, which is reasonably labeled an annotation error (an on going issue with such biological data sets). The most interesting issue is the exon right of center, where the brown cases are high, but the blue cases are essentially unexpressed. This is an event called *alternate splicing* where actually different versions of mRNA are produced from this chromosome region by different people. This is very important to the development of new treatments for cancer, because such phenomena can be targeted by appropriate drugs.

While the alternate splicing present in this gene CDKN2A has been well known for some time, the success in finding it with this type of visualization motivated Kimes et al. (2014) to use this type of idea to scan the whole genome in search of unknown alternate splices. A key challenge was that, as noted in Chapter 12, most automatic clustering methods always find many clusters, whether they represent important biological structure as in Figure 4.8, or not. Hence confirmatory analysis, as discussed in Section 4.2 and Chapter 13 is essential, and was key to the SigFuge method developed in Kimes et al. (2014).

Another important aspect of data representation is *scale* and *normalization* issues. These important parts of OODA preprocessing are discussed in detail in Section 5.2. As seen in several examples above PCA can be a powerful visualization device for finding interesting structure in data. But because PCA is driven by finding directions of maximal variation, it can lose effectiveness in situations where differing variables (i.e. features or traits) have different scalings. In particular, PCA will tend to be driven by those variables with the most variation, while ignoring those with smaller-scale variation. This challenge can be particularly acute in situations where different variables even measure non-commensurate quantities, such as having different units. Approaches to standardization are discussed in Section 5.2.

Another useful response to the tendency of PCA to focus on variation is to replace PCA directions in high-dimensional space with directions aimed at other aspects of the data. The case of directions highlighting differences between subgroups is studied in Section 4.1.4.

4.1.4 Case Study: Pan-Cancer Data

Figure 4.9 shows a PCA scatterplot view of part of the *Pan-Cancer* data set, from Hoadley et al. (2014), who explored many contrasts between 12 cancer types, based on a variety of measurements. That data set was another product of TCGA discussed in Section 4.1.3. The raw data are counts indicating expression of $d = 12478$ genes, again measured using RNAseq. Substantial preprocessing, including log transformation as discussed in Section 5.3 have been done by Hoadley et al. (2014). Studied here is a randomly selected subset of $n = 300$ cases (this number gives clear visualization of the main point about the limitations of PCA) with 50 from each of the cancer types Bladder Cancer (BLCA, magenta), Kidney Renal Cancer (KIRC, blue), Ovarian Cancer (OV, cyan), Head and Neck Squamous Cell Cancer (HNSC, green), Colon Adenocarcinoma (COAD, yellow), and Breast Cancer (BRCA, red). While each of the six cancer types can be clearly seen, there is substantial overlap of the classes in this view. This is because the PCA directions only maximize variance, and they ignore class labels.

Figure 4.10 shows an alternate scatterplot view of the same Pan-Cancer gene expression data shown in Figure 4.9. The symbols and colors are the same, but instead of using PC directions for the axes, the directions used in the projections are designed to deliberately separate pairs of cancer types. Each direction is based on the DWD (Distance Weighted Discrimination) method, trained on pairs of cancer types. DWD was also used in Section 2.4 to visually separate the male and female faces, and is discussed in more detail in Section 11.4. The projection direction used in the first row and column is DWD trained on only the Kidney (blue) versus the Head and Neck (green) cancer types. The projections of the full data set onto that direction (although DWD was trained on just those two) are shown in the upper left and on the same horizontal axis in the other first column plots, as well as the vertical axis of the other plots in the top row. Note that both cancer types stand out as distinct clusters in these views, so DWD has succeeded in separating out the expected biological differences.

Similar excellent separation happens for Colon Cancer (yellow) in the second DWD direction in Figure 4.10, although the Bladder Cancer (magenta) stands out

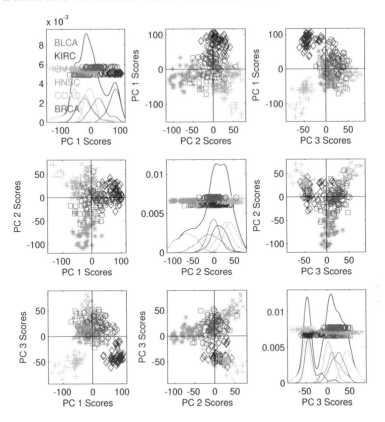

Figure 4.9 *PCA scatterplot matrix view of the Pan-Cancer gene expression data, symbols are tissue samples, colored according to cancer type. Note substantial overlap of several cancer types.*

less well. In the third DWD direction the Ovarian Cancer (cyan) stands out clearly, while Breast Cancer does not. The latter is not surprising because Breast Cancer is well known, see e.g. Perou et al. (2000), to have several subtypes which are quite distinct from each other. Presumably each of these would be clearly different from the others, but because of their diversity the union fails to be very distinct in this sense.

In earlier scatterplot matrix views the axes were orthogonal, as that is a consequence of PCA. However, there is no guarantee of orthogonality otherwise, as can be seen in several off-diagonal panels of Figure 4.10. These views should be considered as projections onto the two-dimensional subspaces generated by the pair of non-orthogonal directions. In each case the horizontal axis shows that direction, and the vertical axis is just its orthogonal complement (in the 2-d subspace). The other line shows the second direction in that subspace. For example, in the first row second column, the nearly vertical line is approximately a transport of

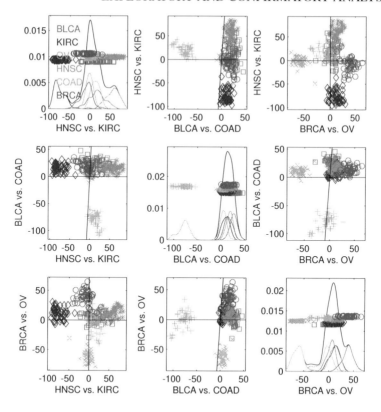

Figure 4.10 *Same Pan-Cancer data from Figure 4.9, with PC directions replaced by directions deliberately aimed at separation of types (highlighted with colors). Shows much better distinction of cancer types, demonstrating that PCA directions may not find all interesting structure in data.*

the line between the means of the dark blue (KIRC) and green (HNSC) clusters. More detail of this type of visualization appears around Figure 6.14.

One might wonder how these particular pairings of cancer types (on which the DWD directions were trained) were chosen. This was done by considering all pairings and deliberately choosing a set of three on the basis of good visual distinction of types. But the main point of these figures is that there can be visualization directions giving much different visual insights than those available from PCA.

Another striking example of PCA providing not the best separation of cancer classes can be found in Liu et al. (2009).

The methods and examples studied in this section provide a somewhat nonstandard way of thinking about high-dimensional data. The currently fashionable notion in much of statistics is that when faced with high-dimensional data, one must use approaches such as *sparsity*, i.e. treating most variables as negligible, to reduce the data to a "manageable dimensionality". While sparsity is a useful approach in some cases and has been tackled effectively using a very large range of methods starting with the LASSO approach of Tibshirani (1996), there seem to be many more OODA contexts where the fundamental sparsity assumptions are far from being reasonably well satisfied. These include almost all of the examples discussed in Chapter 2, and also the rich genetic data discussed in Figures 4.5–4.8. Yet sparsity ideas seem to be currently both over used and over studied in the statistics literature, perhaps because most statisticians tend to think about high-dimensional data in a too *variable centric* way. The OODA viewpoint demonstrated in this section allows taking a more *object, centric* approach, where the primary focus is more usefully placed on the data objects and the relationships between them, not the variables. Of course variables are important, but they should be playing the role of representers of the objects, as opposed to being the focus of the analysis.

However variables often contain useful insights about the *drivers* of the relationships displayed in the above object-oriented views. This is often usefully done through visualization of loadings, i.e. the entries of the direction vectors of each mode of variation. One type of loading visualization was done in the left panels of Figures 1.4 and 1.5, where the curves are multiples of the direction vectors. That curve visualization is less useful in other settings, such as the gene expression data shown in Figures 4.9 and 4.10, where there is no natural insightful ordering of the variables (critical to the display of curves) and often simply too many variables ($d = 12478$ in the present case) for a useful curve display. Some solutions to this are demonstrated in Figure 4.11. The top panel shows indices (horizontal axis) of the sorted loadings (vertical axis) from the DWD direction vectors trained to separate the KIdney Renal clear cell Carcinoma (KIRC) patients from those with Head and Neck Squamous cell Carcinoma (HNSC). Note there are just a few very large both positive and negative values. Those strong loadings are studied in the bottom panel, with bars showing the top twenty (in absolute value) variables that are labeled with gene names. Validation of the relevance of these genes comes from the National Center for Biotechnology Information (2019) website. In particular, the most strongly expressed (in the HNSC direction) gene LYPD3 is commonly expressed in esophagus and skin, while the genes that are strongest in the KIRC direction, CUBN and RBP5, are involved in activity in the intestine and kidney, and kidney and liver, respectively.

Further discussion of various roles played by loadings, especially for PCA, can be found in Section 17.1.2.

4.2 Confirmatory Analysis–Is It Really There?

The visualization methods discussed in Section 4.1 are very good at providing useful insights and at finding population-level structure in data. However an important aspect to keep in mind is that they also have the potential for finding useless artifacts of sampling variation.

This point is illustrated using the Overlapping Classes simulated data sets in Figure 4.12. Here two classes of data were generated in $d = 1000$ dimensions, with $n_1 = n_2 = 50$ data points in each class. It is hard to see much difference

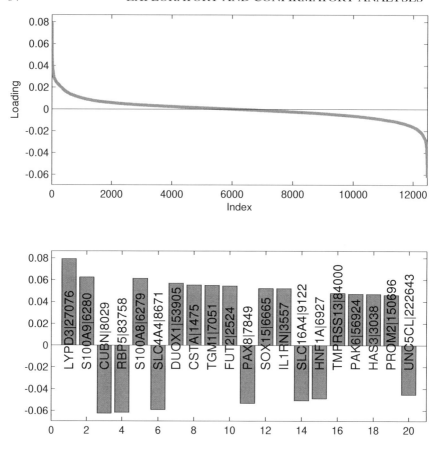

Figure 4.11 *Loadings of first DWD direction (KIRC vs. HNSC) of Pan-Cancer data in Figure 4.10. Top panel shows sorted loadings for all $d = 12478$ genes, bottom panel zooms in on top 20 (absolute value) gene names.*

between the red class (shown as circles) and the blue class (shown as plus signs) in the PCA scatterplot view shown in Figure 4.12. However, an important lesson from the Section 4.1 is that for high-dimensional data, PCA may not find all interesting aspects of the overall distribution because it focuses only on variation in the data.

Figure 4.13 more deliberately targets the class difference between the red plus signs and the blue circles, using the DWD direction which was previously used in Figures 2.9 and 4.10. Projections on this DWD direction appear in the upper left panel. Note that this shows a very clear and distinct separation of the two classes which is visually comparable to some of those seen for the Pan Can data in Figure 4.10. The other two directions are orthogonal PC directions, which are computed using PCA based on the projection of the data onto the $d = 999$ dimensional subspace orthogonal to the DWD direction. This is useful because it

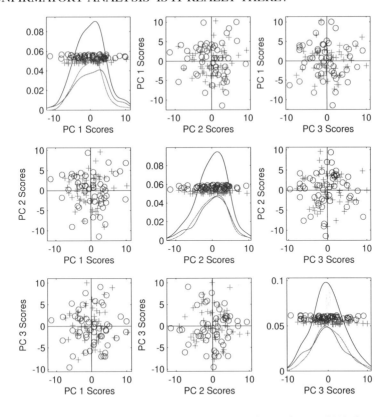

Figure 4.12 *PCA scatterplot view of the Overlapping Classes data in 1000 dimensions. Shows no apparent difference between classes.*

allows the directions used in the scatterplot matrix to be orthogonal, which generally makes the view more interpretable. These issues are discussed in more detail in Chapter 6.

While the DWD separation looks very seductive, it is important to keep in mind that DWD is very efficient at finding directions which separate groups of data in high dimensions. But there is another side to this, which is that DWD can be in some sense too good. That is an issue in the Overlapping Classes example, because both the red and blue classes were simulated from the $d = 1000$ dimensional standard normal distribution, $N_d\left(\mathbf{0}_{d,1}, \boldsymbol{I}_d\right)$, where

$$\mathbf{0}_{d,n} = \begin{bmatrix} 0 & \cdots & 0 \\ \vdots & \ddots & \vdots \\ 0 & \cdots & 0 \end{bmatrix}, \tag{4.2}$$

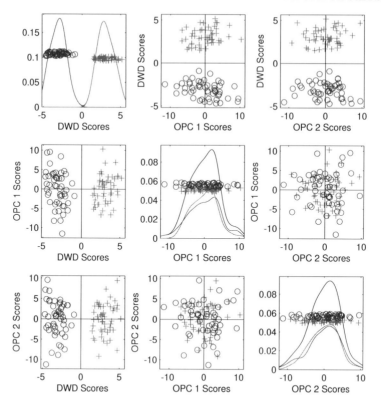

Figure 4.13 *DWD and Orthogonal PC view of the Overlapping Classes data from Figure 4.12. Now shows strong inter-class difference.*

denotes the $d \times n$ matrix of zeros, and where

$$
\boldsymbol{I}_d = \begin{bmatrix} 1 & 0 & \cdots & 0 \\ 0 & 1 & \ddots & \vdots \\ \vdots & \ddots & \ddots & 0 \\ 0 & \cdots & 0 & 1 \end{bmatrix}
\tag{4.3}
$$

is the $d \times d$ identity matrix. This highlights a fact which is discussed in detail in Chapter 14: high-dimensional data can frequently exhibit perhaps surprising behavior. Actually, the specific behavior observed here, and even the amount of visual separation apparent in the DWD direction will be directly explained and even predicted using considerations developed in Section 14.2. Another perhaps surprising aspect of Figure 4.13 is the large standard deviations of the PC scores, clearly far larger than the standard normal value of 1. This will also be explicitly studied and quantified in Section 14.1. The fact that this apparent difference between the red and blue classes is spurious is confirmed in Section 13.1, where it is seen that the DiProPerm hypothesis test (described below) for the difference of the means between these two data sets gives a quite non-significant p-value of 0.82.

Figure 4.13 makes a very important point about data visualization in general. While it can be very useful at finding important population structure in data, it is also quite capable of finding things which are just natural artifacts of the sampling variation (which can appear in unexpected ways). For this reason it is critical to combine any exploratory visual analysis with *confirmatory analysis*, as introduced here and studied more deeply in Chapter 13.

In the more complicated areas of OODA, e.g. many of those illustrated in Chapter 2, confirmatory analysis is still in a relative state of infancy (compared to other parts of statistics). One reason for this is that in some of those areas, such as tree structured or manifold data objects, it can be quite challenging to develop appropriate null probability distributions, which underlie much of classical statistical inference. This has motivated permutation and bootstrap solutions, although careful investigation of their properties remains as a wide-open research area in mathematical statistics.

Some existing confirmatory analysis methods for OODA are discussed in Chapter 13. Section 13.1 discusses a generally useful high-dimensional permutation type of test, called *DiProPerm*. The key steps are:

- Find a DIrection in the data space, such as the DWD direction used in Figures 2.9, 4.10, and 4.13, although any other systematic direction can be used as well.

- PROject the data onto that direction to focus on representative univariate components, e.g. the numbers whose distribution is shown in the top left panel of Figure 4.13. Then summarize the projections with an appropriate summary statistic. A natural choice might be the 2 sample t-statistic. However a surprising result of the careful mathematical analysis of Wei et al. (2016) is that the simple difference of sample means provides a more stable hypothesis test in high dimensions.

- PERMute the data to assess statistical significance. In particular randomly re-assign the group labels (e.g. red and blue for the data in Figures 4.12 and 4.13), recompute the separating direction, the projections and the summary statistic, to generate one element of a simulated null distribution. Repetition generates a simulated null population and comparison with the original summary statistic provides statistical inference such as p-values.

In Section 13.1 it is seen that while it may not always be the most powerful mean hypothesis test, the DiProPerm test is generally useful because it provides direct confirmation (or not) of visually observed effects, such as the difference between the red and blue groups in Figure 4.13. Further examples exploring these issues, together with real data examples highlighting the importance of this type of confirmatory analysis appear in Section 13.1.

While the DiProPerm test provides a very useful reality check for confirming visualized differences between previously defined groups, care must be taken in the comparison of groups discovered say by *clustering*. The useful operation of clustering can be done in an informal visual way, as for the Lung Cancer RNAseq data in Figure 4.7. It can also be carried out in many more mathematically motivated ways, as discussed in Chapter 12. It is seen in Section 13.2 that application of DiProPerm in clustering contexts can be seriously misleading. Yet the method of clustering has led to many important discoveries in data, so it will continue to be an important tool. In parallel to the challenge of spurious visualization illustrated in Figure 4.13, is the question of "which clusters are really there?" as opposed to being spurious artifacts of the sampling variation. An answer to this question, which becomes particularly challenging in the high-dimensional case is the SigClust approach motivated in Section 13.2.

There is one more important aspect of confirmatory analysis in OODA. This is that carefully working from the OODA viewpoint can yield much more powerful and insightful analyses than are available from naive implementation of classical methods. An example of this is the study of osteoarthritis and its impact on knee shape carried out in An et al. (2016). The shape data objects were represented by a set of 60 two-dimensional landmarks, collected from standard x-ray images, analyzed using Procrustes methods as discussed in Section 8.4 and in Dryden and Mardia (2016). Earlier work in this area, such as Gregory et al. (2007) and Nelson et al. (2014), used PCA to summarize the population structure and then did 2 sample t-tests on the resulting sets of scores.

There are 2 ways in which OODA offers improvement over this approach. First is the concept, illustrated in Figures 4.9 and 4.10, that important information in terms of class differences may not show up strongly in any chosen low-rank PCA direction. The second is that the multiple testing requires some type of adjustment, such as a Bonferroni correction or False Discovery Rate calculation, which entails additional loss of power. This issue was shown to be serious in the relatively small scale ($n = 65$) study of An et al. (2016), where an exploration of knee shape in the development versus non-development of osteoarthritis in African American females resulted in a DiProPerm p-value of 0.033, while the first 5 PC based p-values are reported in the first row of Table 4.1. In this case PC3 happens to give a significant result at the 5% level.

However, since multiple tests are considered some adjustment is needed. The second row shows a Bonferroni adjustment based upon the 5 components shown, and the third component no longer gives a significant result. Parallel conclusions follow using the more sophisticated False Discovery Rate adjustment, Benjamini and Hochberg (1995), in the bottom row. Of course the number 5 (of components) is arbitrary, but consideration of more components will result in a less significant result. Also, the separation in the 3rd component of these data may be something of a fluke, as demonstrated by some of the strong cancer type separations shown in Figure 4.10 being clearly not available in the PCA of the same data shown in Figure 4.9. The main lesson here is that thinking in an object-oriented way, i.e. developing the inference in terms of the data objects, instead of doing a naive piecemeal inference on the components (that had been standard in this area), results in a more powerful statistical inference. See Nelson et al. (2017) for related applications.

PC no.	1	2	3	4	5
p-val	0.223	0.505	0.013	0.350	0.204
Bonferroni p-val	1.000	1.000	0.065	1.000	1.000
FDR p-val	0.372	0.505	0.065	0.437	0.372

Table 4.1 *Naive and Bonferroni adjusted p-values for the osteoarthritis data. Shows adjusted results are not significant in contrast to the significant p-value of 0.033 from DiProPerm.*

4.3 Further Major Statistical Tasks

While data visualization, as illustrated in Section 4.1, and confirmatory analysis as discussed in Section 4.2 are important components of OODA, there are also a number of important analytic methods that are used as well. These include:

- *Distance-based analysis.* A number of OODA situations involve data objects which lie in spaces where statistical analysis can be challenging. A straightforward general strategy is to first find a metric on the space and then to compute the matrix of pairwise distances. Chapter 7 discusses various methods for data analysis whose input is only a matrix of distances between data objects. Perhaps most important among these is an analog of PCA called Multi-Dimensional Scaling, studied in Section 7.2. A crucial issue in metric-based analysis is choice of metric, which is essentially a data object representation issue, that is explored in Section 7.3 and other places.

- *Statistics on manifolds.* Chapter 8 discusses data objects lying in *manifolds*, essentially smooth curved surfaces. Relatively simple examples of data objects that are usefully thought of as lying on a manifold include *directional* data, where angles (e.g. wind or magnetic field directions) are the data objects, see Mardia and Jupp (2000) and Fisher et al. (1993); Fisher (1993). More complicated manifold data objects arise in the study of shape, for example the various types of shape data object representations discussed in Section 1.2. Statistical analysis of data objects lying on a manifold remains a controversial topic, as there are a number of ways to approach it, with no clear consensus on issues even as to how population centers should be computed.

- *Tree-structured data objects.* Even more challenging than manifold data are data objects having a tree structure, in the sense of mathematical *graph theory*. This area is studied in Chapter 10, motivated by a data set where each data object is a representation of the arteries in a human brain that was introduced in Section 2.2. As for manifold data, a number of different analytic methods have been proposed, and it is even less clear which approaches are most natural. A perhaps exotic, but quite successful approach has been *Topological Data Analysis*, done in Bendich et al. (2016).

- *Classification* (also sometimes called *discrimination* or *pattern recognition*).

This is a large field and in fact has become a very important component of the field called *machine learning*. A good overview is available in Duda et al. (2001) and Hastie et al. (2009). This area is reviewed briefly in Chapter 11.

- *Clustering*. This is another very large field, with again just some discussion in Chapter 12. A classic reference in this field is Hartigan (1975). Another good source is Kaufman and Rousseeuw (2009). In machine learning clustering is often called *unsupervised learning*, to provide useful contrast with classification being called *supervised learning*, since the goals are related, although in the latter class labels are given, while in the former they are derived from the data.

- *Statistical smoothing*. This is one more field with a large literature and many proposed approaches, often with substantial controversy, as reviewed in Chapter 15. It includes *density estimation*, essentially a smoothed version of histograms, and *nonparametric regression* which is essentially scatterplot smoothing. While smoothing methods are commonly used in exploratory data analysis, less well known is the confirmatory method SiZer, proposed by Chaudhuri and Marron (1999).

- *Robust Methods*. Once again this is a very widely studied area of statistics. The main idea is to develop statistical methodologies which focus on methods with reduced sensitivity to violation of assumptions. Much of that effort has gone toward dealing with outliers, which can be very important in OODA, as discussed in Chapter 16. Major references in this area include Huber and Ronchetti (2009), Hampel et al. (2011), and Staudte and Sheather (1990). See Clarke (2018) for a more recent overview of this area.

- *Data Integration*. This relatively new statistical area is driven by the desire in many research areas to make multiple types of measurements and to integrate those in a meaningful way in statistical analyses. In OODA terms, the data objects are typically multiple vectors, which could be merely concatenated into a single vector, but there is often interest in understanding how these relate to each other. This is commonly done using regression methods, which makes sense when the goal is prediction, but not when the goal is a non-directional understanding of the relationship. The latter is accomplished by methods such as Canonical Correlation Analysis, Partial Least Squares, and the more general JIVE approach discussed in Sections 17.2.1–17.2.3.

CHAPTER 5

OODA Preprocessing

An acronym going back at least to the early days of computer programming was GIGO for "Garbage In–Garbage Out". That principle certainly applies to modern data analysis, yet seems to be all too frequently ignored. This chapter on OODA preprocessing describes some useful ways for understanding sometimes hidden data problems and some remedies that scale in a reasonable way to larger data sets, even those with many variables (i.e. traits). Section 5.1 gives examples demonstrating the importance of a careful study of marginal distributions and how they can be used to guide data object choice. The often useful approach of normalization (usually shifting and scaling of variables, but with some perhaps nonobvious variations) is discussed in Section 5.2. Another data representation point is transformation of variables which is considered in Section 5.3. Finally Section 5.4 studies registration, which is one more data object representation issue that is relevant to image and shape analysis, as well as to phase variation in Functional Data Analysis.

A general term that encompasses all of these issues is *data provenance*, which includes information about the sources and processes that lead to the creation and representation of data, Glavic and Dittrich (2007).

5.1 Visualization of Marginal Distributions

As noted in Section 4.1 an important OODA pre-processing step, which can often help to avoid unpleasant surprises of many types is visualization of marginal distributions. This section recommends use of *marginal distribution plots* to address the challenge of doing this for a large number of variables (i.e. traits or features) in a given data set. The key idea is to select a *representative* subset of the variables to actually look at. The idea of sorting on a one-dimensional summary statistic (e.g. the mean of each variable as in Figures 5.1 and 5.2), is essentially that of Tukey's *scagnostics*, see Wilkinson et al. (2005); Wilkinson and Wills (2008) for good overview and discussion. The difference is that scagnostics use numerical summaries (e.g. correlation) to find interesting scatterplots (two-dimensional views) from a large collection, while in contrast these marginal distribution plots similarly use summaries to understand a large collection of one-dimensional marginal distributions.

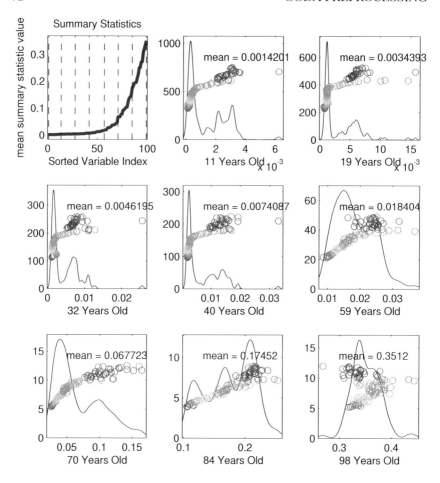

Figure 5.1 *Marginal distribution plot of the Spanish Mortality data from Section 1.1. Variable (trait) means are the blue curve in the upper left panel. Marginal distributions of a representative (equally spaced) set of variables, indicated as vertical dashed lines, appear in the remaining panels. As in Figure 1.2 colors indicate years using a rainbow from 1908 (magenta) to 2002 (red). Shows large variation in variable scaling, many variables have strong skewness and presence of outliers.*

5.1.1 Case Study: Spanish Mortality Data

An example of a marginal distribution plot is shown in Figure 5.1, for the Spanish Mortality data studied in Section 1.1. This is based on the same data matrix that was used in the left panel of Figure 1.1. Recall that the columns of that matrix (the data objects studied there) were indexed by years. The rows of that matrix are viewed as variables (i.e. features or traits) and correspond to ages. The upper left panel in the marginal distribution plot shows the mean mortality of ages, sorted into increasing order. Note that the first half of these averages all appear to be quite small, with much larger values appearing among the second half. This is consistent with the visual impression from the left half of Figure 1.1 that around half of the ages have mortality orders of magnitude smaller than the rest. This set of sorted variable means is also the key to finding a *representative* set of variables (ages in this case) to actually visualize. One notion of representative is to look at an equally spaced subset (among the sorted mean ages), as indicated by the vertical dashed lines. The remaining panels show the 8 marginal distributions of ages corresponding to those 8 lines, using the same format as in the right panels of Figures 1.4 and 1.5. In particular, the circles correspond to the years (i.e. the data objects, colored using the same year rainbow pattern from Figure 1.2) using mortality as the horizontal coordinate, with the vertical coordinate (and color, magenta 1908–red 2002) indicating order in the data set (thus the year). An important point is that this use of order in the data set as vertical coordinates reveals far more structure than would be available from the random ordering of the original jitter plots as discussed in Section 4.1. The black curve is a smooth histogram, i.e. kernel density estimate, as discussed in Chapter 15.

Note that the first two shown ages, 11 and 19, all have very small mortalities on the order of 10^{-3}. The ages in the middle row, 32, 40, and 59, have medium mortalities on the order of 10^{-2}. On the bottom row, all mortalities are larger. An important issue is that data sets having variables with such diverse scales can be problematic for many forms of statistical analysis. This motivates using one of a number of approaches to data adjustment, discussed in detail in this chapter.

These marginal distribution plots show additional challenges to classical analysis methods, such as skewness appearing in most plots, and also the presence of one or more outliers, usually the violet pandemic year 1918 discussed above. These challenges can also be addressed using methods discussed in this chapter on preprocessing.

Given the above described variation across orders of magnitude, log transformation is a natural type of data adjustment to consider for the present data set. Figure 5.2 shows the variable mean sorted marginal distribution plots for the log-transformed mortalities. While there is still natural variation in the means, it no longer spans over several orders of magnitude. This is the reason that the visual impression of variation in the right panel of Figure 1.1 is much more insightful than in the left panel. An added benefit of this transformation is that the skewed distributions above are now transformed into mostly bimodal distributions, which is again very consistent with the fairly rapid overall improvement in mortality, observed in the discussion of Figure 1.4. Note that the impact of the outlying year 1918 is also substantially diminished.

Finding a representative set of variables (ages) by sorting on the variable means was very effective for understanding critical aspects of this mortality data set, as shown in Figures 5.1 and 5.2. As demonstrated in the following chemometric example, other summary statistics can highlight different, and very insightful, notions of representative variables as well. An issue is the number 8, of representative variables shown in Figures 5.1 and 5.2. This was chosen purely for graphical convenience, in the present format. In other situations $15 = (4 \times 4) - 1$ allows

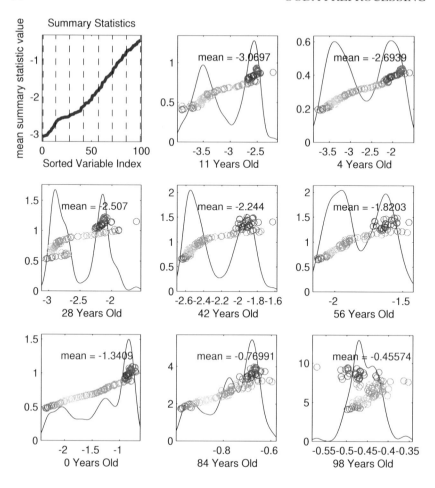

Figure 5.2 *Marginal distributions similar to those shown in Figure 5.1, but for* $\log_{10}$ *Spanish Mortalities. Shows strong beneficial effects of log transformation.*

simultaneous viewing of more representatives. A much larger number results in each marginal distribution being too small for easy viewing.

5.1.2 Case Study: Drug Discovery Data

The next example comes from the area of drug discovery, or more precisely Quantitative Structure-Activity Relationships as discussed in Cherkasov et al. (2014). This particular *Drug Discovery* data set is from Borysov et al. (2016). There are $n = 262$ chemical compounds, that are represented by $d = 2489$ chemical features (i.e. variables or traits). The primary goal is to distinguish *inactive* compounds shown as blue circles, from *active* ones shown as red plus signs in the coming graphics.

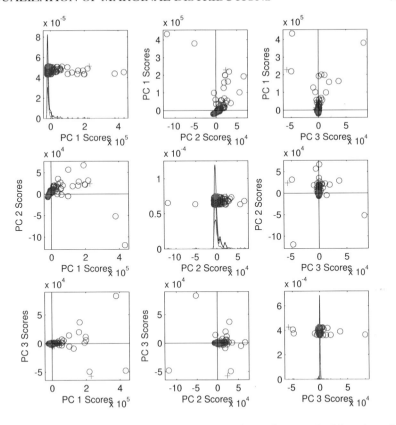

Figure 5.3 *PCA scatterplot of raw Drug Discovery data. This view is driven by a few outliers and gives very poor separation of the active (red) and inactive (blue) compounds.*

Visualizing the data as curves, e.g. for the Spanish mortality data in Figures 1.1 and 1.2, is not insightful as there is no natural ordering of the variables. A PCA scatterplot matrix, using the same format as Figures 4.4, 4.6, 4.7, and 4.9, is shown in Figure 5.3. This view of the data is dominated by relatively few of the data objects. Almost all of the $n = 262$ data points are tightly clustered near the origin, which seems to be where any meaningful differences between the actives and inactives may be found. However, as indicated in Figures 4.9 and 4.10, there can be a large amount of interesting structure in data which is not apparent from merely looking at PC scores. There are many potential causes of such behavior. One of these, that is frequently worth checking, is the behavior of the marginal distributions. For example highly skewed marginal distributions (such as the log normal) can frequently generate such data views.

A number of marginal distribution analyses (based on various sorts of trait summary statistics) for this Drug Discovery data will now be presented to demonstrate the usefulness of that approach. As in Figure 5.1, a reasonable starting point is based on a sort of the sample means, with visualization of an equally spaced set of distributions. The upper left panel shows the sorted variable means as a blue curve. Note that most of the means appear to be around 0 with a few relatively huge values on the far right. Because PCA finds directions of maximal variation, these very few variables are potential drivers of the unfortunate population structure observed in Figure 5.3, and there may yet be useful population structure that will emerge when those variables are properly handled. Note also the rather small downturn in the blue variable mean curve on the far left.

As above, eight representative marginal distributions (that number again chosen merely for convenience of plotting) corresponding to the vertical dashed lines are shown in the remaining panels. These show huge heterogeneity in the variables present in this data set. The first few show no variation at all, i.e. all values are exactly the same. In the first marginal distribution, they are all equal to -999 (this perhaps surprising value is explained below). For the next three variables, all values are 0. The center-right variable is all 0's except for a single 1. The variables on the bottom row are also wildly different from each other, with a discrete distribution on the left, and a clearly skewed distribution, with values that are four orders of magnitude larger, on the right.

That value of -999 is sometimes used to code missing values (in fact this is the case here), perhaps with the idea that it is so different from all the others that it would be easily noticed and properly dealt with during an analysis. However, that idea failed in this data set, because there are some variables that are so much bigger in magnitude (which may have been added to this combined data set by a different analyst). In particular, because -999 is a number, it would be easy to make the big mistake of treating them as meaningful data. This type of effect easily arises in Big Data contexts where there is a lot of merging of diverse data sets in contexts for which no individual has a complete understanding of all aspects. This marginal distribution type of visualization is often effective in discovering such anomalies.

Figure 5.4 makes it clear that a number of variables with no variation (and thus no information about active vs. inactive compounds) can be deleted from the data set with no loss of information. It also indicates that careful attention should be paid to the missing values, coded as -999, and finally that both the relative magnitude and skewness of other variables will need careful consideration.

Figure 5.5 shows the marginal distributions this time sorted on the variables sample *standard deviation* (SD). This summary statistic gives a more clear focus on those variables with no variation, revealing that they are nearly half of the $d = 2489$ variables. This also sustains several important lessons from Figure 5.4 such as there are a few variables that are several orders of magnitude larger, and the distributional shapes are very heterogeneous.

An aspect not yet discussed is that the text below each distributional subplot is the name of the variable (i.e. chemical feature). This can be very useful for identification of important features in data. Note that the variable with largest SD, labeled ww, is different from the variable SRW10, which has the largest mean (bottom right in Figure 5.4). In addition the variable nHBonds, which was seen to have the smallest mean (with each value being -999) in Figure 5.4, does not appear in Figure 5.5. The reason is that in the latter graphics, there are a large number of

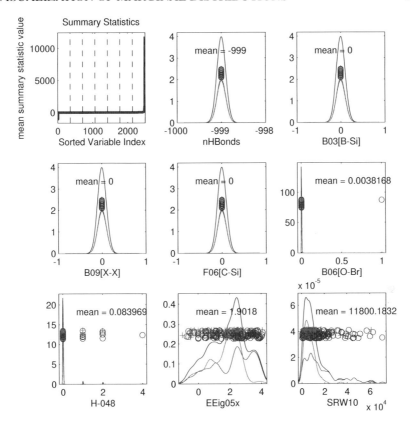

Figure 5.4 *Marginal distribution plot view of the Drug Discovery data, sorted on sample means. Shows great diversity of variation over variables. Many have no variation and a very few are orders of magnitude bigger.*

variables with SD = 0, and the large number of ties are broken by simply using the original variable ordering.

As seen in Figures 5.4 and 5.5, much can be learned from marginal distribution plots using equally spaced (with respect to the variable summary statistic) representative distributions. However, in some situations it is very useful to focus in on particular parts of the collection, often those with the smallest and/or the biggest variable summary measures. For example, with the goal of taking a more careful look into the missing values coded by -999 in the Drug Discovery data, Figure 5.6 again considers distributions sorted by variable mean, but now shows the 8 smallest mean values. This is reflected by all eight of the vertical dashed lines being on the far left (although very hard to see because they are so close to the vertical axis). It reveals that there are six variables that are all missing (i.e. all values are -999) and at least two more with some missings.

Figure 5.7 highlights a different set of variables, this time by sorting on the

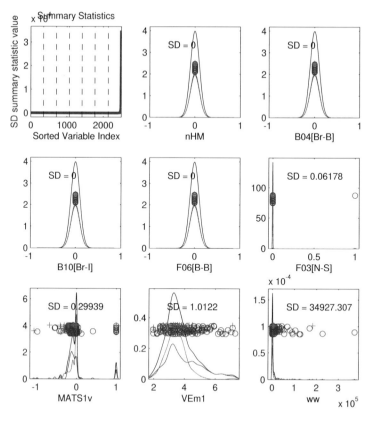

Figure 5.5 *Standard deviation sorted marginal distribution view of the Drug Discovery data. Reveals many variables with no variation.*

minimum value of each marginal distribution. In this situation, this view is less informative than those above, again because of how ties are handled in the sorting. For example, instead of seeing all values at -999 as in several of the above plots, the shown variable, IC4, only takes on -999 once (and does not appear in Figure 5.6). The next 6 variables all have a minimum of 0, so they appear in an arbitrary order, which is not particularly useful for understanding specifics in this data set, although it does happen to show the wide heterogeneity present here.

Based on the above insights, straightforward calculation verified that there were 1315 variables with no variation, and 16 that had at least one value of -999. Removal of these variables resulted in a cleaned data set of $d = 1164$ variables, which is further studied in the following. In other situations, much more care may be needed to deal with missing values. A usually important issue is whether all variables with missing values should be deleted (which is sensible here, since there are so few missings), or else imputed in one of various ways. See Gelman and Hill (2007) and Enders (2010) for good overview of many possible approaches

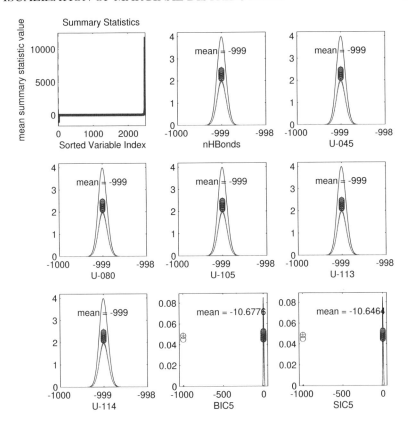

Figure 5.6 *Drug Discovery data explored using the eight smallest mean marginal distributions. Shows six variables are all -999 and others have a few -999s.*

to data imputation. This type of data object choice can be very challenging, and is often most effectively done in consultation with collaborators having domain knowledge.

A PCA view of the cleaned data is not shown here because it looks exactly like Figure 5.3 above. The reason is that both views are driven by the large magnitude variables, so of course eliminating those variables with no variation changes nothing. Also removing the -999s has no visible impact on the PCA scores view because the larger variables are so much larger, as revealed by a careful look at the axis labels in Figure 5.3.

Figure 5.8 studies SD for this cleaned data set. This shows that in addition to a few variables with extremely large variation, there are also some with extremely small variation and seems to indicate the presence of some binary variables (taking on only the values 0 and 1). This very wide range of variation suggests that some type of normalization, say rescaling each variable by its standard deviation, will give a much different result, perhaps revealing other types of structure in the data,

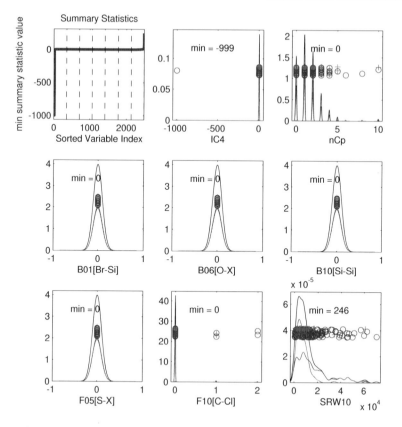

Figure 5.7 *Minimum value sorted marginal distributions for the Drug Discovery data. Again shows wide range of variation, but less informative because of arbitrary handling of many ties.*

as discussed in Section 5.2. Note that the lower right (maximal SD) variable, ww, is the same as the lower right variable in Figure 5.5. The others are all different, because they are equally spaced in a smaller set of variables.

Several of the above plots, e.g. those in the lower right panels of Figures 5.5, 5.7, and 5.8, suggest that skewness is a serious issue for at least some of these variables. This is explicitly studied in Figure 5.9 by sorting the variables this time on skewness. The upper left blue curve indicates more right variable skewness than left skewness, although both types are present. It also once again shows a strong presence of binary variables.

Note in addition that the variables with strongest skewness (typified here by F05|O-Ci| in the lower right panel) take on just a single value of 1, with all other $n - 1 = 261$ values being 0. The large flat spot on the upper right of the blue curve shown in the upper left panel indicates that in fact there are around one hundred such variables. Conventional wisdom is that those variables contain little useful

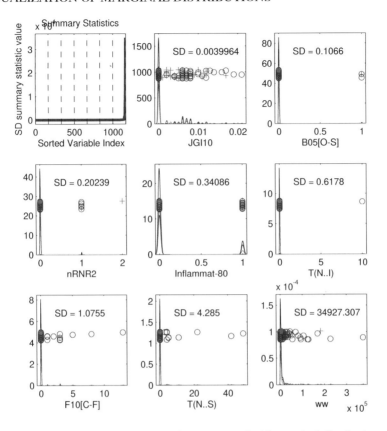

Figure 5.8 *Drug Discovery data, after cleaning viewed with marginal distribution plots sorted on variable standard deviation. Shows widely differing variation.*

information and thus should also be eliminated from consideration. However, the large number of them suggested this issue may be worth a second look. This was done in Borysov et al. (2016) who showed that in this case, there actually is useful information in these variables and incorporating them in data analyses actually gave improved classification.

When there is a combination of discrete and continuous variables in a data set, as suggested in the above analyses, there are other sortings of marginal distributions that are also quite useful. One of these is the number of unique values for each variable, as shown in Figure 5.10. This clearly highlights binary variables (which have only two unique values) by putting them first, with the blue curve in the upper left panel showing nearly 500 such. In addition to a number of clearly quite discrete variables, there are variables such as ZM1 in the middle left that appear to be continuous and yet contain a large number of exact replicates. The upper right part of the blue curve reveals that there are very few truly continuous

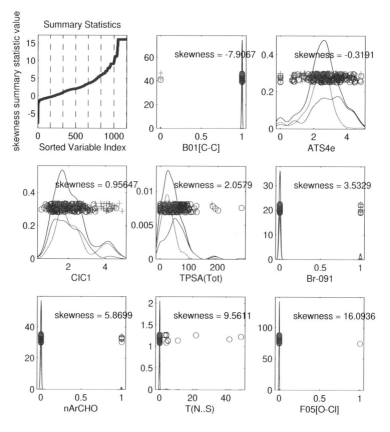

Figure 5.9 *Marginal distribution view sorted on skewness, for the clean Drug Discovery data. Shows many variables with strong skewness, including binary distributions.*

variables, for which the number of uniques is $n = 262$, as for AEigp in the lower right panel.

A different way to study discreteness versus continuity of variables is to sort on *number most frequent* as in Figure 5.11. This measure counts the number of times each value appears in the distribution and reports the largest count. This number is 1 for continuous variables, because all values are different. Note that the shown continuous variable, TIE, is different from the continuous variable AEgip shown in the lower right panel of Figure 5.10. Hence there are at least two truly continuous variables (namely AEgip and TIE), which appear in different orders in the two sorts. The question of how many continuous variables are present could be more carefully studied by looking at the smallest variables in this sort, or the largest number of uniques using the ordering in Figure 5.10. In this case there are only 9 variables which are truly continuous in the sense of having $nuniq = 262$. In this view, the binary variables appear on the right, in the order of how many times the majority values appear. This is another way of seeing that hundreds

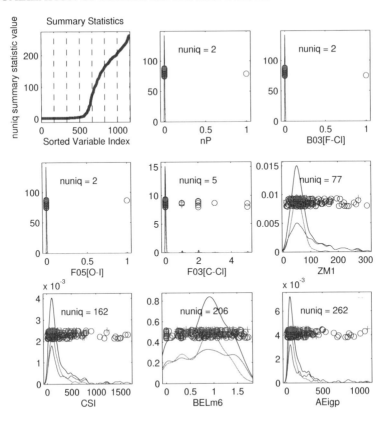

Figure 5.10 *Clean marginal distributions of Drug Discovery data, sorted on number of uniques. Shows wide range, from binary (nearly 500 such) to completely continuous variables.*

of variables have just a very few nonzero values. Because relatively few statistical methods (with the important exception of *zero-inflated models*, going back at least to Lambert (1992)) are designed to handle such a mix of continuous and discrete variables (i.e. traits), the need to use methods such as these visualizations is becoming increasingly important in Big Data contexts.

Given the large number of binary variables, a question arises as to how much information for separating active versus inactive compounds is present in those variables only. To study this, the data are reduced to only the $d = 364$ binary variables, with the resulting PCA scatterplot shown in Figure 5.12. This shows some perhaps surprisingly rich structure in this data with most of the active cases focused in just a few regions and larger regions with essentially no active cases. As noted above, Borysov et al. (2016) gives a more detailed analysis of this binary data set. Improvements in statistical power available from the various versions of this Drug Discovery data considered above (and some others below), for

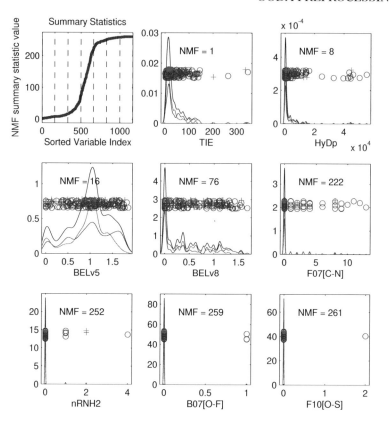

Figure 5.11 *Drug Discovery data, after cleaning. MargDistPlot sorted on number most frequent. Another way of contrasting discrete and continuous variables, again showing a wide range of both types.*

distinguishing the actives from inactives, are given using the DiProPerm hypothesis test in Table 13.1.

In the PCA in Figure 5.12 the binary data (0–1) are treated as real numbers. An alternate analysis, that stays within the binary domain by using entirely Boolean operations is Binary Matrix Factorization, proposed by Zhang et al. (2007b), and further discussed in Section 6.5.

This case study has shown that marginal distribution plots can discover many important aspects of data sets. In many situations, these can be essential in indicating strategies such as variable deletion, scaling (as studied in Section 5.2), and/or transformation (see Section 5.3) that can be very important to arrive at an effective data analysis. A few commonly used summary statistics have been demonstrated here, but many others can be equally revealing in other situations. For example small values of kurtosis can easily find variables with strong bimodal structure. Data sets with endemic outliers can adversely affect conventional moment based

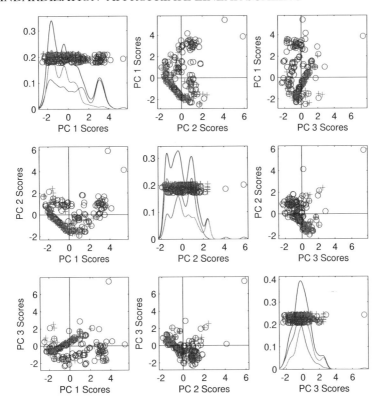

Figure 5.12 *Drug Discovery data, after cleaning. PCA based on binary variables only. Shows these variables contain red-blue separation information.*

summaries, in which case robust summaries, such as the robust skewness-like measure of Bowley (1920) and the robust kurtosis-like measure of Ruppert (1987), can be very useful. An (2017) developed some L-statistics based methods that are quite effective at finding important genes in the context of cancer research.

5.2 Standardization–Appropriate Linear Scaling

The point from Section 4.1, that scaling and normalization issues can be very important in OODA, is explicitly illustrated in Figures 5.13 and 5.14 in this section. As noted there, the need for scaling is obvious when different variables (i.e. traits or features) are not *commensurate*, e.g. when they use different units. But it is also critical when variables have scalings that are different by orders of magnitude (as seen for the Drug Discovery data in Figure 5.5). Since PCA seeks directions of maximal variation, this issue can have a very strong impact. As noted in many classical texts, such as Mardia et al. (1979), Muirhead (1982), Jolliffe (2002), and Anderson (2003), this can be handled by *pre-whitening*, i.e. standardizing each variable by subtracting the mean and dividing by the standard deviation. That operation followed by PCA is equivalent to replacing the usual sample covariance matrix input with the sample correlation matrix in PCA.

5.2.1 Example: Two Scale Curve Data

A toy example designed to explore this issue is the *Two Scale Curves* data set shown in Figure 5.13, whose format is quite similar to Figure 4.1. The $n = 200$ raw data curves in $d = 100$ dimensions appear in the upper left panel, using a rainbow color scheme. Note that the first 20 variables (as indexed on the horizontal axis) exhibit a much higher amount of variation than the remaining 80. The mean in the top center panel is essentially 0, so the mean residuals in the top right are nearly the same as the raw data. The first PC mode of variation in the second row is clearly driven by the first 20 variables, and is a mode reflecting all 20 variables moving up or down together. Similarly the second PC shown in the third row, has a mode of variation which is a contrast between the first and second 10 variables, which is orthogonal to the PC1 mode, and of course reflects less total variation. The last row shows some remaining variation of much smaller scale.

The same Two Scale Curves data set as in Figure 5.13 is re-analyzed in Figure 5.14, shown in the same format. This time the data are pre-whitened by standardizing each variable (mean subtracted, divided by standard deviation). The similar variation of each variable is immediately clear in the input data plot in the upper left. Unlike Figure 5.13 the last 80 variables are now as prominent as the others. Very different are the discovered modes of variation, as now the first two modes focus mostly on structure in the second 80 variables, while the variation that dominated the analysis in Figure 5.13 now shows up only in PC3 and its residuals (which would thus show up in PC4 had that been plotted). The reason that variables 21–100 now drive the analysis is that the per-variable magnitude of the signal is now comparable with variables 1–20 and there are simply more of them, giving more overall variation (they are simulated to be independent, thus the variation goes in essentially orthogonal directions) so these are now the dominant modes in the PCA.

Quantification of these ideas is given in Table 5.1, which contains the percentages of sums of squares of each mode of variation (as illustrated in Figure 3.5), with respect to the residuals about the mean. These are the numbers that are graphically displayed in the scree plots of Figure 3.5. Because the raw data curves in the upper left of Figure 5.13 have almost all their variability on the left side of the range, it is not surprising that part of the range drives two very large PC components explaining almost all the variation in the data (with the first having about 3 times the energy of the second), as seen in the top row. The bottom row shows a more even spread of variation, which is consistent with the visual impression of the standardized data in the top left of Figure 5.14. In particular, the table together with the figure shows that it is variation on the right part of the range which has become dominant, containing about 80% of the variation. The remaining 20% is split at the same 3:1 ratio between the third and fourth modes of variation.

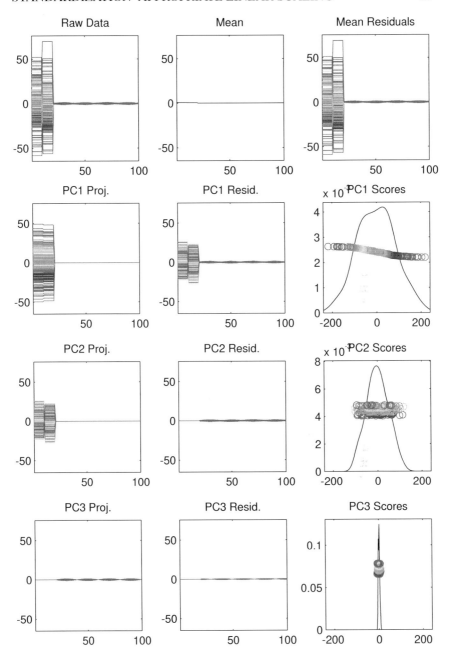

Figure 5.13 *Two Scale Curves toy FDA example, illustrating challenge of differing variable scaling. The first 20 variables have much more variation and thus drive the first two principal components.*

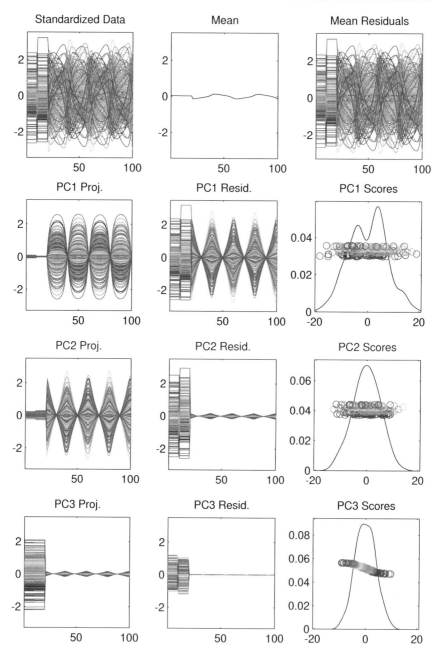

Figure 5.14 *Two Scale Curves data set from Figure 5.13, with pre-whitening based on correlations. Now the last 80 variables drive the variation, i.e. appear in the first two components, leading to much different conclusions. This shows data scaling and normalization have a critical impact on this type of analysis.*

	PC 1	PC 2	PC 3	PC 4
Raw PCA	76%	24%	0.1%	0.03%
Standardized PCA	53%	27%	15%	5%

Table 5.1 *Percent sum of squares explained by each PC component for the Two Scale Curves data in Figures 5.13 and 5.14. Quantifies how raw data components focus on structure on the left, while standardization shifts the focus to the right.*

5.2.2 Overview of Standardization

It is worth considering which of the two very divergent analyses in Figures 5.13 and 5.14 is more appropriate. As noted above, most classical texts on multivariate analysis will recommend doing the analysis based on the correlation matrix (i.e. pre-whitening as in Figure 5.14). This is often a sensible default, especially in situations where different variables are measured in different units. However it is important to realize that in other situations the original data scaling may be most appropriate and thus should be preserved. For example, in the Lung Cancer data in Figure 4.8 pre-whitening by standardization will result in the small exon starting at exonic nt number 500 playing too large a role in the analysis. Clearly this is an important data object choice, deserving careful consideration (and discussion with knowledgeable collaborators) in data analysis. More discussion of the tendency of PCA to focus on large scale variation can be found in Chapter 17.

The effect of pre-whitening on the cleaned Drug Discovery data from Section 5.1 is demonstrated in Figure 5.15. A quantitative measure of the impact is also given in Table 13.1. Unlike the outlier driven PCA shown in Figure 5.3, this view shows much more in the way of interesting relationships between the data objects, i.e. the chemical compounds. In particular, it is clear that there are now very complex (e.g. highly nonlinear) relationships between the active (red pluses) and inactive (blue circles) compounds, which is why Drug Discovery has been a challenging problem over the years. Note for example an indication of small regions where interesting comparisons can be made, which motivates the idea of *activity cliffs*, (regions of abrupt transition between classes) as studied in Maggiora (2006).

There are many other aspects of data normalization that should be kept in mind as needed. For example, while standardization of the variables (i.e. of each row of the data matrix) can be very sensible in some cases as illustrated using the Two Scale Curves data in Figures 5.13 and 5.14, column standardization can be more useful in others. A canonical example of this is genetic molecular measurements, which are based on amplification of DNA or RNA in a way that is not easy to calibrate, resulting in columns of the data matrix which tend to differ by scale factors. For gene expression studies, e.g. Hoadley et al. (2014), this is commonly handled by scaling each column appropriately. While averages could be used for this, that would not allow for expected differing overall expression across cases, so instead normalization to achieve a common third quartile is often used. That

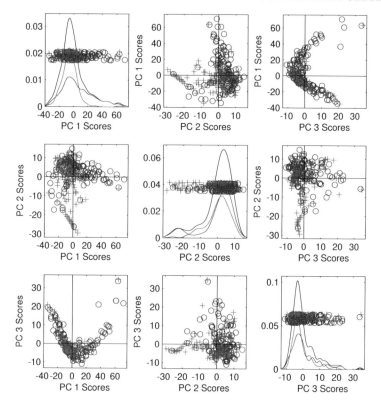

Figure 5.15 *Drug Discovery data, after cleaning. PCA on non-binary variables only. Shows these also contain red-blue separation information.*

choice is driven by the idea that the most actively different genes should not be involved, and the third quartile is sufficiently stable to properly represent overall variation.

In other situations projection of each data object (column vector) to the unit sphere, S^d (recall the notation (3.2)) can be appropriate. This projection is computed as $\frac{x}{\|x\|_2}$, using the notation (3.3). It is useful in situations where vector length contains only nuisance variation, and the interesting variation is in the angles between vectors. See the discussion around Figure 12.6 for more in the context of hierarchical clustering. In such cases, another option is projection to the unit simplex, as discussed in Section 18.3. As noted in Section 18.3, Xiong et al. (2015) found that (for virus detection using DNA data) projection to the simplex gave better results than projecting to the sphere.

5.3 Transformation–Appropriate Nonlinear Scaling

As seen above relative magnitude of variables is an important consideration in OODA. Similarly the distributional shape of the marginal distributions can also have a major effect as seen using the Spanish Mortality data in Figures 5.1 and 5.2. Another example that illustrates this point is shown in Figure 5.16. This is a different subset of the Pan-Cancer data analyzed in Section 4.1.4. Here the focus is on the cancer types Ovarian Cancer, which is labeled as OV in TCGA notation and shown as magenta circles, and Uterine Cancer, labeled UCEC and indicated using green plus signs. Only the 1000 most variable genes, among genes having no missing values, are considered. The full data set (with a few hundred cases of each type) shows a strongly statistically significant difference between these two cancer types for almost any type of analysis, so for good contrast between statistical methods, randomly chosen subsets of size $n_1 = n_2 = 30$ cancer patients of each type are analyzed here for illustration.

The top row of Figure 5.16 shows PCA scatterplot views of the data. Unlike the scatterplot matrices shown above, e.g. in Figures 5.3, 5.12, and 5.15, here each plot shows only the PC2 vs. PC1 scores scatterplot (often the left plot in the second row in matrix views). The upper left panel of Figure 5.16 studies the distribution of raw counts. Note that PC1 is dominated by a single very large case (about an order of magnitude bigger than all others). PC2 is driven by a handful of other cases, but still only a relatively few. While one might hope to see a large difference between the OV and UCEC cases, if it can be seen in this scatterplot, it can only be in the lower left part of the plot, but is very hard to perceive due to over-plotting. For a closer view of potential class differences, the top center panel shows a zoomed in (on the lower left corner) version of that plot. This makes it even more clear that this data set suffers from strong skewness (which can also be easily seen using the Marginal Distribution views described in Section 5.1), with essentially no OV-UCEC difference visible. This does not mean that there is no difference, only that it does not appear in the 2-dimensional subspace of the first 2 principal components.

For such strongly skewed data, a log transformation of each variable is often very useful, as it tends to strongly reduce the influence of data points that are orders of magnitude larger than the others. The top right panel shows the result of the $\log_2$ transformation applied to each variable. That transformation is usual in this field, where the doubling interpretation of that log base is commonly desired. Note that these two modes of variation highlight a clear and strong difference between the cancer types, appearing as mostly the dominant mode of variation (i.e. the PC 1 Scores).

Table 5.2 provides another way of seeing that the $\log_2$ transformation provides a much better scale on which to analyze this data set. In particular, the outlier in the upper left panel of Figure 5.16 is seen to dominate the raw data analysis, with the first PC containing 88% of the total variation about the mean. In contrast, on the $\log_2$ scale the first PC explains 25% while the second explains 10%, which are much more reasonable as there are a large number of diverse biological

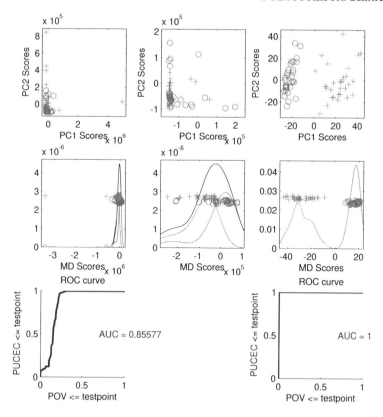

Figure 5.16 *Contrast of Ovarian (magenta circles) and Uterine (green plus signs) Cancer gene expression for the Pan-Cancer data. Top row shows PC1 vs. 2 scores, for raw count data (left, with zoomed in version, center) and log₂ transformed data on the right. Middle row shows corresponding MD projected distributions. Bottom row are ROC curve analyses. Shows log transformed analysis gives much better contrast between cancer types in all ways.*

processes whose presence should be reasonably represented in gene expression measurements.

	PC1	PC2
Raw Data	88.2	4.7
$\log_2$ Data	24.8	9.9

Table 5.2 *Percent of sums of squares about the mean explained by the first two principal components, for the raw gene expression (top) and the $\log_2$ transformed version of the same data. Shows that log transformation gives more sensible distribution of variation in the data.*

The middle row of Figure 5.16 more directly targets the OV vs. UCEC contrast by showing the univariate distribution of projections onto the *Mean Difference* (MD) direction vector, which is just the normalized (to have norm 1) difference between the sample mean vectors. This direction vector is also called the *centroid classifier*, see Tibshirani et al. (2002). See Section 11.1 for much more discussion of the MD direction. The display format is the same as used several times above, e.g. the marginal distribution plots in Section 5.1, using symbols whose x-coordinate reflects the value with y-coordinates simply providing visual separation. Again, the black curve is a kernel density estimate, with the colored curves representing proportional sub-densities for each of the two data types. The middle left panel shows this distribution for the raw counts data. As in the top left panel, the difference is not easy to discern because the view is again dominated by a single large outlier, which obscures how well separated the two groups are. The zoomed view in the center panel shows that actually the MD direction provides decent separation of the classes, with the green plus UCEC cases tending to lie more to the left of the OV magenta circles, although there is substantial overlap. This overlap is quantified using the Receiver Operating Characteristic (ROC) curve of Hanley and McNeil (1982), in the lower left panel. This curve is generated by sliding a cutoff point along the horizontal axis of the left middle panel, and for each such point displaying the proportion of UCEC (green) points that are smaller on the vertical axis versus the proportion of smaller OV (magenta) points on the horizontal axis. The fact that more UCEC points lie to the left is reflected by the curve moving fairly steeply upwards. Once the cutoff point includes all UCEC points the curve remains at height one. The fact that this curve lies mostly toward the upper right of this plot shows that the two populations are relatively well separated. A simple numerical summary of ROC behavior is the *Area Under the Curve* (AUC), which in this case is 0.86.

The middle right panel studies the MD projections for the $\log_2$ transformed version of the data. Given the obvious group separation in the PC 1–2 scatterplot in the upper right panel, it is not surprising that there is a very strong separation between UCEC and OV that is apparent in this projection. The strong visual impression is confirmed in the lower right panel by the ROC plot following first the vertical axis, then the top horizontal line, resulting in an AUC of 1.

A related contrast between the raw count data and the $\log_2$ transformed version is provided using the DiProPerm confirmatory method, described in Section 13.1. That hypothesis test for exploring differences between the UCEC and OV cases, using the mean difference direction and mean difference summary statistics, for the raw counts gave a non-significant p-value of 0.24, while the $\log_2$ counts gave a strongly significant p-value $\ll 10^{-4}$. This is another way of seeing that analyzing this data on the $\log_2$ scale is very well worthwhile.

An important variation of the log transformation is the *shifted log transformation* of the form $\log(\cdot - c)$, where the data are shifted by a constant amount c before application of the logarithm. This is useful both for data which take on 0 or negative values, and in the case of $c < 0$ is also useful as a typically less stringent version of the log transformation, that is useful for data with relatively mild skewness. This works in the same way as the skewness of the log normal distribution is controlled by the mean parameter of the underlying normal distribution. Good automatic choice of the shifted log transformation has been developed by Feng et al. (2016). The usefulness of that method in the case of the Drug Discovery data is explored in Table 13.1.

Another appealing and widely used family of transformations is the Box-Cox family

$$f(x) = \begin{cases} \frac{x^\lambda - 1}{\lambda} & \text{for } \lambda \neq 0 \\ \log x & \text{for } \lambda = 0, \end{cases}$$

proposed by Box and Cox (1964). This is essentially a family of power transformations coupled with a linear transformation. Note that a careful calculation of the limit as $\lambda \to 0$ shows that this is a continuous function of the tuning parameter λ. Hence this provides a broad and flexible way of adapting to skewness in data. One more important general family of transformations is described in Johnson (1949).

5.4 Registration–Appropriate Alignment

As noted in Section 2.1, registration, i.e. alignment issues, are often quite important in many types of OODA. This point is illustrated using an FDA (curves as data objects) toy example in Figure 5.17, which is similar to the Bimodal Phase Shift data of Figure 2.2. The raw data are shown in the left-hand panel. Each curve has two peaks, but there is now substantial variation in both locations and heights of the peaks. In contrast to Figure 2.2, this time the curves are color coded using the height of the left peak, with a rainbow color scheme ranging from magenta (tallest) through green and yellow to red (shortest) with the goal of highlighting the amplitude variation in this case. The varying locations of the peaks create challenges for standard statistical analysis (which ignores the strong phase variation). For example, the (point-wise) mean curve, shown as a thick black dashed line, is not at all representative of the population. In particular, its peaks are substantially lower than any peak in the data set, and the left peak actually appears as two modes. In addition, it will seen be in Chapter 9 that PCA (e.g. as in Figures 1.4, 1.5, 3.4, 4.1, and 4.3) of data sets with such strong phase variation also provides very poor low-rank representations.

The right panel of Figure 5.17 shows the results of a Fisher-Rao registration of these curves as described in Section 9.1. The heights of each curve are the same in both panels, but in the right panel the horizontal axis for each curve has been appropriately warped (i.e. the horizontal axis has been appropriately stretched and compressed separately for each curve) to make the curves align very well. Note that the mean of this set of curves, again shown as a thick black dashed curve is now a quite sensible notion of center, as it lies clearly in the middle of the data set, as well as typifying the general shape of each. As seen in Chapter 9, PCA of such aligned sets of curves similarly provides a quite intuitive and much more

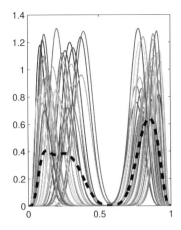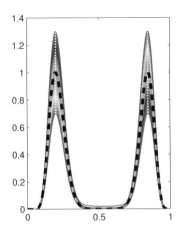

Figure 5.17 *Toy example, of functional data, each curve having two peaks, illustrating usefulness of curve registration. Left panel shows original curves, using rainbow color scheme on height of left-hand peak to highlight amplitude variation. Right panel shows the result of a careful curve alignment. Dashed curve in each case is the sample mean, which is much more representative of the data curves after alignment.*

compact representation (in particular, requiring only a single mode of variation) of data sets with strong phase variation.

For quite similar reasons, registration is also critical in image related tasks. For example, while collecting the Faces data objects in Section 2.4, little attempt was made during the initial photographs to ensure that each face was in the same place in each picture, which presented a challenge that is quite similar in spirit to what is seen in the left panel of Figure 5.17. In particular, various facial aspects (such as eyes, nose, mouth) were seriously misaligned across images, resulting in a far fuzzier analog of Figure 2.9 when the analysis was based on the unaligned data. As noted in Section 2.4, great improvement was made by a simple affine transformation based on landmarks at the center of each eye and the mouth.

Generally in image analysis this type of consideration motivates the study of shapes as data objects. The mathematics that allows convenient and rigorous definition and analysis of such data is discussed in more detail in Chapter 8. Essentially that set of ideas is used to provide a fully automatic approach to curve registration in Chapter 9.

CHAPTER 6

Data Visualization

This chapter considers OODA visualization concepts in more detail. Here the focus is on data objects having a Euclidean feature space representation, where the entire data set is conveniently summarized by a matrix. For the reasons given in Section 3.1, columns of the matrix are taken to represent data objects. A direct and natural visualization of the structure of a matrix of data is a *heat-map* image, whose application in OODA contexts is studied in Section 6.1. *Curve* and *scatterplot matrix* views of data objects, both of which have been used repeatedly in earlier chapters, are developed in further detail in Sections 6.2 and 6.4. Data centering and combined views are considered in Section 6.3.

6.1 Heat-Map Views of Data Matrices

The main idea of a heat-map view of the structure of a data matrix is to construct an image that represents each matrix entry with a colored (or gray level) patch in a rectangular grid. Patterns in the perceived colors frequently give useful insights about the data set represented by the matrix. The basic idea is related to that of digital photography (where the patches are called *pixels* from picture elements), and has been used in many contexts. A good overview of the area, that includes some interesting historical discussion, can be found in Wilkinson and Friendly (2009). This approach to data visualization is commonly used in bioinformatics, where it has proven to be very effective in finding biological insights and discoveries from data, see for example Eisen et al. (1998) and Perou et al. (2000) where this exploratory tool was used in the discovery of clinically relevant cancer subtypes, that have since become key to realizing the improved treatments promised by *precision medicine*. However, heat-map visualization has also attracted skeptics such as Rogowitz et al. (1996) and Borland and Taylor (2007), particularly over color choices and how they relate to human visual perception. A different limitation of heat-map data visualizations is shown in Figures 6.5 and 6.6, where it is hard to see small signals in noisy data. For visualization of a matrix with nonnegative entries a gray level plot, as shown in Figure 6.1 below, can be a good choice because the ordering of colors is clear. A richer color scheme in the nonnegative case is the topographical colors that are classically used in geographic maps: green, yellow, orange, brown, gray, and white. For entries that are both positive and negative, in contexts where knowing which entries are 0 is important, two colors plots such as blue and red are usually more useful, as discussed around Figure 6.5 below. The topographic color scheme can also be extended to this case by including shades of blue suggesting below sea level. In situations where location of the origin is not important, e.g. log scale data such as the Spanish Mortality data in Figures 1.1, 5.16 and a number of other places in this monograph, a rainbow color scheme can also be useful.

It is usually important to keep in mind that ordering of the rows and columns in heat-map visualizations has a major influence on what can be learned. This is illustrated in the Two Clusters data toy example in Figure 6.1. All 3 panels are based on a common set of values. At first sight the left-hand panel appears to be fairly random, with perhaps hard to interpret vertical and horizontal patterns. In the center panel, a clustering algorithm (hierarchical clustering with Euclidean distance and average linkage as discussed in Section 12.2) has been applied to the columns, and they are correspondingly reordered. Despite the high noise level this shows a clear propensity for lighter shades of gray in the right half. The right-hand panel shows the results of similarly clustering and reordering the rows as well, which shows a related pattern between the top and right half of the plot. The contrast between the left and right sides indicates two clusters in the data set (columns) with different cluster mean structure evident in the reordered variables (rows). This ability to show rich structure (even in the presence of very high noise) seems to be why this type of view is commonly used in bioinformatic contexts, where there tend to be strong underlying patterns often corrupted by a relatively high noise level.

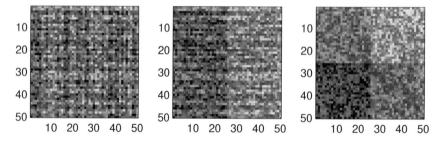

Figure 6.1 *Toy example demonstrating usefulness of hierarchical clustering rows and columns in heat-map image views. Left panel shows a random ordering. Column clustering has been applied in the center panel, and both column and row clustering were used in the right panel, to reveal important systematic "Two Clusters" structure.*

In some situations, the hierarchical clustering itself is of interest, beyond merely enhancing the visualization. In such cases it is common to add dendrograms (as shown in Figures 12.4 and 12.7), revealing the tree of nested clusters for both the rows and columns of the heat-map.

Another issue that is critical to heat-map visualizations is *scaling* as illustrated in Figures 6.2, 6.3, and 6.4. These are three different views of the same data matrix demonstrating the impact of different gray scale choices on which aspects of the data are highlighted. Figure 6.2 shows how the typical first choice of an equally spaced color scale can obscure important structure in the data. In particular, the only thing visible is one small white spot in the heat-map shown in the left panel.

Insight into the cause of this behavior comes from studying the distribution of matrix values, which is shown in the right panel. This uses the same format employed for displaying one-dimensional distributions in many other contexts above, starting with the Spanish Mortality data in Figure 1.4. In particular, the matrix entries lying in the interval $[0, 20]$ are shown as $(2500 = 50 \times 50)$ circles. The curve is a smooth histogram, i.e. kernel density estimate as discussed in Chapter 15. The horizontal coordinate of each circle is the entry of the matrix and the vertical coordinate is the order in the vectorized version of the data where the matrix columns have been successively concatenated into a single long column. Both the circles and the density estimate reveal a highly skewed distribution. The right-hand panel also contains vertical dotted lines which are the boundaries of the 20 equally spaced gray scale regions used in the heat-map in the left panel. The right-hand three-quarters of the plot has very few circles, which reflects very few white or gray pixels visible in the heat-map. But note that the shape of the distribution of the circles suggests there may be additional structure that is not visible in the heat-map because all those points appear in the essentially black region. While equal spacing of color scales is a common default in heat-map software packages, this example demonstrates how that choice may obscure important structure.

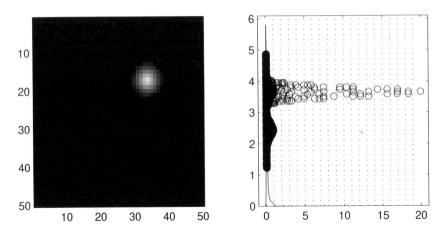

Figure 6.2 *Toy example illustrating gray scale color scaling issues. Heat-map image is shown in the left panel. Distribution of the matrix entries is shown in the right panel, with vertical dotted lines indicating the boundaries of the gray color regions. The heavily skewed distribution indicates poor use of the equally spaced gray levels, resulting in an ineffective heat-map view.*

A simple approach to making better use of the available gray scale is to transform the distribution of matrix entries to have less skewness. The shape of the distribution shown on the right of Figure 6.2 suggests a log transform could be useful. The effect of this, for the data matrix in Figure 6.2, is shown in Figure 6.3. Note that the heat-map image on the left still shows the same white spot as in Figure 6.2, but now it also shows three distinct gray spots. The distribution of the log matrix entries in the right panel clearly shows the impact of the log transformation. The bright spot in the image on the left is reflected by the larger, sparser peak in the matrix values. The less bright but light gray spots in the image appear as the peaks extending about halfway across the range (thus using only about half of the gray levels). There are two such peaks because of the column reorganization operation used to order the matrix pixel values. A careful look at the heat-map reveals additional interesting structure. In particular a regular grid of quite dark gray (versus the black background) points is visible. Because these small dark gray spots appear all across the heat-map, their pixels form a thick band on the left side of the right-hand panel, itself indicating interesting structure in this data matrix.

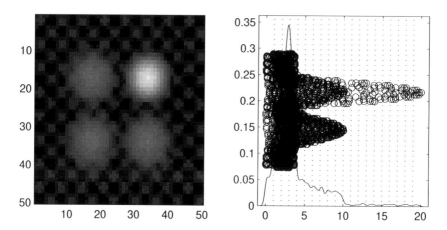

Figure 6.3 *Different heat-map image view of the same data matrix from Figure 6.2 using gray level scale (boundaries again shown using vertical lines in the right panel) based on log of function values. Distribution of log function values shown in the right panel, with vertical lines now showing better use of gray levels, so this color scale reveals more structure in the image.*

While this log scaled heat-map image reveals much more structure in the data matrix than was available from the naive scaling shown in Figure 6.2, there is still room for improvement. For example the entire brighter half of the range (thus the bright half of the gray levels) is devoted to just the single tallest peak in the matrix. An alternate approach, that can be very revealing in some contexts is *quantile scaling* (also known as *histogram equalization*). That color scaling ensures approximately equal numbers of pixels used at each gray level, by placing the vertical lines in the right-hand distribution plots at equally spaced quantiles of the matrix values distribution, as shown in the right-hand panel of Figure 6.4. Note that the corresponding heat-map on the left now strongly reveals the regular background pattern. As with many choices among OODA methods, there is a trade-off of which one should be aware: the clear view of the background comes at the cost of much less contrast between the heights of the major peak and the three other large ones.

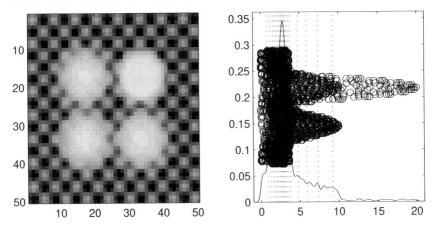

Figure 6.4 *Another heat-map visualization of the same data matrix from Figures 6.2 and 6.3. Highlights different structure in the data using the even spread of the gray scale colors provided by quantile scaling.*

Notice that quantile scaling is the same whether applied to the original data or the log transformed version, because the log is a monotone transformation. However the matrix value distributions in the right panels above are much easier to understand on the log scale (used in Figures 6.3 and 6.4) than on the linear scale of Figure 6.2.

When choosing a heat-map visualization tool, at least having the option for some type of visualization of these heat-map gray level (or color) distributions can be very useful. An attractive option is a histogram, where each bar represents a gray level (or color) with each color used for its corresponding bar. Best insight comes from taking the area of each bar to be proportional to the number of pixels using that color. This is the same data display principle that underlies density estimation as discussed in Section 15.1. Such visualizations of color distribution appear in the lower right panels of Figures 6.9–6.12.

For data matrices taking both positive and negative values where it is important to highlight 0, a common two-color choice codes 0 as black, with shades of red and green indicating magnitude and direction of departures from 0. This color scheme has the disadvantage of being inaccessible to red-green color blind people, which is nearly 10% of the male American and European populations. An alternate choice illustrated in Figure 6.5, codes 0 as white with shades of red and blue for magnitudes and directions. In some fields (e.g. economics or finance) red is frequently used for negative values (suggesting financial loss), while in others (e.g. climatology) it is more natural to use red for positive (corresponding to hotter temperatures). In Figure 6.5 the former scheme is used. Note that the actual coded values are shown using a color bar on the right side of the heat-map.

The main point of Figure 6.5 is to show that while a heat map can provide very useful insights in some situations, it can miss important structure in others. This is done using the Two Class Gaussian data set, which is a matrix with $n = 200$

data object vectors of length $d = 20,000$, which is common in bioinformatics. To enhance visible structure as demonstrated in Figure 6.1, hierarchical clustering using average linkage and Euclidean distance (a common choice for visualization in bioinformatics as noted above) has been applied to both rows and columns. A challenge to heat map visualization of data sets of this size is that the pixel capability of most types of displays tends to be in the very low thousands, which is very inadequate in this case. Hence, most heat-map visualization software packages provide convenient facilities for zooming in on appropriate parts of the display. Here that is done by simply showing the first 200 rows of this much longer matrix. Other subsets show similar apparent pure noise.

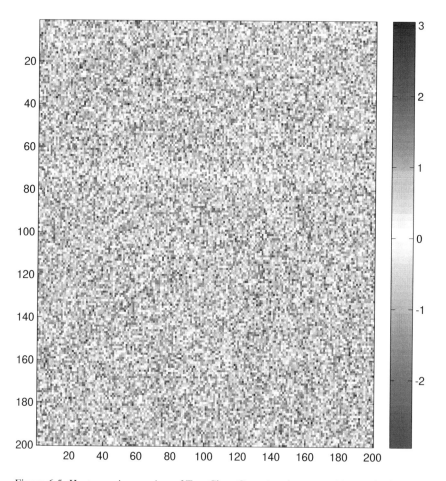

Figure 6.5 *Heat-map image view of Two Class Gaussian data set, taking on both positive and negative values. White indicates matrix entries that are 0, with blue and red for positive and negative values, respectively. It is hard to see much structure, even though rows and columns have been clustered.*

As suggested by the name, there actually is structure in this Two Class Gaussian data matrix. In particular it was generated as the first $n_1 = 100$ columns having all entries distributed as independent $N(0.04, 1)$, with the next $n_2 = 100$ columns drawn independently from the $N(-0.04, 1)$ distribution. While this perhaps small true underlying signal is dominated by the noise in Figure 6.5, it can be seen in other types of visualizations, such as the PCA scatterplot matrix as shown in Figure 6.6. Here the two subpopulations (labeled as magenta circles and green plus signs) are visually distinct, showing that the PCA scatterplot matrix view can reveal structure that is not apparent in the heat-map image.

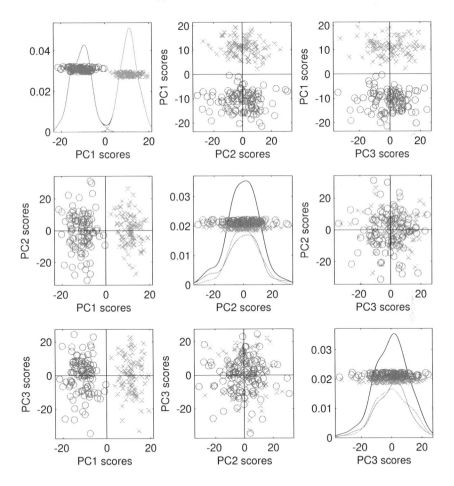

Figure 6.6 *PCA scatterplot matrix view of Two Class Gaussian data set in Figure 6.5. Two subpopulations, indicated with different colors and symbols, are visibly quite distinct, despite the high noise level, in sharp contrast with the heat-map view.*

However, as seen in Section 4.2, one should be skeptical as to whether such visually apparent aspects of a scatterplot represent true underlying structure, or are merely irreproducible sampling artifacts. In this case, the DiProPerm analysis of this Two Class Gaussian data set in Figure 13.1 shows the difference between these groups is strongly significant (p-value much less than 0.001), as it should because there is a systematic difference between the subgroups.

In summary the heat-map visualization tool has proven to be very powerful in the hands of experienced analysts. As discussed at the beginning of Section 6.1, it has a strong track record of finding clusters and their drivers in a single visualization. But as seen in Figure 6.6, it can also miss important structure in high noise situations. One more point to keep in mind is that apparent clusters in a heat-map may not represent reproducible discoveries. A hypothesis test that is well suited for analyzing the statistical significance of clusters (especially in high-dimensional contexts) is SigClust, discussed in Section 13.2.

6.2 Curve Views of Matrices and Modes of Variation

Curve views (standard in FDA) reveal quite different aspects of variation in a sample of data objects, as seen in many situations in earlier chapters. The value of such views includes a clear display of modes of variation, as defined in Section 3.1.4. Some specific examples follow.

- *Figure 1.1, Spanish Mortality data.* These displays demonstrate the value of the log transformation, in terms of making much of the variation in the data available to the analysis.
- *Figure 1.2, Spanish Mortality data.* This curve view graphic, together with the previous figure highlights the value of careful color coding for clear insights about (for example) time structure in the data curves.
- *Figure 1.3, Spanish Mortality data.* Here the focus is on variation about the mean, revealing that many of the age impacts on mortality have been constant over time.
- *Figure 1.4, Spanish Mortality data.* The main mode of variation is an overall improvement in mortality, which is most dramatic for the young, together with an improvement in the keeping of records.
- *Figure 1.5, Spanish Mortality data.* The second mode of variation is a contrast between the 20–45-year-old males with the rest of the population. Recall this mode reflected pandemic, war and automobile effects.
- *Figure 2.1, TIC spectra.* This curve view display shows that alignment can be a challenging problem, that is very effectively addressed by the Fisher-Rao approach to curve registration.
- *Figure 2.2, Bimodal Phase Shift data.* These curves demonstrate how appropriately chosen warping functions can be used to align peaks in curve registration, and to visually reflect phase variation (lower right panel).
- *Figure 4.1, Tilted Parabolas data.* There are a number of curve displays in this analysis which give a clear indication of the variation in the full data, the mean residuals, and several principal components together with their respective

residuals. The left plot in the second row shows that vertical shift is the dominant mode of variation (a common aspect of many functional data sets that is
more deeply explored in Section 17.1.1). That on the third row reveals the less
obvious tilting mode of variation. The bottom row suggests that the remaining
variation is random.

- *Figure 4.3, Twin Arches data.* Here the curve view shows two non-obvious
  modes of variation, with the first component reflecting a peak-valley versus
  valley-peak mode, while the second mode is peak-peak versus valley-valley.

- *Figures 4.5 and 4.8, Lung Cancer data.* The first shows a large amount of variation that is hard to visually parse. The cluster-based brushing (i.e. coloring)
  of these curves shown in the second reveals important modes of variation as
  discussed in Section 4.1.

- *Figures 5.13 and 5.14, Two Scale Curves data.* These views of the data demonstrate the potential impact of data normalization (through standard deviation
  scaling, as done in unit free correlation matrix approaches) on PCA. In particular this choice can highlight completely different modes of variation.

- *Figure 5.17.* Here studying curves that are mostly horizontally shifted clarifies
  the impact of peak alignment on the pointwise mean of a bundle of curve data
  objects.

These examples demonstrate the potential benefits of curve views when the data
objects are curves, especially when used in tandem with PCA, or some other approach to revealing modes of variation. However as with any OODA method, there
are situations where this approach is less useful, for example when there is a large
amount of noise present. This is illustrated using the Two Clusters toy data (shown
in the right panel of Figure 6.1 where both rows and columns have been hierar-
chically clustered) by a curve view in Figure 6.7. Following the convention in this
book, columns of the matrix are the (curve) data objects, so the horizontal axes in
Figure 6.7 show the indices of the rows. The structure visible in the heat-map in
the right panel of Figure 6.1 can also be seen in Figure 6.7, but the interpretation
takes more effort. In particular, the lighter top half of the former appears as higher
values of the mean curve on the left shown in the top center panel of the curve
view. The brighter gray levels on the right half of the heat-map (i.e. one cluster
of the column data objects) appear as the upper set of cyan curves in the lower
left panel of Figure 6.7, and as the cyan cluster in the scores shown in the lower
right panel. The lower central panel shows the mean plus and minus the extreme
curves, which contain all of the structure available in the heat-map. The top right
panel shows the scree plot (introduced in Figure 3.5), revealing that the noise level
is indeed quite high in this example. While both views distinguish the underlying
structure in the data set, in this case the heat-map is much easier to interpret.

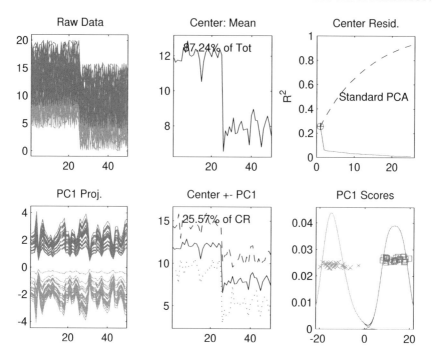

Figure 6.7 *Curve view of the Two Clusters data set from Figure 6.1. Shows that curve view may not be as useful as a heat-map view in high noise contexts.*

In very high noise data sets, such as the Two Clusters data in Figures 6.1 and 6.7, the noise level can have a substantial impact on PCA, in terms of both loadings and scores. This motivated a large amount of research in the FDA community on integrating smoothing with analyses such as PCA. See Ramsay and Silverman (2002, 2005) and Ferraty and Vieu (2006) for good overviews. Additional aspects are treated in Zhang (2014) and Kokoszka and Reimherr (2017). A particularly strong theoretical treatment of the FDA can be found in Hsing and Eubank (2015). A simple approach is to simply smooth the data object curves before doing PCA, see Chapter 15 for further discussion of smoothing methods. One could also do PCA, and then smooth the loadings curves. However generally better is to integrate the smoothing with the PCA, which can be done by various methods as detailed in the above references. Integration of smoothing methods with PCA is also very useful for functional data sets with curves evaluated at sparsely sampled or irregular points. An important method in such situations is the PACE algorithm introduced in Yao et al. (2005). Delaigle and Hall (2016) propose an approach to handling a surprisingly high level of sparse sampling.

Less well understood is that in standard low to moderate noise situations, there is not much gained from explicitly integrating smoothing methods with PCA, because simple PCA frequently has an implicit smoothing effect. This can be seen in the Tilted Parabolas data in Figures 4.1 and 4.2, and in the Twin Arches data in Figure 4.3, where the substantial level of noise visible in the raw data appears in the residual, not in the main components. This appears to be caused by the noise being mostly averaged out when the PCA calculation estimates the smooth underlying loadings vectors.

6.3 Data Centering and Combined Views

The *mean centering* of data is an issue that is very frequently dismissed as trivial. The impact of centering on an OODA seems fairly obvious and routine from a curve viewpoint, as illustrated for example in Figure 1.3. However, when contemplating heat-map views the effects can be surprisingly important and even challenging to intuitively comprehend. In particular, that view motivates the consideration of both horizontal and vertical mean centering of the data matrix. The intuitive impact of these can be surprisingly elusive. That issue has been more deeply investigated in Prothero et al. (2021).

The importance of doing a centering operation (before looking for modes of variation), from the curves as data objects viewpoint (e.g. in FDA), is illustrated in Figure 6.8. This contrasts PCA with a fully uncentered analysis based on a direct Singular Value Decomposition (SVD) of the data matrix. Many aspects of SVD and its relation to PCA are studied in detail in Section 17.1.2.

Figure 6.8 provides a comparison between uncentered SVD and PCA in the context of the simple 2-d Toy Example data from Section 3.1.2. Using the terminology of Section 3.1, the data are shown in a feature space view as the black circles in the top two panels. The top left panel is a view of the SVD approximation of this data set. The red line is the one-dimensional subspace (i.e. line through the origin) that best fits the data. The magenta plus signs are the projections of the data onto that subspace, and are the best rank-one approximation of the data. In particular this direction minimizes the sum of squares of the projected residuals (shown as cyan lines). The first SVD scores are the coefficients of these projections, which appear on the horizontal axis in the left lower panel. The signed lengths of the cyan lines are also the coefficients of the projections onto the second singular vector, i.e. the second SVD scores, which are used on the vertical axis in the lower left panel. Because SVD ignores the center of the data, it fails to efficiently summarize and display the dominant mode of variation in this data set (i.e. major axis of the ellipse of data which is essentially vertical). Instead that notion of main variation is split between both of these modes.

The right-hand panels of Figure 6.8 illustrate the PCA of the same 2-d Toy data set. Recall the difference with SVD is the centering of the analysis at the mean data object (shown as the green plus). This results in the best approximating line, shown in red, being now chosen from direction vectors based at that point. The rank 1 PCA approximations are shown as magenta plus signs, which clearly provide a much better summarization of the data than is provided by SVD, because this direction maximizes the variance of the projections. This direction now appropriately reflects the dominant vertical mode of variation. In particular, these rank one approximations now lie in the middle of the point cloud. This demonstrates

a clear value to data object mean centering. The coefficients of these projections
are the PC1 scores, plotted on the horizontal axis of the scores scatterplot in the
bottom right panel. The cyan lines in the top right panel show the residuals of
this approximation (which of course have minimal sum of squares). The signed
lengths of these are the PC2 scores used on the vertical axis in the bottom right
panel.

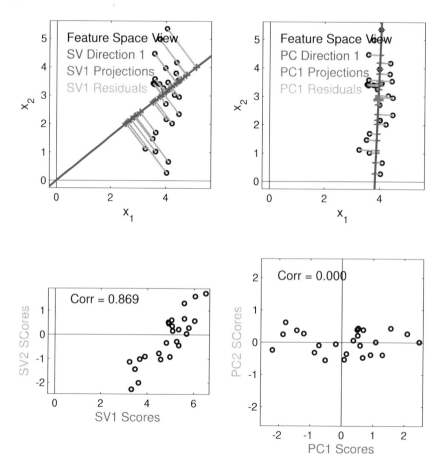

Figure 6.8 *Comparison of SVD and PCA for the 2-d Toy Example data. Top panels are
feature space views of the data objects (black circles) together with the SVD (left) and
PCA (right) approximations. Corresponding scores representations appear in the bottom
panels. Shows the column mean object centering implicit to PCA allows better low-rank
approximation of the data, and also uncorrelated scores scatterplots.*

Figure 6.8 also highlights another aspect of PCA (relative to SVD of uncentered
data) that is worth noting: the scores are uncorrelated. Essentially this is because
when there is correlation, as shown in the lower left panel, the variance of the
larger projection can be made even larger by appropriate rotation. Similarly the

sum of squared residuals (shown as cyan in the top row) can be made smaller by this rotation. It is perhaps surprising, and generally not widely acknowledged, that the lack of correlation in PC scores is a consequence of the column data object mean centering. The mathematics behind these observations is given in the calculation (17.17) in Section 17.1.2.

There are several reasonable notions of centering (i.e. subtracting some type of mean) of a data matrix X (using notation from (3.4)). From the default OODA perspective of column vectors as data objects, a natural definition of center is the *mean vector*

$$\bar{x}_{CO} = \begin{pmatrix} \bar{x}_{1,A} \\ \vdots \\ \bar{x}_{d,A} \end{pmatrix} \in \mathbb{R}^d, \tag{6.1}$$

where $\bar{x}_{i,A} = n^{-1} \sum_{j=1}^{n} x_{i,j}$ for $i = 1, ..., d$. A potential ambiguity (which tends to generate confusion in discussions about data analysis) is whether $\bar{x}_{CO}$ should be called the "column mean", as it is the mean vector of the column vectors, or should be called the "row mean" because it is the vector whose entries are the means of the entries in the corresponding rows of the data matrix X. The OODA resolution of this problem is to use the name *column object mean* for $\bar{x}_{CO}$, hence the subscript of CO. This choice of terminology is deliberately intended to keep the focus of the discussion on the data object column vectors, and not on individual matrix entries.

While column data object vectors are the main focus of OODA, the row vectors that make up the data matrix X are sometimes also of interest, e.g. when considering multi-block methods as in Section 17.2. Using terminology from Table 3.1 OODA terminology for the rows of X is *row trait vectors*. Here the more biological synonym of "trait" instead of "feature" is used because the term "feature vector" already has the common meaning of "data object column vector". The term "trait vectors" for row vectors of the data matrix avoids that ambiguity. Averaging these row vectors gives the *row trait mean vector*,

$$\bar{x}_{RT}^{t} = \begin{pmatrix} \bar{x}_{A,1} & \cdots & \bar{x}_{A,n} \end{pmatrix}, \bar{x}_{RT} \in \mathbb{R}^n, \tag{6.2}$$

which has entries $\bar{x}_{A,j} = d^{-1} \sum_{i=1}^{d} x_{i,j}$ for $j = 1, \cdots, n$.

As noted above deep study of these notions of center, including their representations as projections, can be found in Section 17.1.1. The rest of this section considers how centering by subtracting one or the other (or both) of the column object mean (6.1) and the row trait mean (6.2) impacts a data matrix.

Relationships between heat-map and curve views of a common data set are explored using a *combined view* in Figure 6.9. This is an analysis of the $d = 50$ dimensional Twin Arches data set that was studied in Figures 4.3 and 4.4. That data set was generated with theoretical mean 0 in each coordinate, but due to sampling variation the column object sample mean vector $\bar{x}_{CO}$, shown in the top center panel of Figure 4.3, is only nearly but not exactly equal to $0_{d,1}$. Various views of the input data set are shown in Figure 6.9. The upper right panel is essentially the same as the upper left panel of Figure 4.3 except it has been transposed (so the coordinate axis is now vertical, with curve height on the horizontal axis), and the curves are differently colored. These same curves (actually just vectors) are grouped into a matrix (with corresponding vertical axes) in the heat-map view in the top left panel, using intensities of red (blue) to indicate negative (positive, respectively) curve heights with white for 0. The four clusters revealed in the middle top panel of Figure 4.4 appear as four distinct vertical red-blue color patterns in the columns of the heat-map. Recall from the middle row of Figure 4.3 that two large clusters had arches that followed either a valley-peak or a peak-valley pattern. These appear in the right-hand part of the heat-map, which make it clear that these clusters drove the first PC because there are more column data objects in them. Similarly the last row of Figure 4.3 indicates curves with peak-peak and valley-valley patterns, which appear in the first two smaller clusters of column objects on the left of the heat map. The color scheme of the curves in the top right panel follows the rainbow color bar in Figure 1.2, from magenta through blue, green and yellow to red, in the order of the columns of the matrix. The yellow-red curves are plotted last and follow the valley-peak pattern as shown in the far right cluster of column objects in the heat-map. Second to last are the cyan-green curves with a peak-valley pattern that is similarly consistent with the third set of columns in the top left heat map. The other colors only appear intermittently, as those curves are mostly overplotted by the latter two clusters.

The bottom left panel of Figure 6.9 shows the same data using a parallel coordinates plot of the row trait vectors. Again the horizontal axis is deliberately chosen to correspond to the horizontal axis of the heat map. To make clear the contrast with other panels, a completely different color scheme is employed, which is heat ordered from black through red and orange to light yellow. A key point is that the common variation in the heat map is expressed in quite different ways, although there is an important duality between them. In particular, instead of having four clusters determined by just two peak-valley patterns (as for the column data objects discussed above), from this viewpoint the upper left heat-map shows that the row trait vectors naturally fall into only two clusters determined by more complicated peak-valley patterns. The cluster of rows appearing in the bottom of the heat-map follows a peak-valley-peak blue-red pattern whose curves are highlighted in the bottom left using cooler black and dark red colors. The peak-valley-peak pattern in the curve heights reflects that which is shown with blue-red colors in the upper left panel. The other cluster follows a different peak-valley-peak-valley pattern (again both in the top row cluster in the heat-map as well as in the curves) shown using curves colored with hotter orange and yellow shades in the bottom left panel. The duality between the column object and row trait curve representations is that clusters in one correspond to curve structure in the other.

The bottom right panel of Figure 6.9 shows the distribution of red-white-blue colors used for the pixels in the heatmap. For each color, the height of the bar (and hence its area) indicates the frequency of pixels in the heatmap of that color. The use of this type of graphic was discussed in Section 6.1.

The toy example of how SVD compares to PCA in Figure 6.8 demonstrated the benefits of centering the data by subtracting the column object mean $\overline{x}_{CO}$ (i.e.

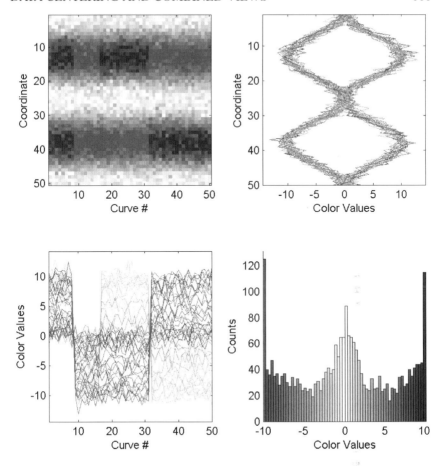

Figure 6.9 *Combined view of the Twin Arches toy data set from Figures 4.3 and 4.4. Upper left is a heat-map image view (red for negative, blue for positive values). Upper right is the columns as data object curves view (using a rainbow color scheme), transposed for good correspondence with the heat map. Lower left shows the corresponding view of the row trait vectors as curves (with a heat ordered color palette). Lower right is distribution of colors in the heat-map.*

the mean of the column vectors). However the heat map view of the data begs the question of why not also center the other way by subtracting the mean of the row trait vectors $\overline{x}_{RT}^{t}$? Note this is the same as subtracting the mean of the entries from each column. Data adjustments of this type are particularly important for various types of bioinformatics data. For example most gene expression measurements involve amplification of quantities such as RNA which essentially involves an uncontrollable random multiplicative factor for each case. When such data are analyzed on a log scale (the usefulness of this is demonstrated in Figures 4.5–4.8), such effects can be mitigated by row trait mean centering, which essentially controls for that random amount of scaling in the original measurements.

The residuals from doing both types of centering, called *double centering* in Section 17.1.1, for the toy data in Figure 6.9 are shown in Figure 6.10. In this case, the column object centering had a negligible impact because as noted above $\bar{x}_{CO} \approx 0$, however the row trait centering does change the visualization. Note that the column object curves view of these mean residuals in the upper right panel is quite different from the column object mean residuals shown in the upper right panel of Figure 6.9. The predominantly cyan-green and orange-red bundles of curves are the same as in the upper left of Figure 6.9, because the means of the entries of those column vectors are all essentially 0 (apparent in both top panels of Figure 6.9). The two new bundles were previously hidden under those bundles (due to being plotted earlier in the process of constructing the upper right panel of Figure 6.9), but have been pulled out by the row trait centering since the mean of their entries is not 0. In particular, each entry of the row trait mean vector $\bar{x}_{RT}^t$ from (6.2) is calculated as the average of the entries of the corresponding heat map column. For the first (left-most) cluster of columns in the heat map, the matrix entries are all between 0 and 10 (apparent in both the upper left heat map and the lower left row curve view of Figure 6.9) so the averages are around 5. The impact of subtracting 5 from each column of the matrix moves the color range [0,10] to [-5,5] in the upper left heat map, and similarly moves the bundle of curves in the lower left panel downwards, while in the upper right panel, this shift of 5 moves only the predominantly magenta bundle of curves to the left, i.e. out from under the other bundles. Similarly, the second cluster in the heat map has column averages around -5, so that cluster has opposite shifts in each panel. In particular in the heat map, the color range shifts from [-10,0] to [-5,5], with a corresponding shift in the lower left panel, as well as pulling the predominantly blue second cluster out from the other bundles toward the right in the upper right panel. A key lesson is that while column object mean centering results in a rigid shift of the full set of column object curves, the row trait mean centering operates on individual column object curves.

As will be seen in Section 17.1.2, for double centered data (both row and column centered), computation of the PCA modes of variation for row trait data objects and for column object data objects are very closely linked. Essentially the loadings for one provide the scores for the other. The first of these PCA modes, for the Twin Arches data of Figures 6.9 and 6.10, is studied using a similar combined view in Figure 6.11. The column object view in the upper right is essentially the transpose of the middle left first mode of variation shown in Figure 4.3, although this color scheme is more interpretable. The reason for this is clear from either panel on the left side. In particular this mode is strongly driven by points in the third and fourth column clusters of the heat map image and essentially ignores the first two clusters which were strongly affected by the row trait mean adjustment. The top left heat map image also illustrates how PCA can be viewed as the best rank-one approximation. In particular, the shown matrix explains as much variation in the full set as possible with a rank-one matrix. Note that the larger number

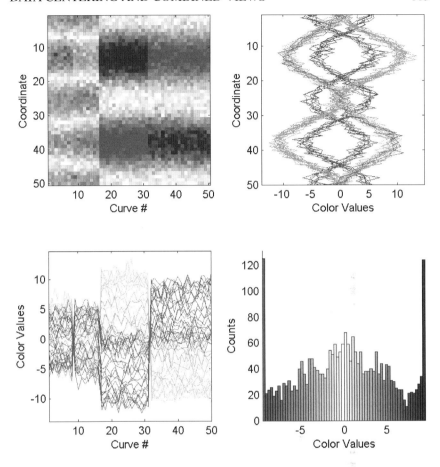

Figure 6.10 *Combined view of Twin Arches data, after both column object and row trait centering (i.e. double centering). Shows interpretation of both types of centering is not always intuitive.*

of nearly white pixels (on the left side of the upper right heat-map) is clearly displayed as nearly zero values in the color distribution plot in the lower right. Also note the much higher vertical scale compared to the color distribution (lower right panel) in Figure 6.10.

The combined view of the second mode of variation based on double centering is shown in Figure 6.12. In contrast to the first mode shown in Figure 6.11, the column object view in the upper right panel appears to be quite different from the mode displayed in the lower left panel of Figure 4.3. But both panels on the left make it clear why. They both reveal that this variation appears in nearly only the first two column object clusters, i.e. in the curves numbered 1–16 on the horizontal axes. At first glance the curves in the top right panel appear to exhibit twice the frequency of those in the lower left of Figure 4.3, but a careful inspection

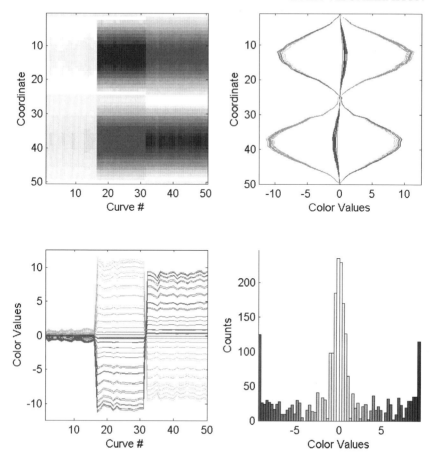

Figure 6.11 *First mode of variation (best rank-one approximation) based on double mean centered Twin Arches data from Figure 6.10. Upper right panel shows the column object mode of variation is similar to that in Figure 4.3. Left two panels reveal this is driven by the large two column object clusters revealed above.*

reveals that this is just the result of separately row trait mean centering each of those two clusters. From the matrix approximation viewpoint, this rank one matrix is the product of a row and a column vector, each of which is orthogonal to the corresponding vector in Figure 6.11, which best approximates the remaining variation. Note that both axes of the color distribution in the lower right panel are quite different from the preceding cases showing even more nearly white pixels.

In the examples shown in Figures 6.9–6.12, basing the analysis on double mean centering gives a very insightful decomposition of the structure of the data set. However, this is not always the case, as illustrated in Figure 6.13. That studies the same Lung Cancer RNAseq gene expression curves shown in Figures 4.5–4.8. The left panel of Figure 6.13 shows the column object mean centered version of

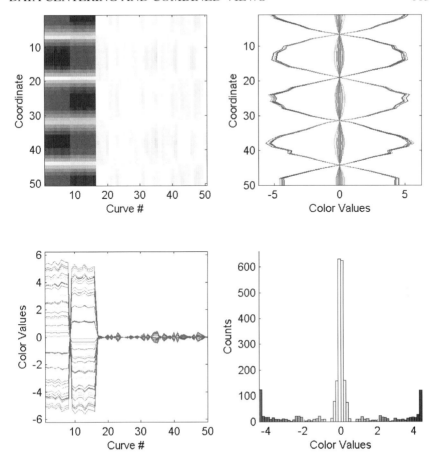

Figure 6.12 *Second mode of variation (rank one approximation) based on double mean centered Twin Arches data from Figure 6.10. The column object mode of variation shown in the upper right panel this time appears different from that in Figure 4.3. Left two panels again explain this via insights about clustering of column objects.*

the gene expression curves, using the colors from Figure 4.8. Note that the essentially unexpressed cases colored red are still clearly distinguished. Furthermore the critical gold versus blue distinction in the exon near the right side is also still very clear. However, there is some loss of insight relative to the very clear explanation available from the Figure 4.8 view. This loss of interpretability becomes much worse in the double centered view shown in the right panel of Figure 6.13. This time the impact of row trait mean centering (with strong potential for moving clusters as demonstrated in the top right panel of Figure 6.12) moves these clusters in a way that strongly impairs seeing the key lessons in this data set. In particular the fact that the (unexpressed) red cases are substantially lower than the others mostly disappears because each curve has separately had the mean of its entries subtracted which obscures important aspects of the data.

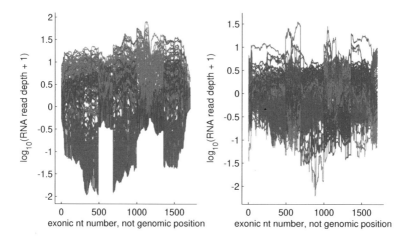

Figure 6.13 *Centered views of Lung Cancer data from Figure 4.8. Column object centering on the left and double centering on the right. Shows how centering can give less interpretable views of data.*

An important lesson from these examples is that substantial thought should be devoted to centering of data. Both column object centering and row trait centering have the potential to either reveal or obscure important insights. Further centering issues, which include various projection viewpoints, are studied in greater depth in Section 17.1.1.

6.4 Scatterplot Matrix Views of Scores

Scatterplot matrix views provide yet another frequently useful type of insight, this time highlighting the *relationships between data objects*. This has also been seen many times in the above examples as detailed here.

- *Figure 4.4, Twin Arches example.* This view provided clear presentation of the important structure in the data set that was challenging to understand from the curve views of this data in Figure 4.3. In particular, it revealed that there are four clusters in the data, whose 1-d projections generated the perhaps surprising effects shown there. That example also illustrated axis scaling issues, where the PC3 axis was rescaled to minimize white space, at the cost of not clearly revealing that those scores were an order of magnitude smaller than the others, i.e. the noise component was generated from an isotropic Gaussian distribution.

- *Figure 4.6, Lung Cancer data.* This PCA scatterplot matrix revealed the important structure of clusters in the data, which motivated the development of a novel method for finding alternate splicing using gene expression data.

- *Figure 4.7, Lung Cancer data.* This brushed version of Figure 4.6 defined a color scheme whose application led to Figure 4.8, which revealed the nature of the clusters in this data set, in terms of alternate splicing. This led to the invention of the SigFuge method, that was effective at finding new alternate splicing events.

- *Figure 4.9, Pan-Cancer data.* Here is a PCA scatterplot aimed at showing the limitations of PCA, using a cancer gene expression data set. Colors and symbols indicate cancer types, and reveal some differences, although also substantial overlap of some of the classes.

- *Figure 4.10, Pan-Cancer data.* This demonstrated how computing scores by projecting the data onto non-PCA directions can give much better visual contrast between cancer types. This used DWD directions (discussed in Chapter 11) trained on pairs of individual cancer classes, which directly target maximal separation of the classes.

- *Figure 4.12, High-dimensional Gaussian example.* Here is another example designed to reveal limitations of PCA. No apparent structure is apparent in this PCA scores scatterplot, especially in terms of comparison of the two classes.

- *Figure 4.13, High-dimensional Gaussian example.* This again shows that the DWD direction, which deliberately optimizes for separation between classes, reveals a strong difference between them. Such views generally improve on PCA for understanding potential class differences. Note that the DWD scores exhibit substantially less variation than is apparent in the PC directions. This is again an axis scaling issue, that could be highlighted at the cost of introducing white space.

- *Figure 5.3, Drug Discovery data.* The PCA view revealed that the raw data were dominated by outliers, which distracted from the important activity differences which were the main motivation of the analysis.

- *Figure 5.12, Drug Discovery data.* Here the impact on the PCA view of data cleaning and focusing only on the binary variables is revealed. Shows much richer structure in the data, and reveals interesting insights about activity differences, shown with color and symbols.

- *Figure 5.15, Drug Discovery data.* This shows that the appropriately cleaned and transformed non-binary variables do a much better job of understanding activity differences.

- *Figure 6.6, Two Cluster example.* Here it is seen how scatterplot matrices can reveal structure in data that is not easily available from heat-map views.

In most of the above scatterplot matrices, the graphics are straightforward because the directions that determine the axes are orthogonal. But this is not always the case, for example the three DWD directions in Figure 4.10. A naive way to handle this is to just plot the scores on the vertical and horizontal axes. An important drawback to that is the scores plot no longer maintains relative positions of the data in the underlying Euclidean space, and thus can give a quite distorted view, especially when the direction vectors are far from orthogonal. This is demonstrated by the toy data set in Figure 6.14. The data lie on four small circles in $\mathbb{R}^2$, with more points on the circles centered on the line from the blue to the cyan circles, shown using a solid line type. Another direction of interest in such data is along the line from the center of the red circle to the yellow one shown with a dot-dashed line type. Recall that for both of those lines, projection of each data point onto them consists of finding the closest point on the line, which could be connected by a line segment that is perpendicular to the given line, e.g. the cyan line segments in the right panel of Figure 3.2 (these are not shown to keep the picture from being too busy). The distance from the origin to each projection is the corresponding score. Such scores are plotted in the other panels of Figure 6.14. The top right shows the red-yellow scores (i.e. projections on the dot-dashed line) on the vertical axis, and the blue-cyan scores on the horizontal axis. There is large distortion in the relative position of the circles (the blue and red circles appear much closer to each other, as do the cyan and yellow) and also strong distortion in the shape of the circles.

Such distortions generally occur from naively plotting scores with respect to non-orthogonal axes. Visualizations that avoid such distortions are shown in the bottom two panels. The bottom left has the same blue-cyan scores on the horizontal axis, but the vertical axis is now the projections (i.e. scores) on the line perpendicular to that axis, which is shown using the dashed line type in the panels on the left. Everything is now a rigid rotation of the quantities in the upper left panel, which thus preserves relative locations. Such graphics were used in the plots above the diagonal in Figure 4.10. The lower right plot in Figure 6.14 illustrates the plots below the diagonal. There the horizontal axis represents the red-yellow direction (still shown with the dot-dashed line type) and the vertical axis is the orthogonal direction (again not shown as yet another line for ease of understanding). As for the lower left panel there is no distortion of the relative positions, and the projection of the other direction of interest (the solid blue-cyan line) is shown to maintain the comparison between the two directions. With this understanding in mind, it could be useful to review the corresponding above and below diagonal plots for the Pan-Cancer data in Figure 4.10.

For comparison of slightly different populations, such as viewing the impact of adding a few new observations, or partially perturbing some data objects, a useful graphical device is to flip between two different scatterplot matrices on a computer screen. This requires use of exactly the same coordinate axes, which can be a challenge for some implementations of PCA. The issue is that PCA is based on an eigen analysis of the sample covariance matrix (or equivalently on a Singular Value Decomposition of the centered data matrix, as noted in Section 17.1.2). While eigen analysis provides a very useful decomposition, there is a natural ambiguity in terms of the direction of the signs of eigenvectors (direction vectors). In particular, for any eigen vectors all entries can be multiplied by -1, i.e. the vector can be rotated $180°$, and it will still be a direction of maximal variation. This can strongly disrupt attempts to flip between two graphics, and perhaps other tasks as well, so it is useful to adopt a convention to minimize the occurrence of this event. A recommended choice is to make the sum of the projections onto the diagonal vector $(1, 1, \cdots, 1)^t$ (denoted $\mathbf{1}_{d,1}$ in (10.1)) positive. This is simply achieved by choosing the direction which makes the average of scores positive.

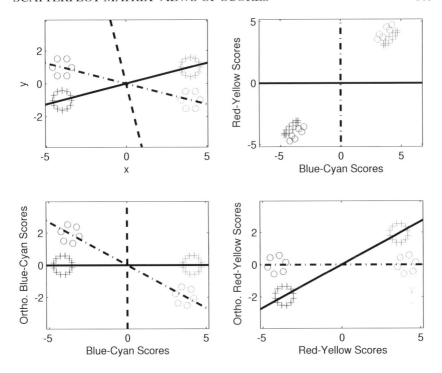

Figure 6.14 *Toy example showing problems with plotting projections onto non-orthogonal lines. Raw 2-d data in upper left shows lines (directions of interest) connecting clusters of interest. Upper right panel is a plot of projections on the red-yellow line (dot-dashed) as a function of projections on blue-cyan line (solid), which introduces large distortions in placement of data. Lower left panel shows projections on the blue-cyan (solid) line, and on the orthogonal line (dashed), which keep relative positions, i.e. are just rotations (and flips) of the original data. Lower right is same for projections on the red-yellow (dot-dashed) line.*

This will fail when the data have been row trait mean centered, i.e. each column has mean 0 (since that inner product gives the sum of the vector entries). In that case a reasonable alternative is to choose the direction which makes the maximum absolute score positive. The subspace generated by the diagonal vector $J_{d,1}$ is studied much more deeply in Section 17.1.

Note that many, but not all, of these scatterplot matrices are based on PCA directions. Some exceptions are the use of DWD (e.g. the Pan-Can data in Figure 4.10) to give directions with better class separation. But there are many other potentially interesting directions, depending on the context, as discussed in the next section.

6.5 Alternatives to PCA Directions

PCA is a well-proven workhorse in finding insightful directions for the computation of scores that reveal structure in data, because the set of directions of maximal variation which it targets frequently highlight such structure. However PCA does not necessarily reveal all types of interesting structure, especially when the latter appears in directions of relatively lower variation. This issue becomes more serious in high dimensions. Hence it is useful to realize that there are many alternatives, which can give important insights depending on the particular context. These include the following. Some of these come directly from multivariate analysis (recall this is viewed as a special case of OODA at the beginning of Chapter 1), but others arise from other considerations.

- *Classification Directions.* The statistical context called classification (also termed discrimination) is discussed in detail in Chapter 11. Binary linear classification methods seek to find a hyperplane that separates two labeled subsets of the data usually in some optimal sense. The normal vector to a separating plane often provides very useful directions for visualization when there is interest in highlighting differences between two subsets of data. The Mean Difference direction (computed as the difference between the means of the two subsets, and discussed in detail in Section 11.1) is a simple and generally interpretable direction for this purpose. Often better visual impression comes from Distance Weighted Discrimination (DWD studied in Section 11.4), because the target of the optimization is more directly related to visual separation than are subset means. Examples of this use of DWD appear in Figures 4.10 and 4.13. In relatively low-dimensional situations, good visual separation of multiple classes can be achieved by Canonical Variate Analysis, see Mardia et al. (1979), which is also called Multiple Discriminant Analysis in Duda et al. (2001). Deeper discussion of related issues can be found in Section 11.4.

- *Independent Component Analysis.* This source of interesting directions in high-dimensional data space seeks directions of maximal non-Gaussianity. The key idea was first developed by Cardoso and Souloumiac (1993) and is discussed in detail in Hyvärinen et al. (2001). Its main motivation is called *Blind Source Separation*, which mathematically models the ability of human hearing to comprehend multiple simultaneous conversations. This is done using two different mixtures (linear combinations) of time series signals. When the coefficients are known, signal separation only needs simple matrix inversion. More challenging is the blind case of unknown coefficients, which Independent Component Analysis solves by finding coefficients resulting in maximal non-Gaussianity. This works because linear combination is essentially an averaging operation that tends to increase Gaussianity. The approach has become very popular in many neuroscience analyses. Such directions are often useful in exploratory data analysis, as they tend to find interesting structure such as directions of multi-modality. The idea of finding directions of interest by maximizing non-Gaussianity goes back at least to the *varimax* subspace rotation from *factor analysis* proposed by Kaiser (1958).

- *Fourier Subspace.* For applications where periodicity plays a central role, Fourier basis vectors can provide useful directions for projection. This idea was used to give an insightful analysis of the Spellman et al. (1998) Yeast Cell Cycle data by Zhao et al. (2004). That noisy data set featured gene expression measured over two cell cycles with a high level of background noise. The noise was high enough that a standard PCA resulted in non-interpretable directions which did not exhibit the periodicity that was the point of the study. A much better noise reduction was done by projecting the data onto the subspace of even Fourier frequencies, which kept only those parts of the data where the first half of the time series is identical to the second half. Good access to aspects of Fourier analysis important to statistics can be found in Brillinger (1981) and Bloomfield (2000).

- *Wavelet Decomposition.* This important variation of Fourier Analysis shifts the focus from approximately periodic signals to those with more local structure, resulting in orthogonal bases with good signal compression (and thus efficient data object representation) for curves with local structure such as jumps and bumps. See Section 3.3 for references and more discussion of wavelets in the context of curve estimation.

- *Known modes of variation.* In some applications particular modes of variation are given as very important. For example in the evolutionary biology work of Izem and Kingsolver (2005), the focus is on modes of variation of central importance to the ability of a population to adapt to environmental changes. Some of these modes were nonlinear, motivating some relatively early research on quantifying variation on manifolds. Theoretical properties of the methods developed there were investigated in Izem and Marron (2007). More recent developments in the area of nonlinear modes of variation are discussed in Chapters 8 and 9.

- *Maximal Smoothness Directions.* As noted several times above PCA is very powerful because of, but can also sometimes be hobbled by, its property of finding directions of maximal variation in a Euclidean feature space. In some evolutionary biology applications, as discussed in Gaydos et al. (2013) and Kingsolver et al. (2015), the major interest is instead on directions of *minimum variation*. This is because they represent ways in which a given population is *unable* to adapt to environmental change, called the *nearly null space*. Such directions can also be easily found by PCA, simply by working with small eigenvalues, instead of the large ones that are the basis of conventional PCA. However, unlike PCA directions, which are typically interpretable because they are driven by important modes of variation, directions of minimal variation are generally far less interpretable as most of them are driven by uninteresting sampling artifacts. To find important interpretable directions of minimal variation, among the nearly null subspace that minimizes variation, those papers propose finding directions of *maximal smoothness*. These are found by eigen analysis of a matrix reflecting *simplicity* (essentially smoothness, but also usually reflecting maximal interpretability) of directions instead of variation as is done in conventional PCA. In particular, Gaydos et al. (2013) propose a method called *PrinSimp*. This studies PCA eigenvectors from a perspective that simultaneously accounts for variation explained as well as simplicity of representation of the null space. Kingsolver et al. (2015) provide a number of interesting applications of PrinSimp. An R software package implementation of PrinSimp is available from Zhang et al. (2014).

- *Non-Negative Matrix Factorization.* In some functional data applications, not only are the input data curves nonnegative, but there is also a desire for all aspects of the analysis to be nonnegative. An important example is chemical spectra as data objects, where the height of the curves represent mass, for example the TIC curves in Figure 2.1. Standard functional PCA can be very unattractive to chemists, because while the first component direction will typically have nonnegative loadings, the orthogonality constraint of PCA entails that all other components will have negative loadings, which chemists tend to find challenging to interpret (because mass is an intrinsically nonnegative quantity). Such considerations are addressed by *Non-negative Matrix Factorization* (NMF) as proposed by Lee and Seung (1999). Simple insight into the relationship between NMF and PCA comes from considering (as detailed in Sections 1 and 17.1.2) the latter as providing an optimal approximation of a mean centered $d \times n$ data matrix X, by a matrix product of a $d \times r$ loadings matrix L (e.g. in the Twin Arches analysis of Figure 4.3 the lower left modes of variation are multiples of the $r = 3$ columns of L) and an $r \times n$ scores matrix S (e.g. in Figure 4.4 various pairs of the $r = 4$ rows of S are plotted against each other). In particular, as seen in Section 17.1 L and S essentially provide minimizers (subject to orthogonality conditions) of the Frobenius norm (defined at (7.9)) of $X - LS$. NMF is based on a similar optimization, but with the constraints that all the entries of both L and S are non negative. As noted say in Fogel et al. (2006) there are other applications, e.g. in gene expression, where an essentially additive, nonnegative decomposition of variation can be viewed as more intuitive than the orthogonal decomposition provided by PCA. As discussed in Section 8.6, a weakness of classical NMF is that it is not *nested*. This issue has been addressed by the Nested Nonnegative Cone Analysis proposed by Zhang et al. (2015).

- *Binary Matrix Factorization.* In some applications, the data matrix X may be *binary*, i.e. each entry may take on only the values 0 or 1. The entries are numbers, so one can analyze the variation using a classical PCA decomposition. But for most purposes, such a decomposition will not be very interpretable, because neither loadings nor scores will be binary. A more interpretable version is offered by *Binary Matrix Factorization* proposed by Srebro et al. (2005) and Zhang et al. (2007b), which has a goal in the spirit of NMF, i.e. an approximation of the form $d\,(X, LS)$, where the matrix product LS is computed with Boolean operations, and d is an appropriate metric on binary matrices such as Hamming (1950) distance which just counts the number of different entries.

Also appealing is the Jaccard (1901) distance which is the proportion of the
number of differences among the total number of ones.

- *Lines Containing Data.* In computer vision, when analyzing images involving
  man-made objects, such as touristic city scapes, many useful methods are based
  on finding lines in an image. This is done by filtering the image to find *points
  of interest*, e.g. points of maximal change in various senses. Lines are then discovered by finding lines containing a reasonable number of such points, which
  are used as building blocks for more complex types of analysis. The RANSAC
  (more precisely Random Sample Consensus) method, proposed by Fischler and
  Bolles (1981), is a useful algorithm for finding such *inlier lines*, featuring a few
  but "very close to collinear" points. These ideas are an interesting inversion of
  the robustness ideas of Chapter 16 where the focus is on outliers.

- *Maximal Data Piling.* A perhaps surprising phenomenon in high-dimensional
  data analysis (when the dimension is larger than the sample size), is that given
  two subsets of the data, there is (with high probability) a set of directions where
  each subset projects to a single point. The direction which maximizes the distance between these points is called the *Maximal Data Piling* direction, which
  is studied in more detail in Section 11.1. This direction is not particularly useful
  for most visualization purposes, but is still worth keeping in mind when contemplating the large variety of directions available in high-dimensional space.

- *Auxiliary Data Directions.* In many situations, visualization in directions that
  incorporate additional data is very useful. The simplest of these may be standard multiple linear regression, when the additional information is in terms of
  a response value Y_i for each data object. More sophisticated approaches, especially in high dimensions include the *Sliced Inverse Regression* of Li (1991)
  and the *Sufficient Dimension Reduction* of Cook and Lee (1999). When richer
  vector-valued data are available, methods such as Partial Least Squares and
  Canonical Correlation Analysis, discussed in Section 17.2, also provide very
  useful directions for visualization.

CHAPTER 7

Distance Based Methods

This chapter is about OODA methods that are based only on distances between data objects. An advantage of such approaches is that they are quite broadly useful, which can be important in exotic data spaces such as manifolds, or tree/graph spaces, where simply developing data analytic methods can be challenging. In particular, the only structure needed on the object space for such methods is the presence of a *metric* (i.e. distance). In some situations, even less may be needed, e.g. perhaps only a *pseudo metric* or even a *dissimilarity*, but these will not be further pursued here.

In this chapter, the symbol δ will be used to denote metrics, i.e. distance functions. As in any elementary analysis textbook, as apparently originally formulated by Fréchet (1906), a metric is a two argument function from some space to the real numbers which is symmetric, nonnegative with the value 0 taken if and only if the two arguments are the same, that satisfies the triangle inequality. Given a set of data objects in an arbitrary space

$$\{\chi_i : i = 1, \cdots, n\},$$

and a distance δ, the corresponding $n \times n$ symmetric *distance matrix* is

$$D = \begin{bmatrix} 0 & \delta(\chi_1, \chi_2) & \cdots & \delta(\chi_1, \chi_n) \\ \delta(\chi_2, \chi_1) & 0 & & \vdots \\ \vdots & & \ddots & \delta(\chi_{n-1}, \chi_n) \\ \delta(\chi_n, \chi_1) & \cdots & \delta(\chi_n, \chi_{n-1}) & 0 \end{bmatrix}. \qquad (7.1)$$

Working with data objects in this type of format is rather common in the machine learning literature, see Cristianini and Shawe-Taylor (2000), Schölkopf and Smola (2002), and Shawe-Taylor and Cristianini (2004) for good overviews. Indeed a central idea in that area is called the *kernel trick*, where one works with a matrix of inner products (closely related to the usual distance in Euclidean spaces), with the important goal of large computational benefits.

As noted in Sections 3.1 and 4.3, a quick and dirty default approach to OODA is to first find a metric, and then simply carry out a distance-based analysis. Such notions of *center* are discussed in Section 7.1. Methods for understanding variation about the center, in the spirit of PCA, using only distances are explored in Section 7.2. Clustering is another set of data analytic methods, that are frequently based only on distances between data objects. Cluster analysis is discussed in Chapter 12.

While the metric first approach gives useful insights of some types, a drawback

DOI: 10.1201/9781351189675-7

is that it tends to only yield scores (displaying relationships between data objects e.g. for the Twin Arches data in Figure 4.4) and *not* full modes of variation as defined in Section 3.1.4. In particular, it does not easily give the one-dimensional indexed sets in the data space that convey all of the insights available from full modes. A more careful use of particular data space structure can result in full modes of variation in many OODA contexts, as discussed in Chapters 8, 9, and 10.

A critical aspect of metric-based analyses is that choice of distance has a major impact on the results. For a particularly dramatic example consider the *discrete metric*

$$\delta_D(\chi, \varsigma) = \begin{cases} 0 & \text{for } \chi = \varsigma \\ 1 & \text{for } \chi \neq \varsigma \end{cases}.$$

This exists for any space, but is useless for OODA analyses, because it contains no information about how the data objects relate to each other. Section 8.7 contains some perhaps surprising examples on the strong impact of metric choice in the context of covariance matrices as data objects.

Generally a good choice of metric is very situation dependent. For example, in the case of Euclidean data objects, say $x, y \in \mathbb{R}^d$, the standard *Euclidean L^2 distance* (a special case of the L^p norms defined in (3.3))

$$\delta_2(x, y) = \|x - y\|_2 = \left(\sum_{j=1}^{d} (x_j - y_j)^2 \right)^{1/2} \tag{7.2}$$

is often very useful. However, when it makes sense to think in terms of polar coordinates, and the important variation happens in the angular direction with mostly distracting noise in the radial direction, a more useful metric can be the *cosine distance*

$$\delta_C(x, y) = 1 - \frac{2}{\pi} \cos^{-1} \left(\frac{x^t y}{\|x\|_2 \|y\|_2} \right), \tag{7.3}$$

which is driven only by the angle between x and y projected onto the unit sphere. The cosine distance is actually a pseudo metric because it does not distinguish between vectors that are multiples of each other. When outliers are a major concern, the L^1 *distance*

$$\delta_1(x, y) = \|x - y\|_1 = \sum_{j=1}^{d} |x_j - y_j| \tag{7.4}$$

is another useful approach because of its natural tendency to down-weight their influence. The L^1 distance is also very useful when sparsity is desired because of a tendency to give optimization results with many 0 entries. In some situations, a drawback of the L^1 distance is a lack of rotation invariance, i.e. a strong dependence on the coordinate system, as shown in Figure 7.4. In situations where the variables have a relationship determined by a non-singular covariance matrix Σ, a unit free metric is the *Mahalanobis distance*

$$\delta_M\left(\boldsymbol{x}, \boldsymbol{y}\right) = \left(\left(\boldsymbol{x} - \boldsymbol{y}\right)^t \boldsymbol{\Sigma}^{-1} \left(\boldsymbol{x} - \boldsymbol{y}\right)\right)^{1/2}.$$

The Mahalanobis distance to the mean is the multivariate generalization of the one-dimensional notion of number of standard deviations from the mean.

7.1 Fréchet Centers In Metric Spaces

An important notion of *center* of a set of data objects $\{\chi_1, \cdots, \chi_n\}$ in an arbitrary metric space S (with distance δ) is the *Fréchet mean*,

$$\arg\min_{\chi \in S} n^{-1} \sum_{i=1}^{n} \delta\left(\chi, \chi_i\right)^2, \qquad (7.5)$$

from Fréchet (1948). The factor of n^{-1} does not affect the minimizer, but provides convenient notation below. This is a direct generalization of the standard sample mean $\overline{\boldsymbol{x}} = n^{-1} \sum_{i=1}^{n} \boldsymbol{x}_i$ in Euclidean space $\mathbb{R}^d$, because it is straightforward to show that $\overline{\boldsymbol{x}}$ is the solution of (7.5) in the case of Euclidean distance (7.2). Insight as to how the Fréchet mean works is given in Figure 7.1.

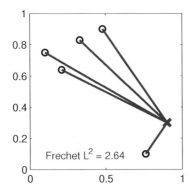 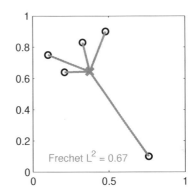

Figure 7.1 *Toy two-dimensional example illustrating the Fréchet mean. Left panel shows a (far from optimal) candidate point as a blue x, with line segments representing the distance to each data point. Right panel shows the optimal Fréchet mean as a green x, with again line segments showing this point is much closer to the data points than the blue candidate on the left. These issues are quantitated in each panel using the Fréchet average of squared L^2 distances.*

A toy data set of $n = 5$ data points in $\mathbb{R}^2$ are shown as black circles in both panels. The blue x in the left panel is a candidate choice of mean. The Fréchet criterion in (7.5) (multiplied by $n = 5$) is the sum of the squared lengths of the blue line segments, whose numerical value for this candidate is seen to be 2.64. The operation of solving the Fréchet optimization problem can be thought of as moving the candidate point to minimize the Fréchet criterion. The solution is shown

128 DISTANCE BASED METHODS

as the green x in the right panel (actually just the familiar sample mean vector $\bar{x}$ in this special case), based on an overall shorter collection of line segments and the smaller criterion value of 0.67.

Depending on the choice of metric, many familiar notions of center can be viewed as Fréchet means. For example on the real line $\mathbb{R}^1$, the median can be written in the form (7.5), by taking $\delta = \delta_2^{1/2}$ (note that it is straightforward to show that the square root of any metric is again a metric). Furthermore on $\mathbb{R}_+ = \{x \in \mathbb{R} : x > 0\}$ the geometric and harmonic means are Fréchet means, based on $\delta(x,y) = |\log x - \log y|$ and $\delta(x,y) = |x^{-1} - y^{-1}|$, respectively.

As noted in Koenker (2006), for any distance δ an important variation is the *Fréchet median*,

$$\underset{\chi}{\arg\min}\, n^{-1} \sum_{i=1}^{n} \delta(\chi, \chi_i), \qquad (7.6)$$

which differs from the Fréchet mean only by replacing the power of 2 by 1. This and other variations of the Fréchet mean have been extensively studied in the field of *robust statistics*, see e.g. Hampel et al. (2011), Huber and Ronchetti (2009), Staudte and Sheather (1990), and Clarke (2018), because both the median and also distances δ which down-weight points farther away, have good properties in terms of reduced sensitivity to outliers. See Fletcher et al. (2009) for interesting applications of this approach in the context of Riemannian manifolds. Deeper discussion of robustness in OODA contexts can be found in Chapter 16.

Direct comparison between the Fréchet mean and median is provided in Figure 7.2. This example features the same toy data set as in Figure 7.1, but this time the center-point is calculated as the Fréchet median based on δ_2. Again a poor candidate point is shown as the blue x, with the optimal solution shown in green. The key difference is that the sum of lengths of the line segments (not their squares) make up the Fréchet sums that are shown.

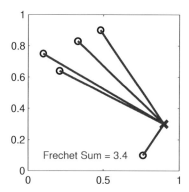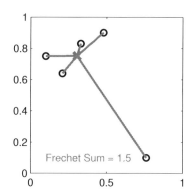

Figure 7.2 *Toy example illustrating the Fréchet median, based on the distance δ_2, for the same data set as in Figure 7.1. Again a sub-optimal candidate appears on the left, with the Fréchet median on the right. This estimate is more robust in the sense of reduced sensitivity to the outlying point in the data set.*

First note the substantial difference between the Fréchet mean and median in this case. In particular, the improved robustness property of the Fréchet median can be seen in terms of the green x being in the middle of the convex hull of the 4 nearest points in Figure 7.2, while the outlier is far enough away to pull it outside in Figure 7.1, because it has much stronger influence when the distances are squared.

The impact of metric choice is again illustrated in Figure 7.3. This is also the same toy data as in Figures 7.1 and 7.2, but now the distance used in the Fréchet median is the L^1 distance δ_1. Because this distance is the sum of absolute distances in each coordinate direction, each is now represented as a horizontal and a vertical line.

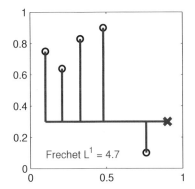 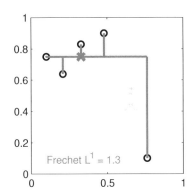

Figure 7.3 *Study of the impact of metric choice using the same toy data from Figures 7.1 and 7.2. Again the left panel shows the Fréchet sum, but now with δ_1, for a poor candidate, and the right shows the optimum, i.e. the L^1 Fréchet median. This version of center also demonstrates reduced influence of outliers.*

This variation of the Fréchet median reveals some perhaps surprising differences with the δ_2 version. In particular, the solution here is the point-wise median (i.e. this can be computed by simply taking the median of each coordinate of the data vectors), which is not true for the δ_2 version of the Fréchet median. Thus the δ_1 variation is easier to compute, while an iterative (but rather fast even in high dimensions) computation is needed in the case of δ_2.

Another basis for the comparison of these notions of center is *rotation invariance*, as illustrated in Figure 7.4. The panels show different rotations of the same toy data set in $\mathbb{R}^2$, together with both the δ_1 and δ_2 Fréchet medians. Because the distance δ_2 is rotation invariant, the greenish blue x shows that the Fréchet median has the same relative position when the data are rotated. But because the distance δ_1 is not rotation invariant, this is not true for that version, shown as the red + sign. This example is deliberately constructed to show that the δ_1 Fréchet median (i.e. the coordinate-wise median, which is here seen to be a naive choice of "multivariate median") can be viewed as a poor notion of center, in the sense that it actually lies on the boundary of the convex hull of the data.

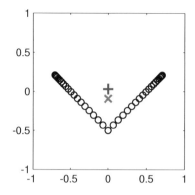 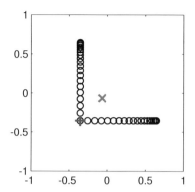

Figure 7.4 *Two-dimensional toy example studying rotation invariance. Data in the right panel are a 45-degree rotation of the data on the left. In each case the δ_2 based Fréchet median is shown as the blue-green x, with the δ_1 Fréchet median appearing as the red plus sign. This shows the latter is not rotation invariant.*

Despite these clear differences, note that both of these medians are direct generalizations of the familiar one-dimensional median in $\mathbb{R}^1$ in the sense that they both reduce to that quantity in the case $d = 1$. As discussed in Section 16.3, actually there are a number of other quite different notions of center in $\mathbb{R}^d$ which also reduce to the median in the case $d = 1$.

Neither the Fréchet mean nor median are guaranteed to be unique. This is typically handled by working with *mean sets* (*median sets*, resp.). For example, on $\mathbb{R}^1$ when n is even, the interval between the central two data points is the median set. In that situation, and in others, sometimes a representative is chosen for the set, e.g. the midpoint of the interval in the $\mathbb{R}^1$ median case. The extreme case of non-uniqueness is the discrete metric δ_D, with respect to which both the Fréchet mean and median sets are the full data space, and thus are of no practical use.

From the mathematical statistical perspective an important issue is just what is being estimated by all of these notions of center. Useful models for this are based on the notion of randomness and probability theory. The convention here and in the rest of this book is that a tilde over notation indicates a random quantity as shown in Table 7.1.

Type of Quantity	Notation for Random Version
random number (scalar)	$\widetilde{x}$
random vector	$\widetilde{\boldsymbol{x}}$
matrix of random variables	$\widetilde{\boldsymbol{X}}$
element of arbitrary space	$\widetilde{\chi}$

Table 7.1 *Notation for random quantities.*

When the data objects are thought of as being a *random sample* (i.e. drawn independently) from a probability distribution on the object space, then appropriate notions of population center are defined using Fréchet criteria based on expected values. In particular, sample Fréchet means as defined in (7.5) can be usefully treated as estimates of the corresponding theoretical (i.e. population) Fréchet means

$$\arg\min_{\chi \in S} E\delta\left(\chi, \widetilde{\chi}\right)^2, \tag{7.7}$$

where $\widetilde{\chi}$ is a random variable with the population probability distribution. Similarly for Fréchet medians. A good overview of many results studying various types of asymptotic convergence of estimates to their population versions, in the challenging contexts of data objects on manifolds and stratified spaces, can be found in Patrangenaru and Ellingson (2015). That book also provides a good overview of affine and projective shape manifolds. For some early related discussion of statistics on stratified spaces see Bhattacharya et al. (2013). Some interesting more recent results can be found in Huckemann and Eltzner (2017) who derive asymptotics for backwards methods as developed in Section 8.6 and Huckemann and Hotz (2016) who explore nonparametric methods on manifolds. See Hotz et al. (2013) for a particularly unusual limiting distribution theory for the Fréchet mean in a stratified space, involving the nonstandard notion of *stickiness*.

There are several synonyms of Fréchet mean. These include *Riemannian barycenter* (really the center of mass in physics, when using the Euclidean distance δ_2), and *geodesic mean* (when δ is a geodesic distance, say on a curved manifold, as discussed in Chapter 8). The term Karcher mean is sometimes also used, with a common distinction being that it refers to only a local minimizer, instead of a global minimizer as in (7.7). However this terminology is not universally accepted, see Karcher (2014). Similarly there are synonyms for the Fréchet median including *geometric median*, *spatial median*, and L^1 *M estimate*.

Also useful is to define a *weighted Fréchet mean* in an arbitrary metric space S with fixed weights w_i, $i = 1, \cdots, n$, by solving

$$\arg\min_{\chi \in S} \sum_{i=1}^{n} w_i \delta(\chi, \chi_i)^2$$

where $\sum_{i=1}^{n} w_i = 1$. The Fréchet mean given in (7.5) has equal weights $w_i = 1/n$, $i = 1, \cdots, n$. An example where the weighted Fréchet mean is useful is the construction of an *interpolation path* between two objects χ_1 and χ_2, where the weights depend on a parameter $\alpha \in [0, 1]$:

$$\arg\min_{\chi \in S}(1 - \alpha)\delta(\chi, \chi_1)^2 + \alpha\delta(\chi, \chi_2)^2. \tag{7.8}$$

Values $\alpha = 0$ and $\alpha = 1$ correspond to the two objects at either end of the path, and the Fréchet mean of the two objects is in the middle of the path given by $\alpha = \frac{1}{2}$. Insightful examples of interpolation paths on the manifold of covariance matrices are given in Figure 7.14, where quite different behavior is observed for each choice of distance. For certain types of distances (intrinsic distances) the interpolation path also provides a *minimal geodesic path* between points on a manifold.

Values of α in (7.8) which are outside the interval $[0, 1]$ lead to an *extrapolation* path. In particular if we choose $\alpha < 0$ we extrapolate beyond χ_1 in an extension of the path from χ_2 to χ_1, and if $\alpha > 1$ we extrapolate beyond χ_2 in an extension of the path from χ_1 to χ_2.

7.2 Multi-Dimensional Scaling For Object Representation

As with notions of center, there are many ways of quantitating variation about the center, based solely on metrics. A very simple one is the *Fréchet variance* which is just the minimum value attained in (7.5).

But generally more useful for data analytic tasks, such as those discussed in Sections 4.1–4.3, are distance-based analogs of PCA. One approach to this is *Multi-Dimensional Scaling* (MDS). MDS has been very popular in the psychometrics literature, and at least the nomenclature is usually attributed to Torgerson (1952, 1958) and Gower (1966). However, the underlying mathematics is substantially older, see Eckart and Young (1936) and Young and Householder (1938). MDS is also called Principal Coordinates Analysis (abbreviated PCoA) in the biological literature.

In its simplest form, MDS starts with a set of n data objects, with a known set of pairwise distances between them summarized as a distance matrix D as in (7.1), and seeks to represent the objects as a set of points $x_1, \cdots, x_n \in \mathbb{R}^d$ for some d, in such a way that the Euclidean distances $\delta_2(x_i, x_j)$ approximate the elements of D as well as possible, in various senses. When the input distance matrix D is itself composed of pairwise Euclidean distances δ_2, typical basic algorithms return x_i as the vector of the first d PC scores. In that sense MDS extends PCA to cases where only distances are known. A toy example is shown in Figure 7.5.

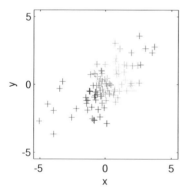 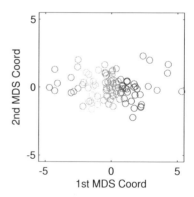

Figure 7.5 *Toy 2-d Gaussian example, illustrating MDS. Raw data is shown using + signs in the left panel. Right panel shows the corresponding MDS scores as circles (with the same coloring), calculated from the Euclidean distance matrix, which are essentially the same as the PC scores for the data on the left.*

The raw data in the left panel are an elongated Gaussian point cloud, colored along the major axis of elongation. To construct the plot in the right panel, the δ_2 distance matrix D was computed, and then the corresponding classical (Torgerson (1952, 1958)) MDS coordinates were plotted in the right panel. Note this looks much like what would be expected from PCA scores for this data set, which is consistent with the above discussion. As discussed in Section 6.4 for PCA scores, each of these coordinates is only determined up to an arbitrary sign flip. In particular the colors suggest a reversal along the long axis.

Figure 7.6 studies how the choice of metric is critical to MDS. The data are called "Equatorial" because they are distributed roughly along the equator of S^2 as shown in the left panel. That is done using a spread Gaussian distribution to determine the longitude together with latitudes from a tight Gaussian. One approach to the distance between data objects is the Euclidean distance in the embedding space $\mathbb{R}^3$. The MDS scores with respect to that Euclidean distance matrix are shown in the center panel. It is not surprising that the distribution looks much the same as the projections of the data into the plane determined by the equator, except they are centered at the mean. This follows from the fact that MDS based on Euclidean distance is essentially PCA (of the data in $\mathbb{R}^3$). A much different geodesic distance is used in the right panel. There distance is measured by arc lengths along the surface. That distance naturally follows the spread of Gaussian points, so that the distance between points in opposite tails is nearly 2π. That results in a far different set of scores, giving a more intuitively useful view of the distribution of the data, that is much more indicative of how they were generated.

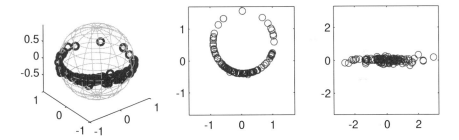

Figure 7.6 *Equatorial data on S^2, for illustrating impact of metric on MDS. Euclidean distance MDS is shown in the center panel, exhibiting a strong horseshoe shape. The right panel displays the MDS using geodesic distance (along the surface of the sphere), which eliminates the metric induced distortion.*

As noted in Section 14.5.2 of Mardia et al. (1979), point cloud shapes such as those in the center panel arise in situations where we can measure distance accurately when objects are close together, but not when they are far apart. This situation occurs surprisingly often in practice. That general pattern was given the name *horseshoe* by Kendall (1970). Diaconis et al. (2008) pointed out that horseshoe effects tend to occur in situations where distant points are less far away than would be expected from linear extrapolation of the set of local distances. Figure

7.6 gives a canonical example of this. Another context where far away points have distances less than expected by extrapolation appears in the study of micro biomes as detailed in Morton et al. (2017). They coined the term *distance saturation* to describe the situation where far distances are less than expected by linear extrapolation. Horseshoe effects, together with a higher order version, are discussed again in Section 9.1.1.

A real data example that may be due to this horseshoe effect can be seen in the Drug Discovery data shown in Figure 5.15. This suggests that an alternate representation of that data set could be useful.

There are many generalizations of the basic MDS idea. An important case in the psychometric literature extends the input of a distance matrix to merely a *dissimilarity matrix*, whose entries are not required to satisfy the metric properties such as the triangle inequality, see Kruskal (1964) for good discussion. More overviews of this large area can be found in Cox and Cox (2000), Borg and Groenen (2005), Buja et al. (2008), and Chapter 3 of Zhai (2016).

For OODA, an important generalization of MDS is to replace the embedding into Euclidean space $\mathbb{R}^d$, with embeddings into curved spaces. This is useful for data spaces that are strongly curved, such as phylogenetic tree space, as discussed in Section 10.1.2.

A weakness of distance and dissimilarity matrix-based methods, is that they only give analyses of the data set at hand, and are challenging to extend to contexts requiring generalization to additional data, such as the classification discrimination tasks considered in Chapter 11. An interesting approach to this is the *out of sample MDS* ideas of Trosset and Priebe (2008). See Section 3.4 of Zhai (2016) for a more detailed overview of out of sample MDS.

As noted above, from the perspective of modes of variation from Section 3.1.4, a perhaps more serious weakness of MDS type analyses is that, while they can provide useful and insightful scores, they fail to provide an analog of loadings that would be needed to give full modes of variation. This means that understanding of the drivers of modes, e.g. as done with loading plots in Figure 4.11 for the Pan-Cancer data, is not available from this approach.

MDS has also been used as a way of representing landmark shapes, using the set of all distances between pairs of landmarks to represent the shape (up to a reflection). This approach has been used by Bandulasiri and Patrangenaru (2005), Dryden et al. (2008), and Bhattacharya (2008). Important choices to be made are: how to average the distance matrices and how to project back to the space of configurations.

Duin and Pekalska (2005) survey another way of handling data presented only in terms of a distance matrix D from the machine learning literature. That is to simply take the columns of D as data objects, i.e. to let the rows of D define the features (using the object feature terminology from Section 3.1). In other words the traits become the distances to the other objects. The impact of this approach on the relationship between this choice of data objects is studied in Figure 7.7, where direct understanding comes from using the same data (and colors for keeping track of individual data points) as in Figure 7.5.

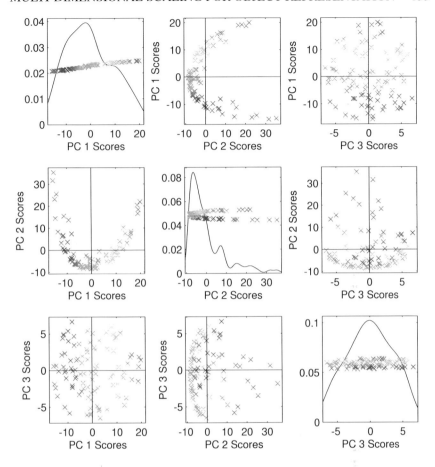

Figure 7.7 *PCA of data using columns of the Euclidean distance matrix as data objects, for the same data and colors as in Figure 7.5. Shows quite strong distortion of the original relationships between data objects.*

This PCA scores scatterplot matrix is in the same format as shown several times in Chapter 4. The colors in the upper left panel, indicate that the first PC scores are much as expected. However, the PC1 vs. PC2 scatterplot in the middle panel of the top row shows structure that appears to be the same horse-shoe effect studied in Figure 7.6. In particular, points at both the magenta and red ends are closer in this data object representation, than they are in either panel of Figure 7.5, which generates a very strong distortion in the relationships between the data objects.

Insight into this distortion again comes from the ideas of Diaconis et al. (2008) as illustrated in Figure 7.8. Think of a data set of numbers equally spaced on $[0, 1]$. The dashed (dot-dashed) line segments represent distance matrix column data objects for the points 0.4 (0.8, respectively), which have distance $\delta = 0.4$ between them. The L^1 distance between the piecewise linear data object representations is the cyan shaded area. To first order, this is $\delta \times 1 = \delta$. However, a closer look shows the area is more precisely $\delta - \delta^2/2$ (half the area shown as a small square box needs to be subtracted). A very similar calculation shows the corresponding L^2 distance is approximately $\sqrt{\delta^2 - \delta^3/2}$. Both of these have the property that the dependence on δ is sub-linear, which fits exactly into the Diaconis et al. (2008) explanation of horseshoe effects. For most purposes these issues seem like a serious weakness of the practice of using columns of a distance matrix as data objects. However in the spirit of kernel methods described in Section 11.2 there may conceivably be situations where this type of distortion of the relative positions of the data objects could be useful.

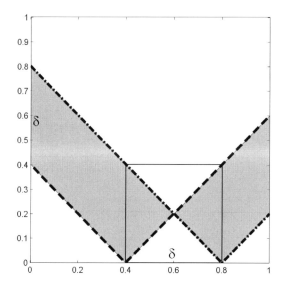

Figure 7.8 *Reveals reason for horseshoe distortion caused by using columns of a distance matrix as data objects, seen in Figure 7.7. Half the area of the box shows how typical distances are sublinear.*

7.3 Important Distance Examples

7.3.1 Conventional Norms

As noted at the beginning of this chapter, a broadly useful family of distances on $\mathbb{R}^d$ are based on the L^p norms, defined in (3.3). The L^p norms are also directly extendable to metrics on function spaces as

$$\delta_p\left(f, g\right) = \left\| f - g \right\|_p = \left(\int_{-\infty}^{\infty} \left| f(x) - g(x) \right|^p \right)^{1/p}.$$

There are a number of other useful norms on the space of matrices, including the *Frobenius norm* which is the square root of the sum of squared matrix entries:

$$\|\mathbf{X}\|_F = \sqrt{\mathrm{trace}(\mathbf{X}^t\mathbf{X})} = \left(\sum_i \sum_j x_{ij}^2\right)^{1/2} \tag{7.9}$$

which is also sometimes known as the Euclidean norm or Hilbert-Schmidt norm. The Frobenius norm is used in Procrustes analysis in Sections 7.3.3 and 7.3.4, in the analysis of covariance matrix data objects in Section 7.3.5, and in the analysis of networks as data objects in Section 10.2.

Other matrix norms include the *operator norm* (the largest singular value), the *one norm* (maximum absolute column sum of the matrix), the *infinity norm* (maximum absolute column row of the matrix) and the *maximum norm* (maximum modulus of all elements). Finally, the *entrywise norm* of the $m \times n$ matrix $\mathbf{X}$ is the L^p norm (3.3) of the vectorized version of $\mathbf{X}$, i.e.

$$\|\mathbf{X}\|_{EW} = \left(\sum_{i=1}^{m} \sum_{j=1}^{n} |\mathbf{X}_{ij}|^p\right)^{1/p}. \tag{7.10}$$

For $p = 2$ the entrywise norm is the Frobenius norm. Entrywise norms are particularly appropriate for computation with high-dimensional, sparse matrices. An example comparing several of these matrix norms is given in Section 10.2.4 when comparing high-dimensional, sparse networks in the context of natural language processing.

7.3.2 Wasserstein Distances

A particularly useful metric on the space of probability measures is the *Wasserstein* (i.e. Kantorovich–Rubinstein) *distance*. In the case of discrete measures this is also called the *Earth Mover's distance* (usually in Computer Science). That name nicely expresses the intuition that this distance essentially quantifies how much probability mass needs to be moved to get from one measure to the other. For this reason it is not surprising that computation of Wasserstein distances in general involves optimal transport formulations. In theoretical studies of the bootstrap a popular criterion is *Mallows distance*, proposed in Mallows (1972) and shown to be equivalent to the Wasserstein Distance by Levina and Bickel (2001). A comprehensive resource for many important aspects of Wasserstein distances is Panaretos and Zemel (2020). Theorem 1.5.1 of that book shows that in the case of probability measures on the real line, the Mallows distance is just the L^2 norm of the quantile functions (defined in Section 3.3.1). Hence the good properties of the Wasserstein metric are very consistent with the conclusion from Figures 3.7 and 3.8 that quantiles are frequently useful representatives of probability distributions as data objects. The ideas of this section suggest that another useful approach to tasks like PCA in that case is MDS with respect to the Wasserstein distance.

As discussed in Section 3.3, the optimal transport aspect of the Wasserstein distance accomplishes in a more rigorous way the goals of the Visual Error Criterion of Marron and Tsybakov (1995). This is seen in Figure 7.9, which shows the main example in that paper (which was motivated by one in Kooperberg and Stone (1991)). The example consists of a target curve (shown as the solid line type) and two candidate estimates (dashed and dotted). The target curve has two distinct bumps, one quite thin, and the other broader. Especially in situations where bumps are critical, Estimate 2 seems preferable as it has similar bumps, although the thin bump is not quite in the right location. On the other hand Estimate 1 has smoothed out that perhaps important feature. Now if effectiveness of the curve estimation is measured using the L^2 norm (the same holds for the L^1 norm), Estimate 1 is closer to the target. This is because the differences between the curves are measured vertically, so Estimate 2 is penalized twice for having its bump in the wrong location, while Estimate 1 only pays the peak penalty once. However, measuring discrepancy using the squared error version of the Wasserstein metric gives the opposite result. In that sense Estimate 2 is closer to the target than Estimate 1.

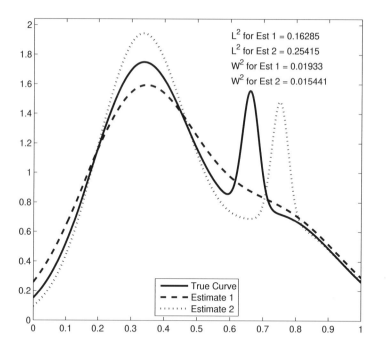

Figure 7.9 *Kooperberg-Stone example showing a curve estimation context that is conceptually challenging to the L^2 norm, but nicely quantified using the W^2 version of the Wasserstein metric.*

A useful application of MDS based on the Wasserstein metric, in the context of FDA curve registration of the Shifted Betas data is shown in Figure 9.4.

7.3.3 Procrustes Distances

An introduction to the shape of landmark configurations as data objects was given in Section 1.2.2. Much more discussion appears in Section 8.4. We now consider suitable distances that measure differences between the shapes of two landmark configurations.

Consider a configuration of k landmarks in m dimensions represented by a $k \times m$ matrix $\mathbf{X}$, where $k > m$. Usually we have $m = 2$ or $m = 3$. Procrustes analysis involves matching a set of such configurations using translation, rotation, and possibly scale using least squares techniques. Procrustes analysis was developed initially for applications of factor analysis in psychology. The name was given by Hurley and Cattell (1962) following the loose analogy with Greek mythology where Procrustes would stretch his victims to fit exactly to a bed if too short, or chop off their limbs if they were too tall. The technique can be traced back to Boas (1905) and Mosier (1939). Other important references include Gower (1975), Kendall (1984), and Goodall (1991), and a summary is given by Dryden and Mardia (2016, Chapter 7).

Ordinary Procrustes Analysis is where one configuration is fitted to another by minimizing the sum of squared distances of landmarks between the configurations, using translation, rotation and possibly scale.

Example: Digit 3 Data

The *Digit 3 data* are available in the `shapes` package in R (Dryden, 2021) and further discussed in Sections 7.3.4, 8.4, and 8.4.2. The landmarks on each digit were located by hand on images of envelopes containing British postcodes by Anderson (1997), and there are $n = 30$ handwritten digits each with $k = 13$ landmarks in $m = 2$ dimensions.

In Figure 7.10 we see two of the handwritten digit 3s in red and blue. The left-hand plot shows the unaligned data. The right-hand plot shows the two digits after they have been centered and the blue digit has been rotated and rescaled to match the red digit as closely as possible using Ordinary Procrustes Analysis.

Example: DNA Molecule Data

The *DNA Molecule data* form part of the study by Dryden et al. (2017) and are further discussed in Sections 7.3.4, 8.4, and 8.4.3. The DNA Molecule data set consists of $n = 50$ DNA molecules, each with $k = 22$ landmarks in $m = 3$ dimensions. More specifically the data set is the TFC molecule in Dryden et al. (2017) (a type of damaged DNA molecule) provided by Charlie Laughton and the observations have been temporally thinned to each 50th observation. The landmarks are located at the phosphorous atoms in the molecule.

In Figure 7.11 we see two of the DNA molecules. Each landmark is marked by small red and blue spheres in the left-hand plot for each molecule. Lines are joined along each of the two strands and also between the strands to indicate the base pairs. The familiar helix shape of DNA can be seen. In this perspective view

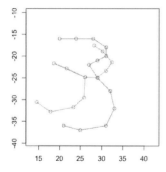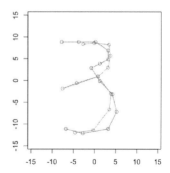

Figure 7.10 *An example of two handwritten digits from the Digit 3 data. The left-hand plot shows the original data. The right-hand plot shows the digits after they have been centered and in addition the blue digit has been rotated and rescaled to match the red digit as closely as possible.*

of the DNA molecule the spheres are smaller for atoms further away and larger for atoms that are closer to the viewer.

In the right-hand plot the blue molecule is translated, rotated, and rescaled to minimize the sum of squared Euclidean distances between pairs of atoms. We say that the blue DNA molecule has been Procrustes rotated onto the red DNA molecule.

We can measure the distance between two shapes using Procrustes analysis. We first transform the configurations to pre-shapes by centering and rescaling to unit norm (see (8.1)), and then optimize over-rotation. The *rotation group* $SO(m)$ is the set of all rotation matrix operators on $\mathbb{R}^m$, which is the set of all $m \times m$ matrices with orthonormal columns (simply summarized as $U^t U = I_m$) and determinant 1. The *partial Procrustes distance* is a shape distance given by

$$\delta_{PP}(\mathbf{X}_1, \mathbf{X}_2) = \min_{\mathbf{\Gamma} \in SO(m)} \|\mathbf{Z}_2 - \mathbf{Z}_1 \mathbf{\Gamma}\|_F, \qquad (7.11)$$

where $\mathbf{Z}_j = \mathbf{C}\mathbf{X}_j / \|\mathbf{C}\mathbf{X}_j\|_F$, $j = 1, 2$ are called pre-shapes;

$$\mathbf{C} = \mathbf{I}_k - \mathbf{J}_{k,1}\mathbf{J}_{k,1}^t / k \qquad (7.12)$$

is the centering matrix; and $\|\mathbf{X}\|_F$ is the Frobenius norm (7.9). Other alternative distances include the full Procrustes distance (8.3) and Riemannian shape distance (8.2) which are described in Section 8.4. For small shape changes there is little difference between these choices of shape distance, and the distances between the pairs of Digit 3s (a large distance) and the pairs of DNA molecules (a small distance) are given in Table 8.1.

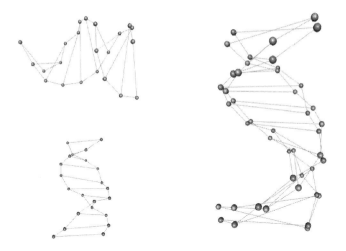

Figure 7.11 *An example of two molecules in red and blue from the DNA Molecule data, each consisting of $k = 22$ landmarks in $m = 3$ dimensions. The left-hand plot shows a view of the two DNA molecules in arbitrary positions. In the right-hand plot the molecules have been matched using Ordinary Procrustes analysis with the blue molecule translated, rotated and scaled to match the red molecule.*

7.3.4 Generalized Procrustes Analysis

In order to compute the mean from a sample of shapes the most commonly used method in practice is Generalized Procrustes Analysis, e.g. see Gower (1975), Goodall (1991), and Dryden and Mardia (2016). The method of Generalized Procrustes Analysis involves translating, rotating, and rescaling the configurations relative to each other so as to minimize a total sum of squares

$$\sum_{i=1}^{n} \|(\beta_i \mathbf{X}_i \boldsymbol{\Gamma}_i + \mathbf{1}_k \boldsymbol{\gamma}_i^t) - \boldsymbol{\mu}\|_F^2 \qquad (7.13)$$

with respect to scale β_i, rotation $\boldsymbol{\Gamma}_i$, translation $\boldsymbol{\gamma}_i, i = 1, \ldots, n$, and an overall mean $\boldsymbol{\mu}$, subject to an overall size constraint (which avoids everything shrinking to the origin), such as

$$\sum_{i=1}^{n} S^2(\beta_i \mathbf{X}_i \boldsymbol{\Gamma}_i + \mathbf{1}_k \boldsymbol{\gamma}_i^{\mathrm{T}}) = \sum_{i=1}^{n} S^2(\mathbf{X}_i), \qquad (7.14)$$

where

$$S(\mathbf{X}) = \|\mathbf{CX}\|_F \qquad (7.15)$$

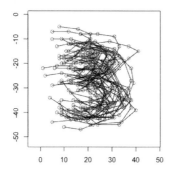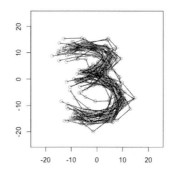

Figure 7.12 *The 30 handwritten digits from the Digit 3 data in red (left) and the digits matched using Generalized Procrustes Analysis in green, where translation, rotation, and scale have been removed (right). This registration gives a more focussed view of the population variation.*

is the centroid size of $\mathbf{X}$. If we write $\hat{\boldsymbol{\gamma}}_i, \hat{\boldsymbol{\Gamma}}_i, \hat{\beta}_i$ for the optimal translation, rotation, scale then the Procrustes mean configuration after matching is

$$\hat{\boldsymbol{\mu}} = \frac{1}{n} \sum_{i=1}^{n} \hat{\beta}_i \mathbf{X}_i \hat{\boldsymbol{\Gamma}}_i + 1_k \hat{\boldsymbol{\gamma}}_i^t. \tag{7.16}$$

The translations $\hat{\boldsymbol{\gamma}}_i$ are simply obtained by object centering. However, the rotations and scales must be obtained by an iterative algorithm. A summary of some algorithms is given by Dryden and Mardia (2016, Section 7.4) and an implementation in R is given by the function `procGPA` in the shapes package (Dryden, 2021).

The shape of $\hat{\boldsymbol{\mu}}$ is also equal to the Fréchet mean (7.5) with respect to the full Procrustes distance (where δ_{FP} is given in (8.3)), which is obtained from

$$\hat{\boldsymbol{\mu}} = \arg \min_{\boldsymbol{\mu}} \sum_{i=1}^{n} \delta_{FP}(\mathbf{X}_i, \boldsymbol{\mu})^2.$$

Methods for studying variation in shape appear in Section 8.4.1. In particular, after alignment by Generalized Procrustes Analysis, Principal Components Analysis and other techniques can be carried out to explore the modes of shape variation.

Case Study: Digit 3 Data

An example of Generalized Procrustes Analysis for the full Digit 3 data (introduced in Section 7.3.3) is given in Figure 7.12. Here there are $n = 30$ handwritten digit 3s with $k = 13$ landmarks in $m = 2$ dimensions. The original digits are viewed in the left-hand plot in red. In the right-hand plot we see the Procrustes registered digits in green, which are obtained using Generalized Procrustes Analysis.

In both plots the shapes of the digits are the same, but in the right-hand plot the unimportant rotations, translations, and scales have been removed.

Case Study: DNA Molecule Data

An example for the DNA Molecule data (introduced in Section 7.3.3) is given in Figure 7.13. Here there are $n = 50$ DNA molecules of $k = 22$ atoms in $m = 3$ dimensions. The unaligned molecules have been pictured in the left-hand panel in red. In the right-hand panel one can see the Procrustes registered molecules in green, which are obtained using Generalized Procrustes Analysis. Here the unimportant rotations, translations, and scalings have been removed in the right-hand plot. Alignment can be thought of as a form of noise reduction which results in distinctly visible landmarks in this case.

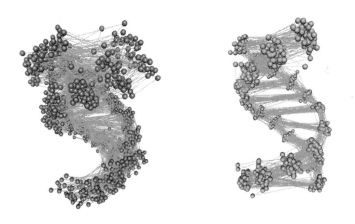

Figure 7.13 *The* $n = 50$ *unaligned molecules from the DNA Molecule data in red (left). The molecules matched using Generalized Procrustes Analysis in green, where translation, rotation and scale have been removed (right).*

7.3.5 Covariance Matrix Distances

Following Dryden et al. (2009), let us compare and discuss some possible choices of distances between covariance matrices S_1, $S_2 \in PSD_k$ (recall the set of symmetric and positive semi-definite $k \times k$ matrices from Section 3.3). The Euclidean (or Frobenius) distance between two matrices is given by

$$\delta_E(\mathbf{S}_1, \mathbf{S}_2) = \|\mathbf{S}_1 - \mathbf{S}_2\|_F = \sqrt{\text{trace}\{(\mathbf{S}_1 - \mathbf{S}_2)^t(\mathbf{S}_1 - \mathbf{S}_2)\}}, \qquad (7.17)$$

where $\|\mathbf{X}\|_F$ is the Frobenius norm (7.9). However, PSD_k is more naturally considered as a curved manifold. This is because when linearly extrapolating beyond the data, for example when using (7.8) with $\alpha < 0$ or $\alpha > 1$ one can leave PSD_k.

This undesirable feature of using a linear extrapolation in a Euclidean space is also seen in the warping example of Figure 9.13, where the extrapolated paths using Euclidean PCA are no longer strictly monotone, sometimes leaving the space of warps.

Another drawback of the Euclidean metric noted by Arsigny et al. (2006, 2007) is that this distance suffers from a swelling effect when averaging *diffusion tensors*, which are 3×3 covariance matrices that represent the local movement of water molecules in white matter fibers of the brain. The swelling effect occurs when the average of two tensors using the Fréchet mean (7.5) has a bigger volume than either of the two individual tensors. An example which displays interpolation paths between common diffusion tensors is given in Figure 7.14 using the Fréchet interpolation (7.8) for a selection of distances. The swelling effect can be clearly seen for the Euclidean distance in the first row.

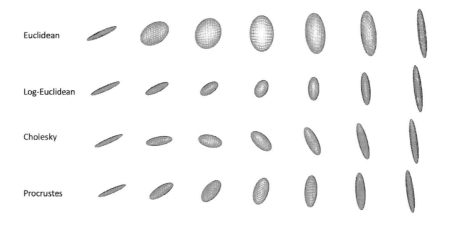

Figure 7.14 *Four different interpolation paths between the two common diffusion tensors (3×3 covariance matrices) in the left-most and right-most columns. Each covariance matrix is represented by an ellipsoid at the 95th percentiles of a multivariate normal distribution with that covariance matrix. The interpolation paths are obtained from (7.8) using Euclidean δ_E (1st row), log-Euclidean δ_L (2nd row), Cholesky δ_C (3rd row), and Procrustes δ_S (4th row) distances.*

These drawbacks can be overcome by using any of several covariance matrix distances, which essentially results in the covariance matrix objects lying in a curved manifold. One such distance is based on the matrix logarithm. We write the logarithm of a positive definite (all positive eigenvalues) covariance matrix $\mathbf{S}$ as follows. Let $\mathbf{S} = \mathbf{U}\boldsymbol{\Lambda}\mathbf{U}^t$ be the usual spectral (i.e. eigenvalue) decomposition with $\mathbf{U} \in O(d)$ (the *orthogonal group* which consists of the *orthogonal matrices*, i.e. those $d \times d$ matrices which have orthonormal columns with determinant ± 1, which consists of both rotation and axis flip operators) and $\boldsymbol{\Lambda}$ diagonal with strictly positive entries. See Section 17.1.2 for further discussion of eigenanalysis of matrices. Let $\log \boldsymbol{\Lambda}$ be a diagonal matrix with logarithm of the diagonal elements of $\boldsymbol{\Lambda}$ on the diagonal. The *matrix logarithm* of $\mathbf{S}$ is given by $\log \mathbf{S} = \mathbf{U}(\log \boldsymbol{\Lambda})\mathbf{U}^t$ and likewise the *matrix exponential* of $\mathbf{S}$ is $\exp \mathbf{S} = \mathbf{U}(\exp \boldsymbol{\Lambda})\mathbf{U}^t$. Arsigny et al. (2006, 2007) propose the use of the log-Euclidean distance, where Euclidean distance between the logarithm of covariance matrices is used for statistical analysis, i.e.

$$\delta_L(\mathbf{S}_1, \mathbf{S}_2) = \| \log(\mathbf{S}_1) - \log(\mathbf{S}_2)\|_F . \qquad (7.18)$$

Another logarithm-based distance uses a *Riemannian metric* in the space of square symmetric positive definite matrices, also known as the affine-invariant metric:

$$\delta_R(\mathbf{S}_1, \mathbf{S}_2) = \| \log(\mathbf{S}_1^{-1/2}\mathbf{S}_2\mathbf{S}_1^{-1/2})\|_F, \qquad (7.19)$$

which has been explored by several authors including Pennec et al. (2006), Moakher (2005), Schwartzman (2006), Lenglet et al. (2006), and Fletcher and Joshi (2007). A disadvantage of both the log-Euclidean and Riemannian metrics is that the covariance matrices must be strictly positive definite: non-full-rank matrices cannot be considered. Advantages are that the log-based distances do not suffer from the extrapolation problem–all extrapolations remain positive definite. Also, the interpolated covariance matrices are much less prone to swelling. In the example in Figure 7.14 the second row shows the interpolation path using the log-Euclidean distance. Note that the volume of the tensor in the middle of the path is smaller than when using the other metrics. The Riemannian metric interpolated path is visually very similar for this example.

We have many other choices of distance for comparing covariance matrices. For the one-dimensional case we can look at the distance between variances (Euclidean distance), log variances (log-Euclidean or Riemannian distances), or square root of variances. The square root scale is a reasonable choice that works well in many applications. But there is non-uniqueness in choosing a suitable root in higher dimensions. One choice is the symmetric square root decomposition where $\mathbf{S}_i = \mathbf{L}_i^2, i = 1, 2$, and $\mathbf{L}_i$ is symmetric, i.e. the symmetric square root matrix of $\mathbf{S} = \mathbf{U}\boldsymbol{\Lambda}\mathbf{U}^t$ is $\mathbf{L} = \mathbf{U}\boldsymbol{\Lambda}^{1/2}\mathbf{U}^t$. The square root distance is given by

$$\delta_{1/2}(\mathbf{S}_1, \mathbf{S}_2) = \|\mathbf{L}_1 - \mathbf{L}_2\|_F . \qquad (7.20)$$

The square root metric is less prone to swelling than the Euclidean metric, but unlike the logarithm-based metrics it can be applied to non-full-rank covariance matrices.

Another distance (which requires full rank matrices) is based on the Cholesky decomposition (Wang et al., 2004), where $\mathbf{S}_i = \mathbf{L}_i\mathbf{L}_i^t$ and $\mathbf{L}_i = \text{chol}(\mathbf{S}_i)$ is lower triangular with positive diagonal entries. The Cholesky distance is given by

$$\delta_C(\mathbf{S}_1, \mathbf{S}_2) = \|\text{chol}(\mathbf{S}_1) - \text{chol}(\mathbf{S}_2)\|_F . \qquad (7.21)$$

The Cholesky decomposition is commonly carried out for efficient numerical computation.

A final choice is the best root in terms of minimizing distance over-rotation and reflection between two root matrices. Such a transformation includes permutations of the rows/columns of the covariance matrix and sometimes it is sensible to not care about this ordering in applications. This motivates the choice of the Procrustes size-and-shape metric (Dryden et al., 2009) between two $k \times k$ covariance matrices $\mathbf{S}_1$ and $\mathbf{S}_2$, which is defined as

$$\delta_S(\mathbf{S}_1, \mathbf{S}_2) = \min_{\mathbf{R} \in O(k)} \|\mathbf{L}_1 - \mathbf{L}_2\mathbf{R}\|_F, \tag{7.22}$$

where $\mathbf{L}_i$ is a decomposition of $\mathbf{S}_i$ such that $\mathbf{S}_i = \mathbf{L}_i\mathbf{L}_i^t$ or $\mathbf{S}_i = \mathbf{L}_i^t\mathbf{L}_i$ ($i = 1, 2$) (the latter representation leads to permutations of rows/columns). This metric is generally quite similar to the square root metric, and has similar properties. If a permutation of the rows and columns is appropriate then the Procrustes metric could be useful, as the permutation matrices are a subset of the orthogonal matrices. This property is particularly useful for comparing unlabeled networks as discussed in 10.2.5.

The last two rows of Figure 7.14 show the interpolation paths for the Cholesky and Procrustes size-and-shape metrics. The paths look rather different due to the extra rotation in the Procrustes method. The metrics are both prone to some swelling in the interpolation, but much less than the Euclidean metric.

A broad family of power metrics was introduced by Dryden et al. (2009). Dryden et al. (2010) selected from such metrics with Box-Cox transformations, and the square root transformation was appropriate in their diffusion tensor application.

A promising metric for interpolating between diffusion tensors has been given by Jung et al. (2015) and Groisser et al. (2017), and involves decomposition of the tensors using rotation and scaling. An advantage with their approach is that there is no swelling effect, and the interpolation paths are visually appealing.

A further example with covariance matrices as data objects is given in Section 8.7 in the context of diffusion weighted MR imaging.

CHAPTER 8

Manifold Data Analysis

As discussed in Section 1.2, at several points in Chapter 2 and in Section 7.3.5, data objects that naturally lie on a curved manifold are an important and challenging part of OODA. The value of analyzing data in this way is first illustrated using the perhaps most straightforward example of data objects on the unit circle in Section 8.1. A brief introduction to aspects of manifold geometry needed for OODA appears in Section 8.2. Section 8.3 highlights a succession of improvements in the area of PCA-like analyses of data objects lying on manifolds and Section 8.4 considers the shape space of landmarks in more detail. Section 8.5 briefly explores the Central Limit Theorem for probability distributions on manifolds. A fundamental concept stemming from these developments is Backwards PCA, as discussed in Section 8.6. Section 8.7 contains a brief overview of covariance matrices as data objects, another area where data analysis on manifolds is an important concept.

8.1 Directional Data

Perhaps the oldest and most deeply investigated area of statistics involving non-Euclidean data objects is called *directional data*. There the data objects are *angles* such as wind or magnetic field directions. The large body of methods developed for analysis of such data are summarized in the monographs by Fisher (1993); Fisher et al. (1993), Mardia and Jupp (2000), and Jammalamadaka and SenGupta (2001). Good recent overview of the area can be found in Pewsey and García-Portugués (2021).

As illustrated in Figure 8.1 it can be very useful to carefully take the non-Euclidean nature of the space of angles into account. Both panels of Figure 8.1 show toy data sets with angle data objects as green dots on the unit circle, S^1 (recall this notation from (3.2)). In the left panel of Figure 8.1, the shown angles are $8°$, $14°$, $342°$, and $350°$. One approach to finding the center (a notion of *mean*) of the data is to simply average the four numbers, and the result of that is shown as the red plus sign. Displaying this average on the unit circle S^1 makes it clear this is a very poor notion of *center* of the set of green dots. A much better notion of center is shown as the blue plus sign, which is the Fréchet mean with respect to the arc length metric (also called geodesic distance), as defined in (7.5). This example shows how taking the unit circle (perhaps the simplest non-trivial example of a curved manifold) structure into account is essential to doing a reasonable statistical analysis of directional data. In general, it is similarly useful to carefully consider curvature of the space in the analysis of data objects lying in arbitrary manifolds.

DOI: 10.1201/9781351189675-8

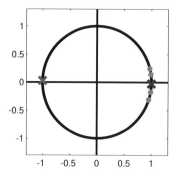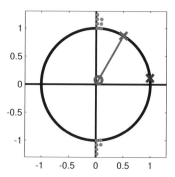

Figure 8.1 *Toy examples of directional data, shown as green dots in S^1 in both panels. Left panel shows the conventional mean of the angles as numbers (red) is a poor notion of center of data, while the Fréchet mean (blue) is more sensible. Right panel compares the extrinsic resultant mean (red) with the intrinsic Fréchet mean (blue).*

Another example of data centering issues can be seen in the right panel of Figure 8.1. That data set (shown as green dots) consists of 7 data angles slightly less than 90° (some shown outside the circle to indicate replication) together with 6 data angles slightly more than 270° on S^1. A notion of mean called an *extrinsic* mean is shown as the red x sign. This is computed by first taking the mean of the green dots (actually the angles on the unit circle) in $\mathbb{R}^2$, shown as the red circle. Next the red circle is projected out to the unit circle to get the estimated extrinsic mean angle (red x sign). In directional data, this notion of mean is often called the *resultant mean* because it is the result of treating each angle as a vector in $\mathbb{R}^2$, finding the vector sum, and then projecting that back to the circle S^1. Advantages of this extrinsic mean, relative to the Fréchet mean (called *intrinsic* because all of its operations happen within S^1 rather than in the embedding space $\mathbb{R}^2$) include straightforward fast computation and better uniqueness properties of the projection. Recall from Section 7.1 that the Fréchet mean is not unique (see Figure 8.3 for an example). An extreme example is an equally spaced data set on S^1 where the Fréchet mean set consists of all the midpoints between consecutive pairs of data points.

A perhaps unattractive feature of the resultant mean is that it does not correspond to one's guess as to where a notion of mean should be, and that it can be rather unstable with respect to small changes in the data. The Fréchet mean with respect to the arc length metric (thus an intrinsic notion of center) is shown using the blue x sign. Many would view this as a more intuitive notion of center in this case. It is also more stable with respect to changes in the data which remain in the 1st and 4th quadrants. This comparison is not unlike comparing the standard Euclidean $\mathbb{R}^1$ mean and median for slightly unbalanced, strongly bimodal data. One more point worth noticing (that does not happen in $\mathbb{R}^1$) is that if the data are shifted slightly to the 2nd and 3rd quadrants, the Fréchet mean is even less stable in the sense of jumping to the complete opposite side of the circle (while the resultant mean moves relatively less far).

This directional data space of angle data objects can also be viewed as another example having the equivalence class (i.e. quotient) structure as the triangle shape space indicated by Figure 1.9. In particular, using polar coordinates every point (except the origin) in $\mathbb{R}^2$ can be represented by an angle and a distance from the origin. Identification of points with the same angle is an equivalence relation, whose resulting equivalence classes are rays from the origin, which are characterized simply by the angles. This is the perhaps simplest example of how spaces of equivalence classes tend to naturally be represented by curved spaces.

8.2 Introduction to Shape Manifolds

As discussed above, a convenient mathematical representation of shapes is based on equivalence classes as data objects. In the case of angles as data objects in Section 8.1, the resulting object space yields the best analyses when viewed as a *curved manifold*.

The general version of this process starts with a group of transformations. For the case of directional data, this group is the set of constant multiples (i.e. scalings) in $\mathbb{R}^2$. For the triangles (or more generally for shape configurations represented by landmarks) depending on the particular application it can be a set with any of translations, rotations, scalings, or perhaps permutation of the vertex labels. Identification of elements which can be reached by a transformation in the group yields an equivalence relation. For directional data, points lying on each ray in $\mathbb{R}^2$ are thus all equivalent to each other. In the case of triangle shapes, all translations, rotations and scalings of a given triangle are identified with each other. The space of resulting equivalence classes (also called *orbits*, which are typically the data objects of interest) is called the *quotient space*. For example angles represent equivalence classes in directional data, while sets of triangles are the shape equivalence class data objects in the triangle case. Generally selection of the group of transformations is a critical issue in OODA.

Quotient spaces are usually *curved manifolds* (e.g. the unit circle in the case of directional data) which are fundamental to differential geometry. These curved surfaces lie in a higher dimensional Euclidean space and are *smooth* in the sense that at each point the surface is approximated by a tangent (hyper)plane in the (limit operation) sense of shrinking neighborhoods. In particular, a *manifold* M is a topological space which locally resembles a Euclidean space near each point. This is illustrated in the left panel of Figure 8.2 which appears as Figure 2.2 in the PhD dissertation Fletcher (2004). It shows how at the point p on the manifold M, that smooth surface is approximated by the tangent plane $T_p M$ on neighborhoods of p.

Further a *Riemannian manifold* is a smooth manifold which has a positive definite inner product defined on the tangent space $T_p M$ at p. The distance between two points on a Riemannian manifold is measured by the length of the shortest path (minimal *geodesic*) in the manifold connecting the points. The Riemannian structure allows defining the useful concept of the *exponential map* at p in Riemannian geometry, also shown in the left panel of Figure 8.2. That is the inverse of the *log map*, which (on neighborhoods of p) maps points on the curved surface of M to points on the tangent plane in such a way that the geodesic distance (to p) along the surface of M is the same as the corresponding distance in the plane T_pM. The exponential-log terminology comes from complex analysis as illustrated in the right panel of Figure 8.2. In particular the first quadrant of the complex unit circle is shown there, which contains the point $e^{i\theta} = \cos\theta + i\sin\theta$, for a given angle θ. Recall that the length of the arc from the point $1 + 0i$ to $e^{i\theta}$ (shown as the thick dashed curve) is θ radians. That is also the length of the tangent line segment from $1 + 0i$ to $1 + \theta i$ (shown as the thick dashed line segment). In this sense the complex exponential maps the tangent line segment to the manifold which is the unit circle so that distances (to 1) on the tangent line are mapped to distances along the manifold (unit circle). Hence the manifold exponential map generalizes that of complex analysis.

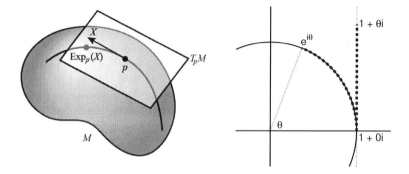

Figure 8.2 *Left panel illustrates the concept of a curved manifold, and its tangent plane approximation at each point, with the exponential map. Right panel shows how the exponential-log terminology comes from complex analysis.*

The PGA (recall Principal Geodesic Analysis from Section 1.2) modes of variation illustrated in Figure 1.11 can be viewed as starting with a tangent plane centered at the Fréchet mean. The data are mapped into that plane using the log map where they are analyzed with conventional Euclidean PCA. Mapping the resulting PC direction vectors back to the manifold using the corresponding exponential map results in the corresponding principal geodesic modes of variation. Reconstruction of corresponding s-rep figures gives the one-dimensional set of object space members that make up each mode of variation as defined in Section 3.1.4.

For a deeper introduction to manifolds and Riemannian geometry, see Chavel (2006) and Lee (2018). A far more in-depth discussion of statistical analysis on Riemannian manifolds can be found in Pennec et al. (2019).

An interesting variant of the data objects lying on a manifold theme of this chapter is *manifolds as data objects* as studied in Kang et al. (2017). This idea has been extended to data objects which are a union of a manifold together with a function on that manifold (motivated by medical imaging applications) in Lila et al. (2016, 2020); Lila and Aston (2020).

Again the focus of this chapter is the statistical analysis of populations of data objects that naturally lie on curved manifolds. However, there is another area where the term manifold intersects with statistics, which is now called *manifold learning*. The main idea is that in the analysis of high-dimensional data, there are situations where the data may lie near a relatively low-dimensional curved manifold, and the goal is to estimate that manifold from the data. An informative example, from Tenenbaum et al. (2000), is a set of digital photographs of the face of a statue from different angles, with different light sources. The vectorized version of such a set of images (as done in Section 2.4) is in some sense low-dimensional, since the 3-d structure of the face is the same in each, and the variation is easily summarized by a few parameters. But this variation is nonlinear in the space of vectorized images, so methods like PCA will not properly find the low-dimensional structure. The field of manifold learning grew out of the *principal curves* idea of Hastie and Stuetzle (1989), in the case of fitting a one-dimensional manifold, i.e. a curve, to data giving a notion of first mode of variation. An interesting issue was how to find a second mode of variation, which can be hard to imagine when trying to generalize PCA, as there is no useful notion of orthogonality in the space of curves. A solution to this problem was provided by LeBlanc and Tibshirani (1994) who proposed finding an appropriate surface, i.e. a higher dimensional manifold, as an appropriate higher dimensional generalization of PCA. A large amount of research in manifold learning, involving many variations, was started by the implementations proposed by Tenenbaum et al. (2000) and Roweis and Saul (2000). See Wang and Marron (2008) for an early approach to estimation of *effective dimensionality*, i.e. the dimension of an approximating manifold in the presence of noise using scale space ideas.

8.3 Statistical Analysis of Shapes

A fundamental aspect of the statistical analysis of shapes as data objects is that the quotient operation mathematically removes irrelevant aspects of the variation from the analysis. As in Section 2.1 such irrelevant variation (e.g. translation, rotation, and scaling in the case of triangle shapes illustrated in Figure 1.9) is usefully called nuisance variation. In that section on curve registration ideas, it was noted that depending on the context either phase or amplitude data objects could be of primary interest or else could merely be thought of as nuisance variation. Less well known is that there is a parallel situation for landmark representations. In particular, the study of plate tectonics and continental drift is also based on landmark data, as studied in Chang (1988) and Royer and Chang (1991). However, in that context an opposite choice of data objects was most sensible. In particular, shape variation was the nuisance, which was removed by essentially the inverse quotient operation. Thus the transformations were the data objects in that analysis.

As can be seen in other statistical areas, there are several rather different ideas as to the "proper" way to statistically analyze data objects lying on a manifold.

An important dichotomy is between extrinsic and intrinsic methods, using terminology introduced in Section 8.1. The main idea of extrinsic analysis is to treat the data objects as points in the embedding Euclidean space ($\mathbb{R}^3$ in the case of the 2-dimensional manifold shown in the left panel of Figure 8.2), do the analysis there, and then project the result back to the manifold. In contrast, intrinsic methods base the statistical analysis on operations within the manifold. One way to achieve intrinsic methods is to work with distance methods, such as the Fréchet mean and MDS as discussed in Sections 7.1 and 7.2, with respect to a metric using distances along the curved manifold surface (hence intrinsic).

This topic was introduced and motivated using bladder-prostate-rectum data objects, in Section 1.2. As discussed there, several rather different representations of shape can be found in the literature, including landmark and boundary representations. Another approach to shape representation takes the data objects to be transformations, typically of some template, such as the Large Deformation Diffeomorphic Metric Mapping of Dupuis et al. (1998) and Beg et al. (2005). A particularly attractive shape representation, based on quotienting over potential reparameterizations to again give equivalence classes (i.e. orbits) as data objects can be found in Srivastava and Klassen (2016). That idea also has provided a compelling approach to curve registration as discussed in Chapter 9. Statistical analysis of any of these shape representations entails working on a curved manifold at some level.

In the rest of this section the focus is on skeletal representations (s-reps), as illustrated in Figure 1.10, because of their strong intuitive appeal and their ability to effectively summarize important aspects of shape as discussed in Pizer and Marron (2017) and Pizer et al. (2020). These data objects naturally lie in a curved manifold because the representation includes a number of angles on S^2. While it is possible to attempt analysis of s-rep data in the ambient space, e.g. to treat points on S^2 as lying in $\mathbb{R}^3$, a major problem is that analytic methods such as PCA tend to leave the space where the data lie. As noted above, one approach to this is extrinsic analysis. The idea is to do the statistical analysis (e.g. the mean or PCA) in the ambient space and then project back to the curved manifold. This approach works well when the data lie in a small region of the manifold where there is not much curvature, so there is little distortion caused by the extrinsic approach.

When the data are distributed more broadly across the manifold, curvature matters more, which provides strong motivation for intrinsic analysis methods. One of these is the Principal Geodesic Analysis (PGA) of Fletcher et al. (2004). As noted in Section 1.2, the main idea of PGA is to think of the standard Euclidean PCA basis as a set of orthogonal lines that (sequentially) best fit the data. In PGA these best fitting lines are replaced by best fitting geodesics (e.g. great circles on S^2) which are a natural analog of lines along the surface of the manifold, that naturally determine modes of variation. A deliberate choice that was made in PGA was to consider only geodesics passing through the Fréchet mean. That restriction allowed straightforward computation of PGA as discussed in Section 8.2, using the corresponding tangent plane. Modes of variation generated by a PGA were shown in Figure 1.11.

More recent research has led to major improvements in statistical methodology that have been realized through improved incorporation of the underlying geometry. Huckemann et al. (2010) made the important observation that the effectiveness of the PGA tangent plane analysis of Fletcher et al. (2004) was strongly tied to the quality of the Fréchet mean as a notion of center-point. Figure 8.3 shows how that can be a serious limitation. The data objects are shown as blue dots distributed along the equator of the ordinary sphere S^2. Note that since they follow a geodesic this data set is essentially one dimensional. The Fréchet means of this data set are shown as red crosses. The Fréchet mean is a mean set (consisting of both north and south poles) in this case, because there is not a unique minimum of the Fréchet criterion (7.5). This is a consequence of least squares penalizing large distances most strongly. In particular, a candidate point along the equator will have some residuals extending halfway around the sphere, while the poles are only a quarter of the way from each data point. If the data only approximately lie along the equator, the Fréchet mean will typically be unique, but it will still be very far from all data points (essentially choosing just one of the north and south poles), and it will similarly be very poorly representative of the data.

Not only are the red Fréchet means in Figure 8.3 an unintuitive notion of data center, they are also exceptionally poor candidates as a point of tangency for a PGA analysis. This is illustrated by the yellow tangent plane in Figure 8.3. The log map projection of the data objects onto the equator, shown as green dots, follow the green circle which is centered at the geodesic mean. This has the unattractive property that it requires two PGA modes of variation to fully represent essential aspects of the data. Such a two-dimensional representation seems inefficient because the data distributed on the equator of S^2 are one-dimensional (in the sense of following a one-dimensional curve in the high-dimensional space). In fact these data objects are actually describable using just the single (nonlinear) mode of variation (as defined in Section 3.1.4) shown as the green circle. That will generally result in much more efficient statistical analysis (e.g. Bayes model fitting for segmentation as seen in Jeong et al. (2008); Jeong (2009); Pizer et al. (2005a); Pizer and Marron (2017)).

The solution proposed by Huckemann et al. (2010) was to consider all geodesics in modeling the data, thus moving beyond the restriction of PGA to only geodesics going through the Fréchet mean. An interesting point here is that the first Euclidean PCA component can be viewed as the line that best fits the data, which will necessarily contain the sample mean, as easily seen using an ANalysis Of VAriance (ANOVA) decomposition of sums of squares (a generalization of the Pythagorean Theorem). However, in non-Euclidean situations (e.g. data objects lying on curved manifolds) this is no longer true, so a conscious decision needs to be made. In particular the use of PGA entails the restriction to geodesics which go through the Fréchet mean. The *Geodesic PCA* proposal of Huckemann et al. (2016) considers all geodesics, resulting in a first mode of variation along the equator for this toy example, which gives a much more appropriate one-dimensional representation (i.e. mode of variation) of this data set.

The toy example in Figure 8.3 of data objects lying on the equator of S^2 may

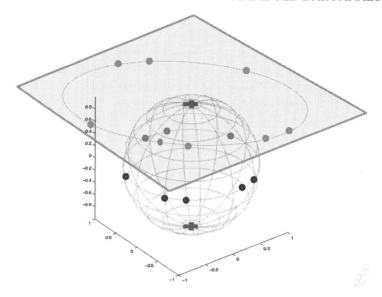

Figure 8.3 *Toy example on sphere S^2 demonstrating poor performance by tangent plane PCA. Data distributed along the equator shown as blue dots, Fréchet means shown as red crosses, with log data shown as green. Note log data lies along green circle in the tangent plane, thus requiring two PGA modes of variation to fully capture the data relationships.*

appear to be artificial, but related modes of variation are actually frequently important to medial and skeletal shape representations, as demonstrated in Figure 8.4. The left panel shows the distribution of a single spoke over a number of realizations from the bladder-prostate-rectum simulator model of Jeong (2009), which captured realistic modes of variation. Note that the data are quite broadly spread over the sphere, as in the above-discussed toy example.

Figure 8.4 also demonstrates a limitation of Geodesic PCA. This is that the spoke variation does not actually follow a great circle (i.e. a geodesic such as the equator in Figure 8.3), but instead follows a small circle (think Tropic of Capricorn on the earth). Hence the Geodesic PCA fit still requires two modes of variation as shown in the center panel. This motivated the *Principal Arc Analysis* proposed by Jung et al. (2011), which generalizes Geodesic PCA to allowing both small and great circle fits to the data as modes of variation. The benefit of this is shown in the right panel of Figure 8.4, where only one mode is needed to fully model this data set. This is an example of where careful exploitation of the curved structure of the manifold gives much more efficient analyses than are available from distance methods as discussed in Chapter 7.

Motivated by landmark-based shape analysis, the idea of using small circles to give an analog of PCA was extended to higher dimensional spheres, S^d for $d \geq 2$, by Jung et al. (2012a), to give a method called *Principal Nested Spheres* (PNS). In the special case of S^2, PNS gives the result shown in the right panel of

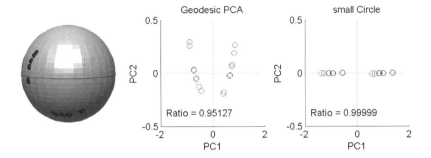

Figure 8.4 *Variation of a single spoke, from a bladder-prostate-rectum simulator model. Left panel shows the blue data objects distributed on the sphere. Center panel shows the results of a Geodesic PCA summary, requiring two modes of variation. Right panel shows the corresponding Principal Arc Analysis, which is much more statistically efficient since only one mode of variation is needed.*

Figure 8.4. One step of the iterative calculation of PNS is illustrated in Figure 8.5, where data objects distributed on S^d are shown as blue dots. The gold circle represents the best fitting sub-sphere of dimension $d-1$. The green dots represent the projections of the data objects onto the gold sub-sphere. Note that the projection (defined as the closest point in the sub-sphere) of each blue dot is calculated with respect to the metric of arc length along the sphere. These lengths are graphically depicted using cyan arcs. The gold sub-sphere is taken to be "best-fitting" in the sense of minimizing the sum of the squares of these cyan lengths. The green dot projections are then the best co-dimension 1 (i.e. one lower dimension) approximations of the blue data objects. Also fundamental to PNS is keeping the signed (depending on which side of the gold sub-sphere) lengths of the cyan arcs as the highest level PNS scores. This process is then repeated iteratively down through dimensions, to generate a full set of PNS scores. The final projections to S^1 play the role of PNS 1 scores. One more aspect of this process worth noting is that the Fréchet mean of the PNS 1 scores is a compelling notion of center called the *PNS mean* or the *backwards mean*. In particular, for the Figure 8.3 toy example of data distributed around the equator of S^2, the backwards mean is the quite reasonable one-dimensional Fréchet mean computed along the equator.

As noted in Pizer et al. (2013); Pizer and Marron (2017); Pizer et al. (2020), PNS has provided major improvements in the Bayes segmentation methods discussed in Section 1.2. For example, representations that required 20 PGA components could be done using just 13 PNS components, meaning very substantial noise reduction and corresponding improved segmentation. This dramatic improvement appears to be mostly driven by the natural prevalence of small circle modes of variation in s-rep data, as illustrated in Figure 8.4. Strong performance of PNS in landmark shape contexts was also demonstrated by Jung et al. (2012a) and is further discussed in Section 8.4.4 for landmark shapes. Also PNS performed well in the study of rotational deformations by Schulz et al. (2015) and

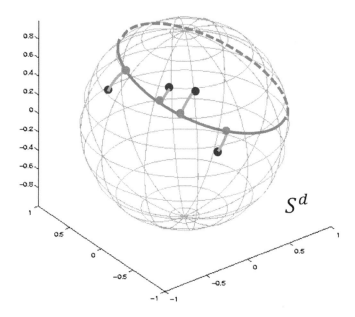

Figure 8.5 *Illustration of one step of Principal Nested Spheres. Raw data shown as blue dots on S^d. Green dots are projections (along the surface) onto the best fit version of S^{d-1} shown as a gold circle, representing the $d-1$ dimensional approximations. Level d scores are signed lengths of the cyan lines.*

in medical imaging by Kim et al. (2019). In addition PNS has been used in the area of Fisher-Rao curve registration as discussed in Section 9.2, which maps curves to a high-dimensional sphere. The benefits of this for blood glucose monitoring data was demonstrated by Yu et al. (2017a). Another important contribution of PNS is that it motivated the development of the *Backwards PCA* viewpoint, discussed in Section 8.6.

While PNS made major improvements in the statistical analysis of s-rep shapes, there is still room for improvement because s-reps do not lie on a single high-dimensional sphere. Instead they lie on a manifold that is a product of spheres called a polysphere. A direct approach to polysphere data, called *Composite PNS* in Pizer et al. (2013), is to use PNS to obtain Euclidean representations for each component sphere, which are then concatenated into long vectors that are analyzed using PCA. While this approach has given useful results as noted above, a weakness is that it makes the perhaps dubious assumption of linear dependence across spheres. A more compelling approach is the Polysphere PCA methodology proposed by Eltzner et al. (2015, 2018). The key idea is to better employ PNS by first distorting the polysphere space into a higher dimensional sphere, and then doing PNS there. This method has been seen to give some improvements, but finding a truly intrinsic approach to PCA for data objects lying on a polysphere space remains an interesting open problem. An important special case, that arises in computational chemistry, is torus type spaces, $\left(S^1\right)^d$, where the data objects are vectors of angles.

8.4 Landmark Shapes

An example of a quotient space of the type discussed in Section 8.3 is the shape space of $k > m$ landmarks in m real dimensions, where m is usually 2 or 3. Landmark configurations were introduced in Figure 1.9 and in Section 7.3.3 and an important method for comparing the shapes of the configurations as data objects is Procrustes analysis, see Sections 7.3.3 and 7.3.4. Examples of landmark shape data were given in those sections, in particular the Digit 3 data in $m = 2$ dimensions and the DNA Molecule data in $m = 3$ dimensions.

Let $\mathbf{X}$ be a $k \times m$ matrix representing a shape object of k landmarks in m dimensions. The shape of the configuration $\mathbf{X}$ is what is left when translation, rotation, and isotropic scale are removed. The geometry of the space is complicated, see Kendall et al. (1999). However, practical progress can be made by first transforming to the *pre-shape*, which involves the easier steps of removing translation and scale by centering and dividing through by centroid size:

$$\mathbf{Z} = \frac{\mathbf{CX}}{\|\mathbf{CX}\|_F} \in S^{(k-1)m-1} \tag{8.1}$$

where $\mathbf{C}$ is the centering matrix (7.12) and $\|\mathbf{CX}\|_F$ is the centroid size (7.15). The pre-shape space is a sphere in $(k-1)m$ dimensions $S^{(k-1)m-1}$, as defined in (3.2).

Rotations of $\mathbf{Z}$ on the pre-shape sphere constitute an orbit or fiber of the pre-shape space $S^{(k-1)m-1}$, which is an equivalence class. Equivalence classes and orbits were discussed in the context of triangle shape objects in Section 1.2.2 and also in Section 9.1.3 for warping functional data objects. Fibers on the pre-shape sphere correspond one to one with shapes in the shape space, and so we can think of a fiber as an equivalence class representing the shape of a configuration. The pre-shape sphere is partitioned into non-overlapping fibers by the rotation group $SO(m)$ and the fiber is the orbit of $\mathbf{Z}$ under the action of $SO(m)$.

The idea behind constructing a distance in the quotient space is to minimize the great circle distance on the pre-shape sphere between points, one in each fiber. This shape distance is the *Riemannian shape distance*, which is the geometrically natural distance inherited from the projection of the fibers on the pre-shape sphere to points in the shape space. The Riemannian shape distance ρ is given by

$$\rho(\mathbf{X}_1, \mathbf{X}_2) = 2 \arcsin\left(\frac{\delta_{PP}}{2}\right), \quad 0 \le \rho \le \frac{\pi}{2}, \tag{8.2}$$

where δ_{PP} is the partial Procrustes distance of (7.11). The interpolation path (7.8) using ρ as the shape distance is the shortest geodesic path between $\mathbf{X}_1$ and $\mathbf{X}_2$, which has length $\rho(\mathbf{X}_1, \mathbf{X}_2)$. D.G. Kendall introduced this Riemannian distance in his seminal paper Kendall (1984).

The projection from the pre-shape sphere to shape space is *isometric*, because

	Figure	ρ	δ_{PP}	δ_{FP}
Digit 3s	7.10	0.5155	0.5098	0.4930
DNA molecules	7.11	0.1464	0.1459	0.1463

Table 8.1 *Shape distances between the pair of Digit 3s and the pair of DNA molecules. This shows that these distances tend to be similar, especially when the shapes are closer to each other.*

distances are preserved. The Riemannian distance is an intrinsic distance in the shape space, and recall that discussion of intrinsic versus extrinsic distance was discussed in Section 8.1. The shape space is a type of quotient space, where the rotation has been quotiented out from the pre-shape sphere using optimization. The quotient space optimization imposes nonlinear constraints on the configuration, and hence a non-Euclidean distance is appropriate.

Recall that an alternative shape distance is the partial Procrustes distance (7.11) which is a type of extrinsic distance between two shapes, where the extrinsic distance is measured in an embedding of the manifold, see Dryden and Mardia (2016, Chapter 4) The partial Procrustes distance is the shortest *chordal distance* measured in the embedding space between two fibers on the pre-shape sphere.

An additional choice of shape distance is the *full Procrustes distance*

$$\delta_{FP} = \sin(\rho) \tag{8.3}$$

which is another extrinsic distance and a variation of the Procrustes distance defined at (7.11). The full Procrustes distance involves optimizing the Euclidean distance in an embedding of the pre-shape sphere over both scale and rotation. The term "full" Procrustes is used as both rotation and scale are used to match pre-shapes, whereas "partial" involves just rotation.

When shape changes are small all three distances (8.2), (7.11), (8.3) are similar. In Table 8.1 we see distances between the pairs of Digit 3s from Figure 7.10 and the pairs of DNA molecules in Figure 7.11. Clearly the shape distances are small for the DNA data, and so all three shape distances are similar. However, for the Digit 3 Data example the shape differences are larger, and thus differ more.

Further discussion of geometrical properties of the landmark shape space is given by Kendall (1984, 1989); Le and Kendall (1993); Kendall et al. (1999); Kent and Mardia (2001); Small (1996); Dryden and Mardia (2016). The geometry is particularly straightforward in $m = 2$ dimensions, where complex arithmetic can be used. In this planar case the shape space is a homogeneous space called a complex projective space $\mathbb{C}P^{k-2}$ (Kendall, 1984; Kent, 1994).

For triangle shapes a toy example was given in Figure 1.9. In the triangle case the shape space is the complex projective space $\mathbb{C}P^1$, which is equivalent to a sphere in three dimensions (Kendall, 1983; Dryden and Mardia, 2016, p.54). A plot of the triangle shape space is given in Figure 8.6, and the equilateral triangles are at the North Pole (anti-clockwise labels) and the South Pole (clockwise labels).

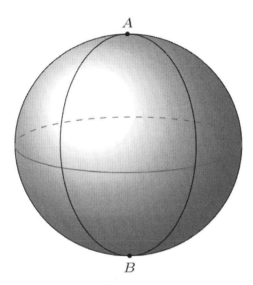

Figure 8.6 *Kendall's spherical shape space for triangles in* $m = 2$ *dimensions. The shape coordinates are the latitude* θ *(with zero at the North pole) and the longitude* ϕ. *The anti-clockwise labeled equilateral triangle is at the North Pole (A) with* $\theta = 0$, *and the clock-wise labeled equilateral triangle is at the South Pole (B) with* $\theta = \pi$. *Longitudes with isosceles triangles are shown in black (* $\phi = k\pi/3, k = 0, \dots, 5$), *and flat triangles at the equator are shown in red at* $\theta = \pi/2$.

Some longitudes with isosceles triangles are indicated, and the equator consists of all flat triangles (where the three points are collinear). If the shape sphere has radius $1/2$ then the great circle distance on the sphere is the same as the intrinsic Riemannian shape distance of (7.19).

For $m = 2$ the shape space is a homogeneous space, which is a nice space to deal with as the structure of the local geometry is the same at each point in the space. However, for $m > 2$ the shape space is not homogeneous in the sense that the local geometry changes depending on the location in the space. Le and Kendall (1993) studied the Riemannian structure of shape spaces in detail, and in particular gave explicit formulae for the average section curvature in the space at each location. Also, for $m > 2$ there are singularities in the shape space when the k points lie in an $m - 2$ dimensional subspace, see Kendall (1989), so we assume that we are away from such singularities. Furthermore the tangent space to the shape space itself is also complicated (Le, 1991). As the shape space is complicated in general, it is convenient to carry out analysis on the pre-shape sphere $S^{(k-1)m-1}$ and then adapt methods so that the analysis does not depend on the rotations of the configurations.

8.4.1 Shape Tangent Space

The tangent space to the pre-shape sphere can be decomposed into two complementary sub-spaces: the *horizontal* tangent space of dimension $(k-1)m-m(m-1)/2-1$ which does not depend on rotation, and the *vertical* tangent space of dimension $m(m-1)/2$ which contains the rotation information. For shape analysis we work with coordinates in the horizontal tangent space to the pre-shape sphere.

Constructing the Procrustes tangent space involves first choosing a pole $\mathbf{P}$, a $k \times m$ matrix, which is assumed to be non-degenerate (Kent and Mardia, 2001) which means that the eigenvalues of $\mathbf{P}^t\mathbf{P}$ satisfy $\lambda_{m-1} > 0$ (i.e. the rank of $\mathbf{P}$ is at least $m-1$). We carry out Procrustes rotation of $\mathbf{X}$ onto $\mathbf{P}$ (where $\mathbf{P}$ and $\mathbf{X}$ have been centered) to give $\mathbf{X}^P = \mathbf{X}\hat{\mathbf{\Gamma}}$, where $\hat{\mathbf{\Gamma}}$ is the optimal rotation from Ordinary Procrustes Analysis in Section 7.3.3.

The partial Procrustes tangent coordinate matrix is then (Kent and Mardia, 2001)

$$\mathbf{V} = \mathbf{X}\hat{\mathbf{\Gamma}} - \alpha\mathbf{P}, \tag{8.4}$$

where $\alpha = \cos\rho(\mathbf{X}, \mathbf{P}) > 0$, with $\rho(\mathbf{X}, \mathbf{P})$ the Riemannian distance between $\mathbf{X}$ and $\mathbf{P}$, and $\mathbf{X}$ is also non-degenerate. We can write

$$\mathbf{X} = (\alpha\mathbf{P} + \mathbf{V})\hat{\mathbf{\Gamma}}^t, \tag{8.5}$$

where $0 \leq \rho < \pi/2$, which is helpful for projecting back to the configuration space.

Multivariate statistical methods can then be carried out in the tangent space on the vectorized coordinates of $\mathbf{V}$, for example PCA, and $\mathbf{P}$ is taken as the Fréchet mean, obtained from Generalized Procrustes Analysis (see Section 7.3.4). We compute the vectorized Procrustes tangent space coordinates, evaluate the sample covariance matrix $\mathbf{S}_V$ and carry out an eigendecomposition of $\mathbf{S}_V$. For detailed discussion of PCA and variants see Chapter 17. The eigenvectors of $\mathbf{S}_V$ are the PC loadings, and the eigenvalues are the variances explained by each of the PCs. In order to visualize the modes of variation revealed by the PCs we back-project using (8.5) to configurations that are located at multiples along each PC, for example at ± 3 standard deviations along each PC. We consider an example of shape PCA in Section 8.4.2 for the Digit 3 Data.

8.4.2 Case Study: Digit 3 Data

In Section 7.3.3 we introduced the Digit 3 data and carried out Generalized Procrustes Analysis in Section 7.3.4. In Figure 8.7 we see a plot of the Fréchet mean obtained from Procrustes analysis in the middle column. We also carry out PCA of the partial Procrustes tangent coordinates of (8.4). In the left and right columns modes of variation are displayed using configurations drawn at ± 3 standard deviations along the 1st (top row), 2nd (middle row), and 3rd (bottom row) principal component directions. This matrix of scatterplots is essentially the transpose of that used to display modes of variation in Figure 1.11. In this example the

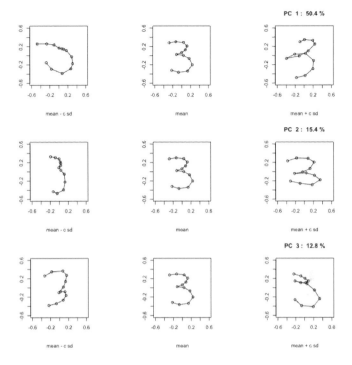

Figure 8.7 *The first three PCs for the Digit 3 data. Each row represents a mode of variation with the Fréchet mean in the middle and configurations at ±3 standard deviations along each PC on the left and right. We see that the first mode reflects the length of the middle prong in the digit, the second mode reflects tall thin digits versus short fat digits, and the third mode reflects an asymmetry in slanting up at one side versus the other.*

percentage of variability explained by these modes of variation are 50.4%, 15.4%, 12.8%, and so more than three-quarters of the shape variability is explained by the first three PCs. There are often complicated combinations of features present in each mode of variation which may not be simple to interpret. In this case the perhaps dominant aspects are: the first mode reflects the length of the middle prong in the digit, the second mode reflects tall thin digits versus short fat digits, and the third mode reflects an asymmetry in slanting up at one side versus the other.

An alternative view of these same modes of variation is shown in Figure 8.8. The piecewise colored line shows the mean (center column of Figure 8.7), with colors used to indicate the ordering of the landmarks from 1 to k. The variation in each mode is displayed by including the mean plus 3 standard deviations (right column of Figure 8.7) as the black piecewise line, and the mean minus 3 standard deviations (left column of Figure 8.7) as the gray piecewise. The line segments connecting the landmarks show the paths of each landmark through the mode

Figure 8.8 *PCA for the Digit 3 data with the mean (shown as the colored piecewise line) and the first (left), second (center), and third (right) modes of variation. Mean ±3 standard deviations are shown as the black and gray piecewise curves.*

of variation (recall from Section 3.1.4, that is a set of data objects that is one dimensional in some sense).

8.4.3 Case Study: DNA Molecule Data

In Section 7.3.3 we introduced the DNA Molecule data and carried out Generalized Procrustes Analysis in Section 7.3.4. We also now carry out PCA using the shape tangent coordinates. In Figure 8.9, the colored piecewise lines are plots of the mean shape configuration obtained from Procrustes analysis using translation, rotation, and scale, with colors indicating landmarks as in Figure 8.8. The larger spheres are located at the landmarks, with the perspective view making the closer spheres larger, and those further away smaller. In addition variation is reflected in a fashion similar to Figure 8.8, with vectors drawn to configurations which are 3 standard deviations along the 1st (left), 2nd (middle), and 3rd (right) modes of variation. In this example the percentage of variability explained by the first three modes of variation are 26.8%, 19.7%, 14.2%, and so just over 60% of the shape variability is explained by the first three PCs. It is difficult to interpret the modes of variation here, but features such as bending and twisting of the molecule may be present, as well as more localized features. The first mode reflects change at the opposite ends of each strand (at red and cyan atoms), and the second mode has orthogonal changes at the same points.

In our examples up to this point, the focus has been on shape, where all of translation, rotation and scale (size) have been removed by the quotient operation. Yet in some situations size matters. For example the part of biology called *allometry* studies relationships between size and shape. An approach to this is a scatterplot matrix comparing data object size, quantified as the centroid size defined in (7.15), with the first 3 shape PC scores shown in Figure 8.10. It can be seen that each PC score has some (strongest for PC1) correlation with centroid size. Also the remaining (shape) scores are unstructured which is consistent with lack of correlation between cases caused by the data thinning.

Figure 8.9 *PCA for the DNA Molecule data with the mean and first three modes of varia-tion. Most modes seem to be various contrasts of variation at the red end versus the cyan fold point.*

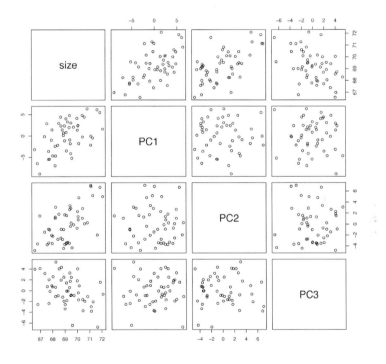

Figure 8.10 *A scatterplot matrix plot of the centroid size and first three PC scores for the DNA Molecule data. We see that these PC scores are all correlated with centroid size.*

A more extensive analysis of the mean and covariance structure for the DNA Molecule data for six different types of DNA which are undamaged or damaged has been carried out by Dryden et al. (2017). Other related work for investigating variability in time-series of DNA molecules includes *time-orthogonal principal components* (Dryden et al., 2010).

8.4.4 Principal Nested Shape Spaces

PCA in shape tangent space is appropriate when the data object variability is not too large and the variability has linear structure. An alternative is an extension of the Principal Nested Spheres (PNS) methodology of Section 8.3 (Jung et al. (2012a)), adapted to Principal Nested Shape Spaces (PNSS) by Dryden et al. (2019a). The method of PNSS involves fitting a nested sequence of pre-shapes to the Procrustes registered data. In order to speed up computation an initial PCA can be carried out, e.g. by first reducing to 20 PCs as in the following examples. Thus a small amount of noise is removed by linear PCA, and then the main computational effort is spent in fitting a sequence of pre-shape spheres of successively lower dimension. As for PNS one has the option to fit either a great sphere (with radius 1) or a small sphere (with a smaller fitted radius) at each level. The lowest levels of PNSS are the primary interest, and it is desirable that these explain high percentages of shape variability. The very final zero-dimensional level (PNSS0) corresponds to the backwards mean shape (an alternative to the Procrustes mean) and the first PNSS mode of variation (PNSS1) is a circular arc in shape space with PNSS1 scores indicating the position along this arc. In order to visualize the effects of PNSS we can back-project the PNSS0 to a configuration, and then draw arcs to configurations along each PNSS mode of variation.

In Figure 8.11 we see the results from PNSS modes of variation carried out on the DNA Molecule data of Section 7.3.4 using a format similar to that of Figure 8.9. The back-projections to configurations show the effects of each PNSS mode of variation. Again the rainbow color is the PNSS mean. The black configuration is +1 standard deviations along the first component, the gray configuration is -1 standard deviations along each component for DNA, and the path along each component is plotted as an arc. Note that the DNA data has some very small radii spheres/circles fitted in the PNSS sequence, indicating that they are rather far back. For PNSS the small circular arcs for the DNA reflect the fact that the distance between the atoms on either strand remains nearly fixed in the data set. These plots should be compared to the linear shape PCA plots that were given in Figure 8.9. The nonlinearity of these nodes of variation is reflected by the small arcs for the atoms being curved here, in comparison to being straight lines due to the linear analysis there.

The percentage variation in the first 5 PNSS scores and PCs can be seen in the top rows of Table 8.2. It is clear that the nonlinear circular variation captured by PNSS1 explains a much higher percentage of shape variation (69.8%) than the linear PCA method (26.8%), because the mode of variation is allowed to follow

Figure 8.11 *DNA Molecule data. PNSS with the PNSS mean and first three modes of variation at 1 standard deviations along the arcs.*

Data	Figure	1st	2nd	3rd	4th	5th
DNA molecule	PCA	26.8	19.7	14.2	7.2	4.6
	PNSS	69.8	7.3	4.1	3.0	2.0
Digit 3	PCA	50.4	15.4	12.8	7.5	4.3
	PNSS	59.3	13.2	9.1	6.7	3.9
Digit 3	PCA	43.6	18.4	13.2	9.0	4.9
w/o outlier	PNSS	64.1	12.1	8.9	4.9	3.3

Table 8.2 *The percentages of variation explained by the first five modes of variation using PCA and PNSS for the DNA Molecule and Digit 3 Data. This shows that the PNSS component explains higher percentages of shape variability than PCA in the first few components.*

a curved path. See Section 9.2 for a parallel phenomenon in the context of curve registration.

We now apply the PNSS method to the Digit 3 data of Section 7.3.4. The percentages of variability compared to PCA are given in the middle row of Table 8.2. In this example there is not such a major difference between the percentages of variability explained. The back-projected PNSS mean and first three modes of variation are given in Figure 8.12. These plots should be compared to the linear shape PCA plots that were given in Figure 8.8, and the structures of PNSS and PCA are quite similar here. The small arcs reflecting variation of each landmark are perhaps surprisingly roughly linear here, suggesting this nonlinear analysis is not far from that of Figure 8.7. In this data set there is a large outlier (the first observation) and the first PNSS and PC modes of variation are both in the direction of the outlier. This outlier (and the paucity of data between it and the bulk of the data) appears to be the cause of the nearly linear modes of variation in Figure 8.12.

To explore this, an alternative analysis without the outlier is given in Figure

Figure 8.12 *PNSS for the Digit 3 Data with the mean (rainbow-colored) and first three modes of variation (gray and black configurations) at ±2 standard deviation along the arcs.*

Figure 8.13 *PNSS for the Digit 3 data with the outlier removed with the mean (rainbow-colored) and first three modes of variation (gray and black configurations) at ±2 standard deviation along the arc. PNSS1 is on the left. Here the PNSS1 variability is clearly nonlinear.*

8.13 with the percentages of variability given in the last rows of Table 8.2. The first PNSS mode of variation now represents more variability (64.1%) than the first PC (43.6%) and the nonlinear aspects of variability are now more clearly seen in the back-projection for the first PNSS in Figure 8.13. In particular the first mode now once again uses a curved path to explain much more variation.

8.4.5 Size-and-shape space

In applications we have a choice of data objects, and in shape analysis one particular choice is whether to retain size in the analysis or whether to remove it. For example with the DNA Molecule data in Section 7.3.4 we removed location, rotation, and scale, but an alternative is to choose to retain the original scale of each molecule and just remove translation and rotation. In this case the space of interest is the *size-and-shape space* (see Dryden and Mardia, 2016, Chapter 5).

The size-and-shape space is again a quotient space, but the normalization by size to obtain the pre-shape is not carried out. The Riemannian size-and-shape metric is based on the Frobenius norm of the residual matrix after carrying out Ordinary Procrustes Analysis using translation and rotation only, and so the distance between $\mathbf{X}$ and $\mathbf{P}$ is

$$\delta_{SS} = \|\mathbf{X}\hat{\mathbf{\Gamma}} - \mathbf{P}\|_F = \sqrt{S(\mathbf{X})^2 + S(\mathbf{P})^2 - 2S(\mathbf{X})S(\mathbf{P})\cos\rho(\mathbf{X}, \mathbf{P})}, \quad (8.6)$$

where S is the centroid size of (7.15) and ρ is the Riemannian distance of (8.2).

The Fréchet mean size-and-shape is given by the sample mean of the configurations after carrying out Generalized Procrustes Analysis using translation and rotation only. The size-and-shape tangent coordinates are given by the residual matrix after Procrustes rotation, and specifically this is given by (8.4) but with $\alpha = 1$. Again multivariate statistical methods can then be carried out on these tangent coordinates, and for the DNA Molecule Data the analysis is similar to the shape space analysis.

The data object choice of whether to scale or not is up to the data analyst. Figure 8.10 took an allometry approach by plotting pairwise views of the first three shape PC scores for the DNA Molecule data (where size was removed), together with the centroid size as a separate variable of interest. An alternative approach is to retain the original sizes of the objects throughout, and in that case we work directly with size-and-shapes.

8.4.6 Further Methodology

Statistical shape analysis has undergone extensive methodological developments, with a large array of applications. Some recent examples include Riemannian smoothing splines applied to peptides (Kim et al., 2021); regression models for human movement data (Dryden et al., 2020); and measurement error modeling and Bayesian regression for face landmarks (Du et al., 2015; Dryden et al., 2019b). Good overviews of contemporary developments in statistical shape analysis include Grenander and Miller (2007); Younes (2010); Bhattacharya and Bhattacharya (2012); Hamelryck et al. (2012); Bookstein (2014); Dryden and Kent (2015); Patrangenaru and Ellingson (2015); Dryden and Mardia (2016); Srivastava and Klassen (2016); Turaga and Srivastava (2016); Zheng et al. (2017); Pennec et al. (2019).

8.5 Central Limit Theory on Manifolds

An important property of the mean $\bar{x}$ of a random sample $\tilde{x}_1, \cdots, \tilde{x}_n$ in Euclidean space $\mathbb{R}^d$ is the Central Limit Theorem (CLT). That Gaussian limiting distribution property is shared by many other notions of center such as the median. As defined in Section 7.1 a random sample means independently drawn from a common probability distribution. The CLT states that under rather broad conditions on the underlying probability distribution (e.g. merely the existence of some moments) the limiting distribution of $\bar{x}$ as $n \to \infty$ is Gaussian in the sense that

$$n^{1/2}(\bar{x} - \boldsymbol{\mu}) \xrightarrow{d} N(\mathbf{0}_{d,1}, \boldsymbol{\Sigma}), \quad (8.7)$$

where μ and Σ are the population mean vector and covariance matrix, using the notation $\mathbf{0}_{d,1}$ for a vector of zeros from (4.2). This CLT is why the Gaussian distribution plays such a central role in statistical inference as taught in elementary courses, and also is why the Gaussian distribution is often called "the Normal distribution" (hence the symbol N in (8.7)). For extension of the CLT to manifolds, a fundamental aspect is that the CLT convergence in distribution happens on neighborhoods that shrink at the rate $n^{-1/2}$. Connecting this idea with the tangent plane definition of manifolds illustrated in Figure 8.2 has resulted in a number of analogs as surveyed in Chapter 5 of Patrangenaru and Ellingson (2015). A good classical reference for CLTs on manifolds is Terras (1985), which has been updated to Terras (2013, 2016). Early work on large sample theory for intrinsic and extrinsic means on manifolds was given by Bhattacharya and Patrangenaru (2003, 2005). An important CLT for PNS (the Principal Nested Spheres method developed in Section 8.3) and related methods has been established by Huckemann and Eltzner (2018). An intriguing variation that appears to happen only on positively curved manifolds is the case of *smeariness* as defined by Eltzner and Huckemann (2019), where limiting Gaussians with a rate slower than $n^{-1/2}$ have been discovered.

While the issue is not critical to analogs of the Central Limit Theorem, because that happens on shrinking neighborhoods, it is natural to contemplate analogs of the Gaussian distributions on manifolds, for example this could be a reasonable approach to the development of a meaningful notion of covariance matrix on manifolds. The Gaussian analog issue can be approached in several ways. An approach that is explained well in Terras (1985) is via the heat diffusion equation, whose Green's function is the Gaussian density in Euclidean space. While this is natural for isotropic diffusion (essentially spherically symmetric covariance) it is not clear how to construct it in a non-isotropic way (i.e. to develop aspherical covariances). Another approach is via normalized *random walks*. In Euclidean space $\mathbb{R}^d$, an n step random walk that is normalized by $n^{-1/2}$ converges (as $n \to \infty$) to a Gaussian distribution. Extending this to a random walk on a manifold is another approach to a Gaussian distribution. Handling anisotropy after the first step is a challenge that has been elegantly addressed using the moving frame idea of Sommer (2020).

When either of these approaches to an analog of the Gaussian distribution is followed on the unit circle S^1, the result is the wrapped normal distribution,

$$\sum_{j=-\infty}^{\infty} \frac{1}{\sqrt{2\pi}\sigma} e^{\frac{-(x-\mu-2\pi j)^2}{2\sigma^2}},$$

see page 50 of Mardia and Jupp (2000). While this is similar to the normal distribution when the scale parameter σ^2 is relatively small (so all but one term of the sum are negligible), this approach entails a substantial price in that the distribution is not an exponential family because of the summation. Hence, likelihood methods are challenging for numerical implementation, for statistical inference, and for mathematical analyses. For these reasons, most of the classical approaches to directional data are based on variations of the von Mises distribution, which has the simple exponential family closed form

$$\frac{1}{2\pi I_0\left(\kappa\right)}e^{\kappa\cos(x-\mu)},$$

for centerpoint μ and scale parameter κ (where I_0 is the modified Bessel function of order 0), and thus straightforward likelihood properties. A contrast with the Gaussian is that the von Mises distribution does not seem to be the limiting distribution of any analog of the CLT. However, it is the conditional (on norm 1) distribution of a bivariate isotropic Gaussian distribution. Furthermore, for smaller values of its spread parameter (larger values of κ) it is very close to the wrapped normal (and to the corresponding tangent plane Gaussian). Many more related examples, asymptotics and discussion can be found in a series of papers by Hotz (2013); Hotz and Huckemann (2015); Huckemann and Hotz (2014).

Much more discussion of extrinsic versus intrinsic issues, together with many other aspects of manifold data including a wide array of theoretical results, are in Patrangenaru and Ellingson (2015). That book goes beyond manifold data object spaces to *stratified spaces*. Typical stratified spaces are richer data spaces than manifolds that essentially consist of a union of manifolds of differing dimension that are appropriately attached together. An important example is the case of covariance matrices as data objects, discussed in more detail in Section 8.7. As noted there the set of covariance matrices of the same rank is usefully treated as a manifold. But a given set of data covariance matrices may contain members of different ranks, which thus lie in manifolds of differing dimensions. These manifolds are naturally connected to each other through limiting operations where some eigenvalues tend to 0. Stratified spaces also play an important role in the analysis of tree-structured data objects considered in Chapter 10. In parallel to the above CLT discussion, even stranger versions of the CLT occur on stratified spaces, such as the *stickiness* first observed by Hotz et al. (2013) (also mentioned near the end of Section 7.1), where the rate of convergence can be much faster than the conventional Euclidean CLT.

8.6 Backwards PCA

PCA is typically introduced as a *forwards* decomposition, where one starts with the mean (best 0-d least squares approximation), then finds the 1-d best fitting subspace (best fit line) and continues sequentially building up higher dimensional approximating subspaces. But PCA can also be developed in a *backwards* way, starting with a hyperplane whose dimension is the rank of the data, and iteratively finding a decreasing sequence of best fitting subspaces. The terminology of forwards and backwards here is consistent with that used in classical variable selection for linear models where forwards (start with small models and sequentially add variables that explain the most variation at each step) and backwards methods (start large and sequentially delete variables explaining minimal variation) were popular before the advent of sparse methods. A detailed study of the backwards PCA idea can be found in Damon and Marron (2014).

Little attention has previously been paid to this forwards versus backwards issue in the context of PCA, perhaps because in the case of Euclidean data they give the same result. As noted in Section 17.1.2, both decompositions are simply calculated from either a Singular Value Decomposition (SVD) of the object mean centered data matrix, or equivalently from an eigen analysis of the sample covariance matrix. The forwards (backwards) decomposition of the data matrix consists of projections onto the increasing (decreasing, respectively) sequence of affine spaces generated by the eigenvectors. As noted in Section 8.3 this equivalence is usefully thought of as a consequence of the ANOVA decomposition of sums of squares (Pythagorean Theorem) that underlies the eigen spectral analysis, which no longer holds in non-Euclidean contexts.

In such non-Euclidean cases, the Pythagorean Theorem no longer applies, so backwards methods are no longer the same as forwards. Damon and Marron (2014) review these ideas noting that the backwards approach seems to frequently give more useful methodologies. An intuitive basis for this observation was developed through viewing PCA in terms of a *nested series of constraints*. That idea can be understood through considering the SVD of a data matrix, described in detail in Section 17.1.2. SVD is studied here instead of the usual mean centered PCA for notional simplicity (otherwise subtraction of means obscures the main point). Such an SVD results in a sequence of subspaces of $\mathbb{R}^d$ having rank $k = 1, \cdots, r$ (where r is the rank of the data matrix)

$$ S_k = \left\{ \boldsymbol{x} : \boldsymbol{x} = \sum_{j=1}^{k} c_j \boldsymbol{v}_j, c_j \in \mathbb{R} \right\}, $$

where the $\boldsymbol{v}_j$ are the eigenvectors and the c_j play the role of scores. Now to find the next smaller subspace S_{k-1} one could simply change the index above the summation to $k - 1$, but much more insight comes from writing

$$ S_{k-1} = \left\{ \boldsymbol{x} \in S_k : \boldsymbol{x}^t \boldsymbol{v}_k = 0 \right\}, $$

where $\boldsymbol{x}^t \boldsymbol{v}_k$ denotes the Euclidean inner product. This highlights the fact that S_{k-1} is easily calculated from S_k by incorporating a single linear constraint. The general success of backwards methods is interpretable in terms of it being relatively easy to sequentially find constraints, as done by backwards methods. On the other hand, to compute a forwards analog of PCA one needs to know the full sequence of constraints in advance, and then sequentially relax them, which takes much more effort.

Many variations of PCA have been proposed for a wide variety of purposes. Some of those that focus on non-Euclidean contexts are discussed here, while others are in Section 17.1. A common theme in many non-Euclidean extensions is an unfortunate tendency to move away from nested approximations. In particular, while one can usually have a rank k approximation for desired k, there is typically no relationship between say the rank k and $k - 1$ approximations, making multi-scale (in the sense of rank) interpretation quite challenging. Nested versions of PCA analogs are also essential for developing modes of variation that yield insightful visualizations of population structure, as detailed in Section 4.1, and also used above in many places. A major strength of backwards approaches to non-Euclidean PCA is that they easily address both of these concerns, using the fundamental idea of defining the method through a nested series of constraints.

In very high dimensions, because backwards approaches are stepwise, they can be computationally slow relative to conventional linear calculations of PCA, which are done with fast Singular Value Decompositions (SVD) as discussed in Section 17.1.2. This can be mitigated by doing an initial SVD with a high enough rank to only eliminate noise and retain almost all the signal followed by using the backwards method to give good compression of the remaining signal part of the data.

In addition to PNS and PNSS, contexts where backwards ideas have had (or apparently could have), a useful impact include:

- *Non-Negative Matrix Factorization.* NMF was introduced in Section 6.5. Recall it is an approximation of the (uncentered) data matrix by a product of loadings and scores matrices subject to the constraint that all matrix entries are nonnegative. There is a substantial literature on various ways to implement NMF (which reference the original Lee and Seung (1999) paper). However most of them are non-nested in the sense that one can request any rank of approximation, but the results are completely different when a different rank is requested. An exception is *Nonnegative Nested Cone Analysis* (NNCA), proposed by Zhang et al. (2015). NNCA specifically uses a backwards implementation, starting with the nonnegative cone generated by the data, and successively reducing the dimension through a sequence of linear cone constraints, keeping that level of scores at each step.

- *Manifold Learning.* As discussed above, this approach is very useful for high-dimensional data sets that tend to lie quite close to, or perhaps even on, a lower dimensional curved manifold. A drawback to all current approaches is that they are not nested, so there is no relationship between a rank k and rank $k-1$ representation of the data, and thus no decomposition into modes of variation. An interesting open problem is to find a nested version of manifold learning, where one iteratively finds a rank $k-1$ approximation as a submanifold of the one of rank k. This would make the most sense by minimizing the sum of squared geodesic distances at each step. This would result in a more natural analog of PCA scores distribution plots than the current PCA typically performed on unrolled versions of the k dimensional manifold.

- *Robust PCA.* Robust methods that are generally useful in OODA are discussed in Chapter 16. One overall approach is to use methods based on the L^1 norm. Such methods typically draw robustness from behavior similar to the median as discussed in Section 7.1. Motivated mostly by computational considerations, Brooks et al. (2013) have proposed a backwards approach to L^1 based PCA.

- *Principal Flows.* Another backwards approach to PCA for data objects lying on a manifold is the Principal Flow approach of Panaretos et al. (2014). This method results in nongeodesic components with an interesting ability to locally track maximal variation.

An important contribution to the discussion of backwards versus forwards PCA is the *Barycentric Subspace Analysis* idea of Pennec (2018). In addition to providing an interesting general framework for manifold PCA, that paper points out that both backwards and forwards methods are essentially greedy searches, which one expects can be improved by solving an overall optimization, although that comes at the price of increased complication. An interesting open problem is how much gain is available from a globally optimized overall method, relative to a simple backwards approach. As noted in Section 8.5, an approach to asymptotic analyses in this direction can be found in Huckemann and Eltzner (2018).

8.7 Covariance Matrices as Data Objects

Covariance matrices as data objects appear in several applications. The one that has received the most research effort has been *diffusion tensor imaging*, started by Basser et al. (1994). In that area connectivity within the brain is studied using 3-d images, where at each voxel the distribution of diffusion of water molecules is summarized in terms of a covariance matrix, as shown in Figure 8.14. These 3-d covariance matrices are represented by shaded ellipsoids representing a contour of the corresponding Gaussian distribution in Figure 8.14, after Dryden et al. (2009). In fluid regions of the brain this covariance is essentially a multiple of the identity matrix, reflecting the ability of the molecules to diffuse freely in various directions. In regions with fibers, such as neuronal brain connections, one eigenvalue of the covariance matrix is much larger than the other two, indicating only one degree of freedom of diffusion along the direction of the fiber. Covariance matrices as data objects also play an important part of the speech sounds as data objects discussed in Section 2.3. Additional examples appear in the survey of *Object Oriented Spatial Statistics* by Menafoglio and Secchi (2017).

A number of approaches to data analysis of covariance matrices have been proposed. Since matrices are just arrays of numbers they could be vectorized into long vectors and then analyzed with standard linear methods. But that approach generally leads to severe complications because linear analyses such as PCA tend to leave the space of positive semi-definite matrices. As noted in Chapter 7 PCA distance methods provide a useful approach to this challenge. An interesting aspect of covariance matrices is that there are quite a few metrics of interest, with quite different properties as seen in Section 7.3.5. The log-Euclidean metric (7.18) was popularized by Arsigny et al. (2006, 2007). Fletcher and Joshi (2004) point out benefits of a Riemannian metric (7.19) approach and Dryden et al. (2009) advocate the Procrustes metric (7.11). Properties of the metrics have been revealed by studying widely varying interpolation paths generated by different metrics, for example in Dryden et al. (2009) and Pigoli et al. (2014a), and some examples were given in Figure 7.14.

A measure calculated from a covariance matrix that is very commonly used in

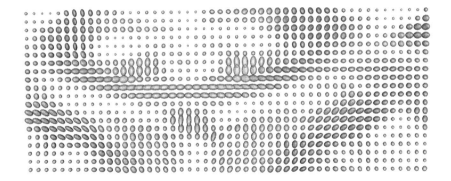

Figure 8.14 *Example of a coronal slice of diffusion tensors estimated from a human brain. Tensors are represented by ellipsoids. Those stretched in a single direction indicate fibers running through that location.*

diffusion tensor imaging is *Fractional Anistropy*

$$FA = \left\{ \frac{k}{k-1} \sum_{i=1}^{k} (\lambda_i - \bar{\lambda})^2 / \sum_{i=1}^{k} \lambda_i^2 \right\}^{1/2}$$

where $0 \le FA \le 1$ and λ_i are the eigenvalues of the diffusion tensor matrix (in which case $k = 3$), or alternatively a covariance matrix. Note that $FA \approx 1$ if $\lambda_1 >> \lambda_i \approx 0, i > 1$ (very strong principal axis of the confidence ellipse) and $FA = 0$ for isotropy. Basser and Pierpaoli (1996) developed the use of FA in the context of diffusion tensor imaging. High values of FA correspond to white matter fibers which connect different parts of the brain. The study of the arrangement of these fibers is a key objective of diffusion-weighted MR imaging.

An alternative to FA for finding voxels indicating fibers is to use the full Procrustes shape distance δ_F (8.3) to isotropy. This gives the *Procrustes Anisotropy* of Dryden et al. (2009)

$$PA = \left\{ \frac{k}{k-1} \sum_{i=1}^{k} (\sqrt{\lambda_i} - \overline{\sqrt{\lambda}})^2 / \sum_{i=1}^{k} \lambda_i \right\}^{1/2}, \tag{8.8}$$

where $\overline{\sqrt{\lambda}} = \frac{1}{k} \sum \sqrt{\lambda_i}$. So $0 \le PA \le 1$, with $PA = 0$ indicating isotropy, and $PA \approx 1$ indicating a very strong principal axis.

A final isotropy measure based on metrics δ_L or δ_R is the *Geodesic Anisotropy*

$$GA = \left\{ \sum_{i=1}^{k} (\log \lambda_i - \overline{\log \lambda})^2 \right\}^{1/2},$$

where $0 \le GA < \infty$ (Arsigny et al., 2007; Fillard et al., 2007; Fletcher and Joshi,

Figure 8.15 *The anisotropy measures FA (left), PA (middle), GA (right) for the diffusion tensors, from Figure 8.14. This shows the differences in displays using the three measures, with arguably PA having better contrast within the (relatively bright) corpus callosum.*

2007), which has been used in diffusion tensor analysis in medical imaging with $k = 3$.

To compare the measures of anisotropy consider the diffusion tensors obtained from diffusion-weighted images in the brain that were plotted in Figure 8.14. In Figure 8.15 we see a slice from the brain with the 3×3 tensors displayed. This image is a coronal view of the brain, and the corpus callosum and cingulum can be seen. At first sight all three measures appear broadly similar, but the GA image has fewer brighter areas than PA or FA. Dryden et al. (2009) argue that PA is slightly preferable here, offering more contrast than the FA image in the highly anisotropic region–the corpus callosum. The data were provided by Paul Morgan and for related discussion see Tench et al. (2002).

Important work on the statistical analysis of covariance matrices and covariance operators as data objects can be found in Dryden et al. (2009), Pigoli et al. (2014a,b), and Aston et al. (2017). Analysis of diffusion tensor data, using local polynomial smoothing methods can be found in Yuan et al. (2012), and a varying coefficient model approach is given in Yuan et al. (2013).

The above works demonstrate that it has been very useful to understand covariance matrices as data objects lying on a curved manifold. However, an even more appropriate mathematical context is a stratified space. As noted in Section 8.5 that is a connected set of manifolds of different dimensions. Stratified spaces are appropriate for covariance metrics of varying rank. For each given rank r, the natural data space is a manifold whose dimension is $r\left(r + 1\right)/2$. These manifolds are naturally connected across rank through limiting operations where eigenvalues tend to 0.

While geodesics reveal interesting aspects of metrics, they are also useful for interpolation between covariance matrices. Also of interest for interpolation purposes are the non-geodesic paths of Jung et al. (2015) and Groisser et al. (2017) which better correspond to intuitively expected interpolation for diffusion tensors.

Additional useful analytic methods for covariance matrices as data objects, including hypothesis testing and even kriging can be found in Secchi et al. (2013), Pigoli et al. (2014b, 2016), and Cabassi et al. (2017).

CHAPTER 9

FDA Curve Registration

As discussed in Sections 2.1 and 5.4, amplitude and phase variation can play an important role in FDA. As noted there and demonstrated in Figure 9.1, many curve registration methods work well on the data sets for which they were designed, yet typically require intensive manual tuning for application to a different data set. OODA ideas, including the use of equivalence classes as data object representations (which were quite useful for quantifying "shape" variation in Sections 1.2, 8.2, and 8.3), are seen in Section 9.1 to provide a mathematically rigorous and hence broadly applicable general solution to this challenge. In particular, that section discusses the Fisher-Rao methodology of Srivastava et al. (2011), which was never published in the statistical literature apparently due to the mathematical overhead involved. Section 9.2 shows how the Fisher-Rao approach can be enhanced using Principal Nested Spheres as introduced in Section 8.3.

Figure 9.1 is a subset of Figure 1.2 of Srivastava et al. (2011), that was constructed by Wei Wu. In each row, the left-hand column shows a set of curve data objects which present different challenges to peak alignment. The data in the top row is similar to the Shifted Betas in Figure 10.2 below. The middle two rows consider other types of shift variation. That in the bottom row comes from Wu et al. (2014). The middle three columns show analyses of each data set by a method from the literature with publicly available software. In particular, the second column uses the method of Liu and Müller (2004), the third column is based on the self modeling warp implementation of Gervini and Gasser (2004), while the fourth column shows the moment-based approach of James (2007). These give various degrees of alignment of the peaks, which are quite example dependent. It is important to note that with sufficient manual tuning, most of these can give good results. The last column shows the corresponding analysis using the Fisher-Rao curve registration method described in Section 9.1. For the reasons given in that section, unlike the other methods shown in Figure 9.1, simple default settings give high quality peak alignment in a fully automatic way.

DOI: 10.1201/9781351189675-9

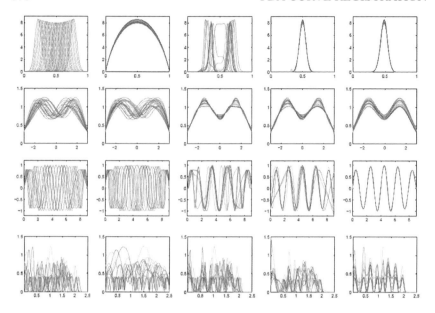

Figure 9.1 *Comparison of several curve registration methods (columns) on four data sets (rows). Shows the Fisher-Rao approach (last column) provides the most stable fully automatic alignment of peaks and valleys in curves.*

9.1 Fisher-Rao Curve Registration

Further motivation for Fisher-Rao curve registration comes from the Shifted Betas example in Section 9.1.1. The impact of misalignment on the sample mean was shown in Figure 5.17. Alignment issues even more strongly affect PCA as seen in Figure 9.2 below. The basics of warping functions that generally underlie curve registration are discussed in Section 9.1.2. Details of the Fisher-Rao approach appear in Section 9.1.3.

9.1.1 Example: Shifted Betas Data

The Shifted Betas raw data object curves are shown in the top left panel of Figure 9.2. These curves are $n = 29$ Beta(α, β) probability densities, with $\alpha = 20 - \beta = 4.4, 4.8, \cdots, 15.6$. There is a very strong phase mode of variation, which is approximately tracked by the means of these distributions, $EX = \frac{\alpha}{\alpha+\beta}$, and coded with a rainbow color scheme. The name Shifted Betas is not entirely appropriate as these densities all have support $[0, 1]$. Nevertheless it does convey the clear left to right shifting of probably mass. The colors clearly indicate a single nonlinear mode of variation. The same colors are used at many other points in this chapter. Because this single mode is nonlinear it can admit insightful decomposition into other (sometimes linear) modes. For example there is a clear amplitude (linear) mode of variation in terms of heights. There is also a nonlinear mode of variation in the widths since as probability densities each curve has area 1 underneath so change in height entails change in width. The presence of these modes of variation (in addition to the obvious phase mode) suggests that an intuitive analysis of this data set should show 3 modes of variation. The sample object mean is shown in the top center panel. For the reasons given in the discussion around Figure 5.17 this mean is not representative of the data objects. In particular the object mean is only a very broad peak, that is much lower than any individual, as highlighted using the same vertical axis as the top left panel. The mean centered residuals are shown in the top right panel, which are then decomposed into modes of variation by PCA in the remaining rows.

The left plot in the second row shows the first PCA mode of variation. This is disappointingly unlike either the phase shift or the peak height modes discussed above. Instead it only shows that the magenta and blue curves are higher on the left and lower on the right, while the yellow and red curves exhibit opposite behavior. The residuals in the center are harder to interpret. However, notice that use of common vertical axes, as for the Tilted Parabolas in Figure 4.1, visually suggests that these residuals represent a substantial share of the variation, in fact 39.7% of the sum of squares about the mean. The color scheme of PC1 scores in the right panel at least suggests that this mode is mostly driven by the shift in peaks, sweeping from left to right.

The third row shows the PC2 mode of variation (28.4% of the sum of squares). The scores on the right seem to be a doubling of the frequency of the sweep through the colors, going right to left and back. The colors of the mode of variation on the left show only the later plotted green to red curves, because these overplot the (hence invisible) magenta to green curves. The 3 bump general pattern can be interpreted as the second order correction of the crude 2 bump approximation of PC1. The next such correction in approximation seems to be the 4 bump pattern shown in PC3 (8.4% of the variation) in the bottom row left. These increasing bumps approximations are very reminiscent of the behavior of orthogonal polynomial basis systems over increasing degree. The corresponding scores in the bottom right again represent an increase in frequency of back and forth sweeps.

The main point of this example is that just a single mode of phase variation is very poorly captured by PCA, with the signal power of that mode being spread widely across the PC spectrum. Using terminology from Marron et al. (2015), PCA is a *vertical* (amplitude) analysis method, which only weakly provides insights about *horizontal* (phase) variation.

It is interesting to study the impact of the systematic back and forth swings of scores using the scatterplot matrix view (this device was used e.g. to visualize the Twin Arches data in Figure 4.4). This is done in Figure 9.3, where the first 3 diagonal plots are the same as in the right column of Figure 9.2. The sequential sweeping effects of the scores over different PC modes generates some interesting patterns. Perhaps initially hardest to interpret may be the PC2 vs. PC4 panel because the green to red points again exactly overplot (and thus obscure) the magenta to green points. Particularly attractive is the knot tied in the PC3 vs. PC4 panel. Together these scores plots indicate that the scores follow a one-dimensional curve

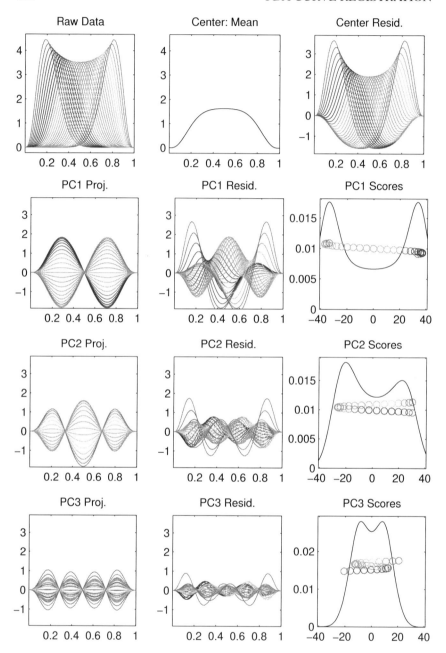

Figure 9.2 *Shifted Betas example, illustrating the impact of phase variation on straightforward Euclidean FDA using PCA. Format similar to Figure 4.1, with data, object mean and mean residuals in the top row. Modes of variation show PCA does a very poor job of displaying the visually obvious shift in peaks as the dominant mode. This motivates alternative approaches to understanding horizontal variation.*

that bends around through the scores space. This is a sense in which the data can be thought of as a single mode of variation. That is consistent with the visual impression of the bundle of curves in the upper left panel of Figure 9.2, and with the fact that those are essentially a one-parameter family of curves. This is another way to understand that while PCA can provide good decompositions of vertical variation as seen at many points above, it has its limits in the presence of horizontal variation, which motivates the coming ideas.

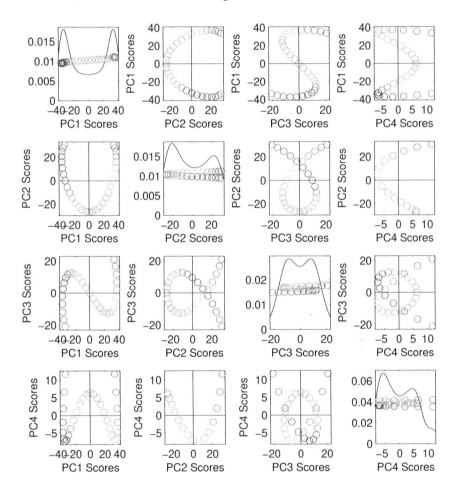

Figure 9.3 *Corresponding PC scores scatterplot matrix for the Shifted Betas data of Figure 9.2. Shows very systematic patterns that are common in applications of PCA to data with a strong component of pure phase variation.*

The second panel of the top row of Figure 9.3 is an example of the horseshoe effect studied in the context of MDS in Section 7.2. The explanation (from Diaconis et al. (2008)) given there applies here, because curves that are farther from each other have much less overlap, they are less far apart than might be expected from linear extrapolation of the shorter distances. This same issue lies at the heart of the shifting support example given in Morton et al. (2017) in a micro-biome context. The distinctive patterns of the scores in Figure 9.3 seems to be a type of higher order horseshoe effect, that could be seen in other examples when higher order components are viewed. Another situation where such patterns have emerged is the time-varying chemical spectra studied in Marron et al. (2004). These appear often enough to suggest the existence of a general mathematical theory (perhaps something akin to Sturm-Liouville theory) behind this phenomenon. That might be a generalization of the concept of *harmonics* in Fourier analysis.

The upper left panel of Figure 9.2 suggests that these generalized horseshoe effects are caused by the L^2 norm doing a poor job of quantifying the shifting probability mass that is a very important aspect of the clear mode of variation. Because these are probability densities, this is another manifestation of the issue discussed using a family of Gaussian densities in Section 3.3.1, that densities often provide a poor representation of probability distributions as data objects. The approach to this that was taken there was to represent the data objects using quantile functions. In the case of one-dimensional distributions as data objects, that was seen in Section 7.3.2 to be equivalent to using an optimal transport based MDS using the Wasserstein metric. Both approaches gave a less distorted representation of this major mode of variation. A Wasserstein MDS appears in the left panel of Figure 9.4. Note that common axes are used to reflect the fact that there is much more variation in the MDS 1 scores than in MDS 2. In particular, this Wasserstein decomposition of the variation seems to be much more informative than the PCA shown in Figure 9.3. The colors suggest that MDS 1 is driven by the clear shifting of mass in the upper left panel of Figure 9.2. The existence of a smaller, but still visible second mode of variation suggested by MDS 2, is consistent with the fact that the peak heights in the density start high, decrease toward the middle, then increase again. Unfortunately it is not simple to visually verify these ideas using loadings (as for the Spanish Mortality data in the left panels of Figures 1.4 and 1.5, or for the Pan-Cancer data in Figure 4.11) because the MDS type of analysis only provides scores, not full modes of variation. One approach is to plot individuals at key points of the MDS scatterplot. For example the first MDS mode can be understood from the magenta and red curves (recall corresponding colors), shown in the top panel of Figure 9.2, as a shift mode of variation. The second mode is a contrast between the light green curve and those other two indicating variation in peak height and width.

The right panel of Figure 9.4 gives a further contrast of PCA with the Wasserstein MDS analysis of the Shifted Betas data, using scree plots (defined in Section 3.1.3) based on the first 4 components. Because the numbers involved vary over several orders of magnitude, the $\log_{10}$ of the proportion of variation explained by each of the first 4 components is shown. The green curve clearly shows how conventional Euclidean PCA spreads the energy of the signal broadly across the spectrum, as discussed near Figure 9.2 above. On the other hand, the magenta Wasserstein scree plot shows the great bulk of the energy appears in MDS 1, with

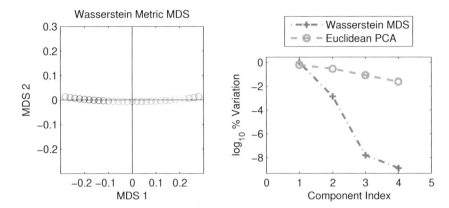

Figure 9.4 *Wasserstein MDS analysis of the Shifted Betas example. Left panel shows scores giving much more intuitive decomposition of the variation than the PCA shown in Figure 9.3. Scree plots (on $\log_{10}$ scale) in the right panel also shows the MDS gives a much more efficient representation from another viewpoint,*

almost all the rest in MDS 2. The remaining very small energy in the other two components seems to be at the level expected from errors in the MDS optimization process. Thus MDS with respect to the Wasserstein metric gives a far more insightful decomposition of the Shifted Betas data. Its optimal transport motivation is clearly useful to keep in mind when facing serious registration issues. This seems to have been done in the microbiome literature where Fukuyama et al. (2012) discuss connections between the Wasserstein distance and the Unifrac metric. However, it still has the drawback of not providing full modes of variation, because of a general lack of loadings. The latter consideration provides motivation for the Fisher-Rao methodology in the rest of this section.

9.1.2 Introduction to Warping Functions

Warping functions are a very useful concept for understanding approaches to curve alignment and to amplitude phase decompositions of the type illustrated using the Bimodal Phase Shift data in Figure 2.2. For numerical convenience these are typically defined on compact intervals. Allowing for linear transformation, these intervals can be taken without loss of generality to be $[0, 1]$. In this chapter, the data objects are taken to be functions $f : [0, 1] \to \mathbb{R}$. The function $\gamma : [0, 1] \to [0, 1]$ is a *warp function* when it

- Is *onto* [0,1] (i.e. surjective),

- Is strictly increasing, hence *one to one* (injective). These last two properties result in γ being invertible (bijective),

- Is a *diffeomorphism*, meaning both $\gamma(x)$ and its inverse are differentiable.

Let Γ denote the set of all such warps. As noted in previous sections, useful intuition comes from thinking of warps in terms of local stretching and compression of the horizontal axis, as illustrated in Figure 9.5. The horizontal axis in each panel shows an equally spaced grid on $[0, 1]$. The function $\gamma(x)$ is shown as a curve from $(0, 0)$ to $(1, 1)$. The thin lines indicate how $\gamma(x)$ induces a stretching and compressing of the equally spaced grid on the vertical axis. The upper left panel shows the diagonal line, i.e. the *identity warp*, $\gamma_I(x) = x$, which leaves the spacing between grid points unchanged. The version of $\gamma(x)$ in the upper right panel is convex, which results in a compression of smaller values and stretching of larger values. The lower left panel shows a concave $\gamma(x)$ which results in the opposite type of stretching and compression. The $\gamma(x)$ in the lower right has both convex and concave parts, resulting in compression at both ends, with stretching in the middle.

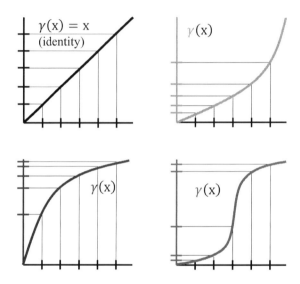

Figure 9.5 *Illustrations of warping functions* $\gamma(x) : [0, 1] \rightarrow [0, 1]$, *showing how they are usefully thought of as stretchings and compressions of the horizontal axis.*

9.1.3 Fisher-Rao Mathematics

There were several major contributions to the curve registration literature made by Srivastava et al. (2011). The presentation here is aimed at maximal understanding of this important set of ideas, which is done with various types of approximation. In particular several function space technicalities are deliberately ignored here. See that paper for the fully rigorous version, which uses concepts such as generalized functions and related approximations for full mathematical precision.

The first contribution of Srivastava et al. (2011) was to distinguish amplitude and phase variation using a quotient structure parallel to that employed in shape analysis in Section 8.3. In particular, the set of all warps is a group of transformations that plays a similar role as the translation - rotation - scaling group of transformations in landmark-based shape analysis. Following that idea, two curves are *identified* when one can be reached from the other by a warping function, i.e. $f_1(x) \sim f_2(x)$ when there is a warp $\gamma(x) \in \upsilon$ so that $f_1(x) = f_2(\gamma(x)) = f_2 \circ \gamma(x)$. That equivalence relation induces a quotient structure, i.e. a set of equivalence classes which are usefully thought of as an insightful type of data objects, playing a role parallel to the shape equivalence classes in Section 8.3. In this chapter, given a function $f : [0, 1] \to \mathbb{R}$, let its orbit be $[f] = \{f \circ \gamma : \gamma \in \Gamma\}$, i.e. the set of all warps of $f(x)$. Any member of $[f]$ is called a *representer* of that equivalence class. Intuitive data analytic understanding of orbits will come from careful choice of representers.

Additional insights into warp equivalence classes come from the examples shown in Figure 9.6. The left panel shows a set of curves that are all in the orbit of a single curve $f(x)$ with three bumps. Not only does each curve share all 3 bumps, each bump also has the same height. Furthermore the bumps are in the same order, so they can reach each other through smooth compressions and extensions in the spirit of the lower right panel of Figure 9.5. Thus they are all elements of $[f]$. A much different example appears in the right panel of Figure 9.6. The curves there all have peaks of differing heights, so none of them can be warped into any other, i.e. none of these are warp-equivalent. Thus $[f_i] \neq [f_j]$ for $i \neq j$ in this example.

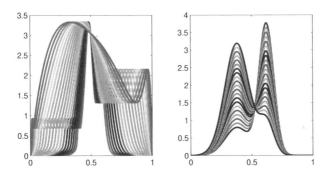

Figure 9.6 *Toy example of some elements of a single warp equivalence class (left panel), and of bimodal curves none of which are warp-equivalent (right panel).*

The notion of warp-equivalence classes allows a formal, mathematically rigorous, definition of the amplitude of a functional data object:

Definition: The *amplitude* of a function f is its warp equivalence class $[f]$.

The left panel of Figure 9.6 indicates that any amplitude data object representer of f contains all of the vertical aspects of f, such as the heights of all peaks, and depths of all valleys, in addition to the relative relationship between peaks and valleys. A parallel definition of phase data object is a deeper concept, and requires development of some intermediate notions.

The second fundamental contribution of Srivastava et al. (2011) to the curve registration literature was the proposal of the Fisher-Rao approach to quantifying the difference between amplitudes (i.e. orbits). A fundamental aspect of that is *warp invariance*. Such metrics on functions induce a proper metric on the quotient space of orbits. The key idea is illustrated in Figure 9.7, which is Figure 3.3 of the PhD dissertation Lu (2013) that was reproduced as Figure 2 in Yu et al. (2017a). This demonstrates the inconsistency of a common approach to the estimation of warping functions, which is to find the warp $\gamma(x) \in \Gamma$ to minimize

$$\|f_1(x) - f_2 \circ \gamma(x)\|_2. \tag{9.1}$$

In Figure 9.7 the functions f_1 and f_2 (called the Step Function toy example) are essentially step functions as shown in the top left panel (actually these are just parallel coordinate plots on a dense grid, so each has one very steep jump between consecutive grid points). The warp function $\gamma(x)$ that minimizes (9.1) is shown in the top center panel, as the blue (dashed) function that essentially shifts the jump point in f_2 until it coincides with the jump in f_1. Because f_1 is unwarped in the computation of (9.1), a red diagonal line, representing the identity warp γ_I is also shown in the top center panel. The bottom center panel gives insight into the minimum value of (9.1), in terms of the area of the rectangle shown with a light blue green color (actually the L^2 norm is the height times the square root of the length). The right column shows the result of minimizing the version of (9.1) where f_1 and f_2 switch roles, so that the jump in the red f_1 is shifted over to the jump in f_2, using the red warping function in the top right panel (with the blue dashed diagonal line indicating $\gamma_I(x)$, i.e. no warping for f_2 in this case). This results in a minimizing value of (9.1) represented using a pink rectangle in the lower right panel. Notice this L^2 minimum value is much different than in the center column showing that criteria of the form (9.1) cannot provide a metric (i.e. distance as discussed in Chapter 7) on the quotient space of amplitudes. In particular, Figure 9.7 reveals that an attempt to base a quotient space metric on (9.1) would fail due to a lack of symmetry, i.e. $\delta([f_1], [f_2]) \neq \delta([f_2], [f_1])$. The lower left panel is explained in detail below.

A natural approach to developing a metric on the space of amplitudes (i.e. the quotient space) is to start with a metric δ on the original space of functions that is *warp invariant* in the sense that

$$\delta(f_1, f_2) = \delta(f_1 \circ \gamma, f_2 \circ \gamma), \tag{9.2}$$

for all warps $\gamma \in \Gamma$. That property of a metric is usefully understood as implying that the equivalence classes are all "parallel" with respect to γ as illustrated in Figure 9.8. In particular, moving through both equivalence classes by the same warp, leaves the distance unchanged. The property (9.2) ensures that δ induces a well-defined metric on the warp-equivalence quotient space as

$$\delta([f_1], [f_2]) = \min_{\gamma \in \Gamma} \delta(f_1 \circ \gamma, f_2). \tag{9.3}$$

In particular the warp invariance (9.2) entails that $\delta([f_1], [f_2])$ does not depend on the representers f_1 and f_2.

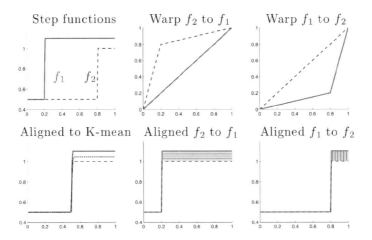

Figure 9.7 *Step Function example illustrating failure of classical approaches to provide a metric on the space of warp equivalence classes. Data objects in the upper left, with aligning warps in the other upper panels. Lower right panels shows impact of warps on the L^2 norms (which are quite different). Lower left shows the impact of the intermediate Fisher-Rao warp estimates $\gamma_{1,K}$, $\gamma_{2,K}$.*

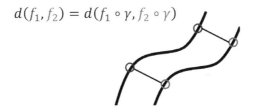

Figure 9.8 *Illustration of how a warp invariant metric results in parallel quotient space orbits.*

The Fisher-Rao analysis of amplitude and phase variation is motivated by the area called *information geometry*, that was started by Rao (1945), and more fully developed by Amari (1985), which studies parametric likelihood estimation from a geometric perspective. The Fisher-Rao approach achieves warp invariance by approaching differentiable functions f through their *Square Root Velocity Function* (SRVF)

$$q_f(t) = sign\left(f'(t)\right) \sqrt{|f'(t)|}, \tag{9.4}$$

i.e. a signed version of the square root of the derivative. The SRVF transformation can be viewed as representing the variation in f in the sense that

$$f(x) = f(0) + \int_0^x q_f(t)\, |q_f(t)|\, dt,$$

where the integrand is similarly thought of as a signed version of the square of

q_f. For a given warp $\gamma \in \Gamma$, integration by substitution shows that the SRVF representation is very useful for curve registration because it gives the L^2 norm a type of warp invariance in the sense that

$$\|q_{f_1 \circ \gamma} - q_{f_2 \circ \gamma}\|_2 = \|q_{f_1} - q_{f_2}\|_2.$$

The Fisher-Rao metric on the quotient space is nearly the L^2 norm on SRVFs, but an alignment of q functions is still needed. In particular, inducing a metric on the space of amplitudes as in (9.3) gives the *amplitude distance* between functions

$$\delta_A(f_1, f_2) = \min_{\gamma \in \Gamma} \|q_{f_1 \circ \gamma} - q_{f_2}\|_2. \tag{9.5}$$

It is straightforward to show that δ_A induces a proper metric on the space of amplitudes, and also that for $f_1, f_2 \in [f]$, $\delta_A(f_1, f_2) = 0$.

Given a set of functions $f_1, \cdots, f_n$, Fisher-Rao analysis starts with the Fréchet mean (called the *Karcher mean* in Srivastava et al. (2011), despite the objection of Karcher (2014) discussed in Section 7.1) of the amplitudes $[f_1], \cdots, [f_n]$ with respect to δ_A. In principle, using (7.5) this gives

$$\overline{[f]} = \arg\min_{[f]} \sum_{i=1}^{n} \delta_A([f], [f_i])^2.$$

But $\overline{[f]}$ is only defined up to warps, leaving open the question of the choice of best representer. For now, let $\overline{f}_C$ denote a *candidate* representer of $\overline{[f]}$, and let its SRVF be $q_{\overline{f},C}$. For $i = 1, \cdots, n$ let

$$\gamma_{i,C} = \arg\min_{\gamma \in \Gamma} \left\| q_{f_i \circ \gamma} - q_{\overline{f},C} \right\|_2$$

denote the corresponding warps of the f_i into $\overline{f}_C$.

The distance δ_A could also form the basis of an MDS (from Section 7.2) version of PC scores on amplitudes. However, direct computation of δ_A is quite challenging, so the approach detailed below is numerically preferable, and also results in fully interpretable modes of variation.

The third contribution of Srivastava et al. (2011) is a parallel definition of phase distance, which integrates appropriately with δ_A to give the Fisher-Rao approach to curve alignment, which is based on also applying the SRVF idea to the warp functions. In particular, given a differentiable warp $\gamma \in \Gamma$, define its SRVF

$$\psi_\gamma(t) = \sqrt{\gamma'(t)}.$$

The definition of ψ_γ does not need the *sign* function used in (9.4) because γ is strictly increasing. A very important property of the SRVFs of warps is

$$\|\psi_\gamma\|_2^2 = \int_0^1 \psi_\gamma(t)^2 dt = \gamma(1) - \gamma(0) = 1, \tag{9.6}$$

i.e. the SRVFs of warps ψ_γ all lie on the unit sphere in function space. A direct analog of the arc length distance on S^d (which is the arc cosine of the inner product of the corresponding vectors in $\mathbb{R}^{d+1}$) is the distance between warps

$$\delta_P (\gamma_1, \gamma_2) = \cos^{-1} \left(\int_0^1 \psi_{\gamma_1}(t)\psi_{\gamma_2}(t)dt \right).$$

Given $f_1, \cdots, f_n$, a candidate center $\overline{f}_C$ and corresponding candidate warps $\gamma_{1,C}, \cdots, \gamma_{n,C}$, result in a candidate notion of the *phase distance* between functions

$$\delta_{P,C} (f_1, f_2) = \cos^{-1} \left(\int_0^1 \psi_{\gamma_{1,C}}(t)\psi_{\gamma_{2,C}}(t)dt \right).$$

Again an analog of PC scores could be derived by MDS applied to the distance $\delta_{P,C}$, but a derivation of full modes of variation, as well as sensible choice of the candidate $\overline{f}_C$ and efficient computation are described below.

A natural sense of average representer of $\overline{[f]}$ is the Karcher mean function $\overline{f}_K$, which is the element of $\overline{[f]}$ with the property that the corresponding warps $\gamma_{i,K}$ of each f_i into $\overline{f}_K$, are centered at the identity warp, $\gamma_I(x)$. Recall $\gamma_I(x)$ was shown as the red diagonal line in the top middle panel (and the blue dashed line in the top right panel) of Figure 9.7. In particular, $\overline{f}_K$ is the member of $\overline{[f]}$ which gives warps whose SRVFs have Fréchet mean

$$\psi_{\gamma_I} = \arg \min_{\psi} \sum_{i=1}^{n} \left\| \psi - \psi_{\gamma_{i,K}} \right\|^2.$$

Choosing the candidate representer $\overline{f}_C$ to be $\overline{f}_K$ in the above calculations results in the final version of the phase distance between functions δ_P.

The fourth major contribution of Srivastava et al. (2011) is an iterative algorithm for numerical approximation of these quantities, based on dynamic programming ideas. This is a special case of the general approach to registration problems described in Srivastava and Klassen (2016). Efficient software can be found in Tucker et al. (2013), with a corresponding R version on CRAN in the package `fdasrvf`.

That computation results in a useful amplitude and phase decomposition of a data set of curves $f_1, \cdots, f_n$ into a couple of useful variations on the theme of data object.

- A Karcher mean function, $\overline{f}_K$, which is a useful version of overall center. In the Step Function Example of Figure 9.7, the Karcher mean is shown as the black dotted curve in the lower left panel. Because that curve takes into account both δ_A and δ_P it contains both phase and amplitude notions of center of the input curves in the top left panel. For the Shifted Betas data in Figure 9.2 the Karcher mean is in the right panel of Figure 9.9.

- A set of warps, $\gamma_{1,K}, \cdots, \gamma_{n,K}$ of each f_i into $\overline{f}_K$. These represent the horizontal variation in the data, and can be thought of as *phase data objects*. Such curves comprised the phase mode of variation of the Bimodal Phase Shift data in the lower left panel of Figure 2.2. For the Shifted Betas data set in Figure 9.2, these warps are shown in the top left of Figure 9.10, where these in turn are decomposed into further insightful modes of variation.

- A set of aligned curves, $f_1 \circ \gamma_{1,K}, \cdots, f_n \circ \gamma_{n,K}$, that represent amplitude variation. Similarly these are usefully considered to be *amplitude data objects*. For the Step Function example, these are the colored curves in the bottom left panel of Figure 9.7. Such curves for the Bimodal Phase Shift data appear in the upper right panel of Figure 2.2. The top left panel of Figure 9.9 shows these aligned curves in the case of the Shifted Betas example.

As noted above, PCA of the amplitude data objects results in insightful vertical modes of variation, which is illustrated in Figure 9.9, based on a Fisher-Rao analysis of the Shifted Betas data in Figure 9.2. The aligned curves, $f_1 \circ \gamma_{1,K}, \cdots, f_n \circ \gamma_{n,K}$, are shown, using the same colors as in Figures 9.2 and 9.3, in the top left panel. As in those figures, most panels show only the green to red curves, as the others have disappeared due to overplotting. Their existence does appear in the scores plot in the bottom right panel, which follows a similar pattern to the PC2 scores in Figure 9.2. That pattern makes intuitive sense in view of the vertical down then up motion of the heights of the peaks in the original data set in the top left of Figure 9.2. The PC1 residuals (bottom center panel) are essentially 0 (the inexactness reflects the discrete approximation used in the numerical algorithm which is also visible in the upper right panel), showing that there is only one major vertical mode of variation in this data set.

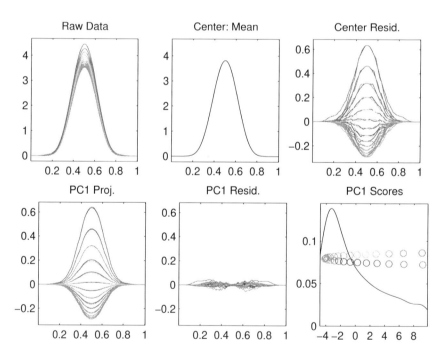

Figure 9.9 *Vertical analysis of the Shifted Betas data from Figure 9.2. PCA of the aligned curves reveals primarily just one vertical mode of variation.*

The nonlinear horizontal objects, which are the set of warps $\gamma_{1,K}, \cdots, \gamma_{n,K}$, can be decomposed into further modes of variation using a PCA as done in Figure 9.10. The warps (for the Shifted Betas example) are shown in the upper left panel, again using the same common colors. Note that the shifting peak mode of variation is nicely reflected in the family of warps which goes from something like the convex warp in the upper right of Figure 9.5 to the concave warp in the lower left. It is not surprising that the mean of the warps is $\gamma_I(x) = x$, as that is the criterion that was used to pick the Karcher mean as a representer of the Fréchet mean equivalence class $\overline{[f]}$. The first mode of variation (center left panel) of these curves appears to be some circular arcs, that are challenging to interpret, although the color scheme, and the scores in the center right panel, show it is clearly about the strong mode of shifting peaks apparent in the original data. The first PC residuals in the center panel are similarly hard to interpret. The bottom row shows that practically all of the remaining variation appears in the second PC component. Note that the PC2 scores (bottom right) follow the same pattern as, i.e. are strongly correlated with, the vertical PC scores in the bottom left of Figure 9.9. This makes sense because the original curves are probability densities (thus having area one underneath), so lower curves are wider. However, this notion of width is hard to understand in terms of warps. The PC2 residuals in the bottom center plot reflect the fact that there cannot be a pure shift mode of variation in densities supported on $[0, 1]$.

The interpretation problems with the warp modes of variation in the left column of Figure 9.10 are addressed in Figure 9.11. That gives a more interpretable version of the families of warps, by taking each curve, turning it back into an actual warp by adding γ_I to it, and then applying that warp to the Karcher mean $\overline{f}_K$. The results of this, for each $\gamma_{i,K}$, once more using the common colors are shown in Figure 9.11. Note that all of these modes are simply warps of the Karcher mean. The left panel very clearly shows that the first mode of variation is a shift in the peak location. The center panel shows that the second mode is the widening that is due to the varying heights together with the curves being densities. The very strong connection of this mode with the dominant phase mode in Figure 9.9 is shown by comparing those scores with the bottom right panel of Figure 9.10. The third panel of Figure 9.11 shows only the last red curve, because it overplots all the others, since there is essentially no variation left for the third component, which is consistent with the PC2 residuals in the bottom center panel of Figure 9.10 being essentially 0.

Fisher-Rao analysis has been quite useful in a number applications. These include the examples following Marron et al. (2014b), including Tucker et al. (2014), Lu et al. (2014a); Lu and Marron (2014a,b), Wu and Srivastava (2014), Kurtek et al. (2014), Staicu and Lu (2014), and Xie et al. (2014), as well as the modeling of seasonal variation in electricity prices in Chen et al. (2019). Thinking about probability distributions as data objects as in Section 3.3.1, in situations where the focus must be on densities and not quantile functions (shown to be preferred in the analyses of Figures 3.7 and 3.8), Fisher-Rao analysis can also be very useful.

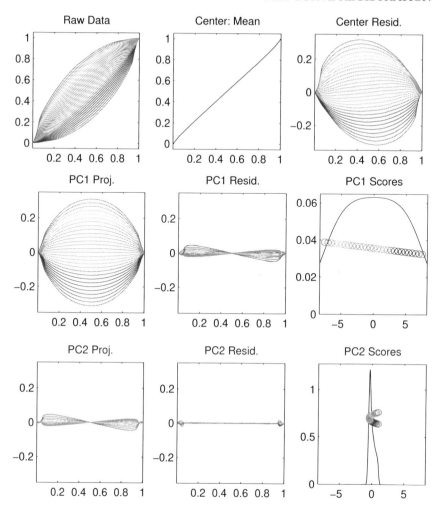

Figure 9.10 *PCA of warps of the Shifted Betas toy data set from Figure 9.2. Shows two modes of variation, whose scores are clear, but loadings are hard to interpret.*

However, the Fisher-Rao analysis is not the only warp invariant approach to the analysis of warp-equivalence quotient spaces. See Section 3.2 of the PhD dissertation Yu (2017) for introduction to a broader class of metrics based on *general root derivative* transformations. Some improvements in the sense of consistency (as $n \to \infty$) are proposed there. However, the Fisher-Rao approach remains a workhorse method because of its excellent properties in terms of peak alignment in curve registration, as shown in Figure 9.12.

The value of the L^2 norm applied to SRVFs, as in (9.5), for peak alignment is demonstrated in Figure 9.12, based on an example constructed by Xiaosun Lu. That is based on two Gaussian curves (not densities as they are just

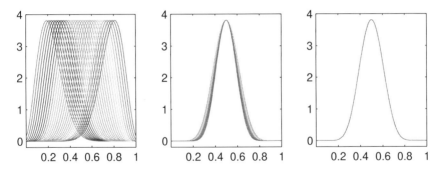

Figure 9.11 *Horizontal modes of variation for the Shifted Betas example, shown by using the modes of variation in Figure 9.10 (added back to δ_I) as warps applied to the Karcher mean. Accurately and intuitively reflects two modes of horizontal variation as expected from the raw data in the upper left of Figure 9.2.*

vertical rescalings), f_1 (dashed) and f_2 (solid), shown in blue in the bottom center panel. The left side of the top panel compares $2\|f_1 - f_2 \circ \gamma_\alpha\|$ (solid black curve) with $\|q_{f_1} - q_{f_2 \circ \gamma_\alpha}\|_2$ (dash-dotted curve) for a family of warps indexed by $\alpha \in (-1, 0]$. The factor of 2 is used to put these on approximately the same vertical scale. The family of warps (one member of which appears as the solid red piece-wise line in the middle left panel) is piece-wise linear, with a strong vertical jump of size α over a very small interval near the center. Those warps squish the interval $\left(\frac{1-\alpha}{2}, \frac{1+\alpha}{2}\right)$ into a very small interval, so that $f_2 \circ \gamma_\alpha$ (shown as the solid curve in the bottom left) suffers what has been called a "pinching effect". The red line in the top panel shows the value of α that minimizes $2\|f_1 - f_2 \circ \gamma_\alpha\|$, and the corresponding warp γ_α is the solid red curve in the middle left panel. It is clear from comparing the red curves f_1 (dashed) and $f_2 \circ \gamma_\alpha$ in the bottom left panel that they are visually much closer together than are the blue curves (corresponding to $\alpha = 0$) in the bottom center panel, which is quite concerning for peak alignment methods based on the L^2 norm between functions. This comparison is quantified in the top panel, where the solid curve is lower at the red line than at the blue line). However, note that the norm of the corresponding SRVFs $\|q_{f_1} - q_{f_2 \circ \gamma_\alpha}\|_2$ (dot-dashed curve) has the opposite behavior. This is because the red solid curve has very steep derivatives near the center of the plot, so the L^2 norm of the SRVFs strongly penalizes against such pinching.

The right side of the top panel does a slightly different comparison, this time of $2 * \|f_1 \circ \gamma_\alpha - f_2\|_2$ versus $\|q_{f_1 \circ \gamma_\alpha} - q_{f_2}\|_2$ for a different family of piece-wise warps γ_α, for $\alpha \in [0, 1)$. That warping family has a flat spot of length α in the center, which stretches (instead of compressing as for $\alpha < 0$) the width of the peak. The warp is applied to f_1 instead of f_2 in this case because that gives smaller norm values. The solid norm of the functions in the top panel shows a local minimum indicated by the green line. That local minimum is intuitively understood by noticing the green curves in the bottom right panel are again closer than shown in the blue panel. Once again applying the norm to the SRVFs results

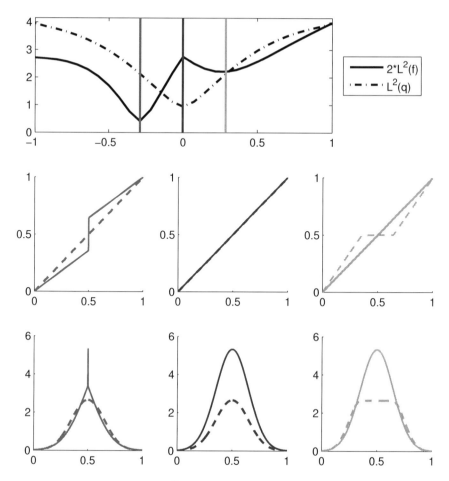

Figure 9.12 *Top row shows the L^2 distance between original functions f_1 and f_2 (solid) versus SRVFs q_1 and q_2 (dot-dashed) for a family of pinching warps (red) indexed by $\alpha \in (-1, 0]$ in the left-hand column and a family of stretching warps (green) indexed by $\alpha \in [0, 1)$ in the right column. The middle column shows the $\alpha = 0$ identity warp in blue and the original f_1, f_2 below. Shows how the L^2 norm applied to SRVFs gives a much more natural notion of peak alignment.*

in a smaller value at $\alpha = 0$ (blue curves) this time because the warped dashed curve has a too small derivative near the center.

Wagner and Kneip (2019) point out that a weakness of the Fisher-Rao numerical approach is that when $\overline{f}_K \equiv 0$ the method will fail. Essentially the dynamic programming algorithm requires a non-constant template mean for convergence. They propose a method that does work in such cases, although it is not as efficient at peak alignment in situations where that is a priority. Other approaches based on the SRVF include Bayesian functional alignment, where prior distributions are placed on the warping functions. Cheng et al. (2016) use a Dirichlet prior on the simplex and Lu et al. (2017) set a Gaussian process prior on the sphere for the warp functions.

9.2 Principal Nested Spheres Decomposition

While the interpretation of the Fisher-Rao approach to curve registration has already benefited in a major way from OODA concepts such as the warp equivalence class quotient space, in some situations there is also technical benefit from other OODA ideas. In particular, the analysis of the warping functions done in (9.6) shows that the warps $\gamma_{1,K}, \cdots, \gamma_{n,K}$, which quantify horizontal variation in a Fisher-Rao analysis, all have corresponding SRVFs that lie on the surface of a high-dimensional sphere in function space. The decomposition into modes of variation in Figure 9.10 was essentially a Euclidean PCA done in the tangent space centered at the Karcher mean $\overline{f}_K$. As noted in Section 8.3, this can result in the problem of a distorted analysis which can be substantially improved using PNS (recall Principal Nested Spheres from Section 8.3) on the corresponding SRVFs. This motivated the proposal of Yu et al. (2017a) to couple PNS with a Fisher-Rao analysis to understand phase variation.

In many situations, there is not a major difference between the tangent plane analysis and PNS. This seems to be because for many data sets SRVFs do not tend to spread very broadly around the sphere (so the tangent plane approximation is frequently adequate). This may be due to the fact that for $\gamma(t)$ increasing, $\psi_\gamma(t) > 0$, and hence γ lies in the positive orthant of the sphere, which in finite dimensions is $\{x \in \mathbb{R}^d : \|x\|_2 = 1, x_1 > 0, \cdots, x_d > 0\}$. However a toy example showing the tangent plane approximation can be a serious problem is shown in Figure 9.13.

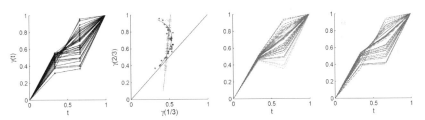

Figure 9.13 *Far left panel shows a family of piece-wise linear warps $\gamma(t)$. Center left shows representations of each warp in terms of heights at $t = \frac{1}{3}, \frac{2}{3}$, together with projections onto Euclidean PCA. Center right is reconstruction of Euclidean projections with cyan curves highlighting invalid warping functions. Far right shows PNS projections, which are all valid warps.*

Figure 9.13 first appeared in the PhD dissertation of Lu (2013), and is Figure 5 of Yu et al. (2017a). The far left panel contains a set of warping functions $\gamma(t)$, which are piece-wise linear, all having just two breakpoints at $t = \frac{1}{3}$ and $\frac{2}{3}$. Thus each warp is entirely characterized by $\left(\gamma(\frac{1}{3}), \gamma(\frac{2}{3})\right)$, and is shown as a small black circle in the center left panel. Because warps must be increasing, i.e. $\gamma(\frac{1}{3}) < \gamma(\frac{2}{3})$,

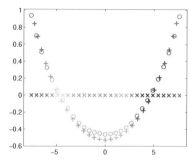

Figure 9.14 *Shows benefits of a PNS analysis of the warps for the Shifted Betas data in the top left panel of Figure 9.10 whose projections onto the 2-d PCA subspace appear as circles. The x signs are the rank 1 PCA approximations. Much better approximation comes from the projections of the rank 1 PNS approximation shown as plus signs.*

all of these circles are above the diagonal black line that represents $\gamma(\frac{1}{3}) = \gamma(\frac{2}{3})$. The results of a direct Euclidean PCA on the family of warps is also overlaid on the center left panel. The red line shows the first eigen direction, and the symbols are the projections of the warps onto the first mode of variation. Red plus signs are those projections that are actually warps (i.e. are increasing functions), while the cyan circles indicate those of the projections that are actually outside of the set of warps (because they are no longer increasing functions). The impact of replacing the Euclidean PCA of the warping functions with a PNS of the corresponding SRVFs is shown in the right panel. Note that each projection onto the first PNS component remains a valid warp. Another view of the effect of PNS is the blue curve shown in the center left panel. That is a plot of $\left(\gamma(\frac{1}{3}), \gamma(\frac{2}{3})\right)$ for a dense grid of points lying along the PNS one-dimensional representation. It shows how appropriately managing the curvature of the space keeps that analysis in the range of functions of interest. A similar issue occurred when studying covariance matrices in Section 7.3.5 where linear extrapolation using the Euclidean metric can lead to negative eigenvalues, hence leaving the space of covariance matrices (which are all positive semi-definite). Recall that challenge was similarly addressed by using a curved manifold for representing covariance matrix data objects.

Figure 9.14 shows another example of the improvement available from PNS of the Fisher-Rao warp functions over the simple tangent plane PCA shown in Figure 9.10. That is based on the Shifted Betas data in the top left panel of Figure 9.2 which was used with the same coloring to illustrate many points in Section 9.1. The input to the PNS is the set of Fisher-Rao warps, $\gamma_{1,K}, \cdots, \gamma_{n,K}$, shown in the top left panel of Figure 9.10. Figure 9.14 shows various projections onto the rank 2 Euclidean PCA subspace. The circles are the (rank 2) projections of the warps onto this space, with the horizontal axis showing the first PC eigen direction, and the vertical axis the second. The x signs show the first PCA mode of variation (shown in a different way in the middle left panel of Figure 9.10), whose constraint of being linear makes them a poor approximation of the clearly nonlinear data set. That strong distortion is addressed by the plus signs, which are the projections of the first PNS approximation of this data set. That mode of variation is essentially nonlinear with respect to this data view. It clearly provides a far better approximation of the data. A careful look at the horizontal and vertical scales shows this effect is not terribly strong (also reflected by the rather small variation in the center panel of Figure 9.11) although still considerable. A more dramatic example (which is based on even stronger phase variation) appears in Section 2 of Yu et al. (2017a).

A real data example, where PNS gives a big benefit over tangent plane PCA can be found in Section 3 of Yu et al. (2017a), in the context of real time monitoring of blood glucose. Additional examples appear in Staicu and Lu (2014) and in the analysis of the Juggling data by Lu and Marron (2014a) shown in Figure 2.3.

CHAPTER 10

Graph Structured Data Objects

In the mathematical field of graph theory a *graph* is defined as $G = (V, E)$, where V is a set of *vertices* (also known as *nodes*) and where E is a set of *edges* (that connect the vertices), i.e. $E = \{[v, w] : v \neq w \in V\}$. In a *directed* graph, the edges are thought of as pointing from say v to w, so $[v, w]$ is taken to be an ordered pair (v, w). Edges in *undirected graphs* ignore directions, and so $[v, w] = \{v, w\}$, just an unordered set. In some situations, it useful to add *attributes* to the graph structure, which are numbers or vectors (or even more general objects) assigned to each vertex and/or to each edge. An example of vertex attributes are the centers and radii of the spheres that compose the artery trees shown in Figures 2.5 and 2.6.

Directed graphs are natural when tabulating ancestry, yielding the terminology w is a *child* of v when $(v, w) \in E$, and v is similarly a *parent* of w. A *tree* is a special case of a graph, which is directed, has a starting vertex called the *root* which has no parent, and is *acyclic* in the sense that every other vertex has exactly one parent and is a descendant of the root. A vertex with no child is called a *leaf*. Brain artery systems as introduced in Section 2.2 and studied further in Section 10.1 are nearly trees, except for an anatomical structure called the Circle of Willis (which seems to ensure that blockage of a carotid artery will not result in loss of half the brain). To give reasonable correspondence, the data objects studied here are unions of the 3, 4, or 5 (depending on the subject) actual trees which flow out of the Circle of Willis. The leaves in the Brain Artery data occur when each artery becomes too small to appear in the MRA images discussed in Section 2.2 (at about 1 mm in diameter). Hence, we do not consider the circulatory system as the closed-circuit that it actually is, because neither the capillaries (between the arteries and the veins) nor the veins themselves show up in the MRA. Thus the Brain Artery data is treated here as a set of trees.

There is a large literature on network (i.e. graph) analysis from various viewpoints. See West (1996), Kolaczyk (2009), and Harary (2015) for much more introductory material. Much of the intersection with statistics is about the variation involved in constructing a single graph. Here we take a different approach by thinking of graphs as data objects and studying sample / population properties. Section 10.1 gives an overview of the literature on the arterial trees. Other examples of the analysis of tree-structured data objects in biomedical applications include airways as data objects in Feragen et al. (2011, 2013); Amenta et al. (2015). Other approaches to data sets of tree-structured objects include the *tree kernel* idea discussed in Vert (2002) and Yamanishi et al. (2007). Early work in this area was done by Banks and Constantine (1998).

DOI: 10.1201/9781351189675-10

Section 10.2 studies the analysis of samples of networks as data objects using *graph Laplacian* representations, which are a subset of the space of covariance matrices.

10.1 Arterial Trees as Data Objects

A deep and challenging set of tree-structured data objects is the Brain Artery data introduced in Section 2.2. A number of quite different approaches have been taken to quantifying the variation in that data set, including *combinatoric* approaches overviewed in Section 10.1.1, a *phylogenetic* point of view as taken in Section 10.1.2, the *Dyck Path* approach studied in Section 10.1.3, and a representation using *persistent homologies* in Section 10.1.4. These methods are compared in terms of performance on the Brain Artery data in Section 10.1.5.

10.1.1 Combinatoric Approaches

The first versions of PCA-like visualizations of these data objects are called combinatoric. In particular, they ignore all attributes such as branch length, thickness, and physical location, and instead only take into account the tree topology (the original nodes and edges). These analyses are in Wang and Marron (2007) and Aydin et al. (2009), which were also the first studies to explicitly use the terminology OODA. Those early analyses involved embedding the trees which naturally lie in a three-dimensional ambient space, into a binary two-dimensional data object representation, as shown in Figure 10.1. Two arbitrary choices of branch location involving either branch thickness or number of descendant branches (called *descendant correspondence* in Section 10.1.3), were considered and gave different results. The most statistically significant results from Aydin et al. (2009) appear in the top row of Table 10.1 in Section 10.1.5, suggesting that the tree topology alone does not provide enough information. Instead it is important to consider at least some tree attributes, as done in the following sections.

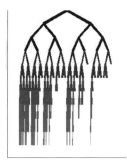 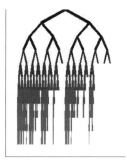 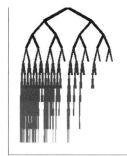

Figure 10.1 *Examples of 2-d embeddings of the Brain Artery data as data objects, for 3 different subjects.*

Two challenges with early versions of the brain artery tree data were the linking of tree branches into a tree structure and the starting point of each tree. These issues have been addressed through careful data objects choices. Linking was initially done in Aylward and Bullitt (2002) using a thresholding operation combined with manual intervention, and the starting point was arbitrarily chosen by the MR operator. In subsequent analyses, arteries were more accurately linked using a visualization device invented in Aydın et al. (2011). As noted above, the starting point issue has been addressed by only including arteries flowing out of the Circle of Willis (a readily identifiable anatomical feature).

Wang et al. (2012) deeply investigated the relationship between age and artery tree structure and found some unexpected age behavior, by inventing an analog of kernel smoothing with a tree-structured response variable. See Alfaro et al. (2014) for another combinatoric approach to PCA of the Brain Artery data.

10.1.2 Phylogenetics

Another approach to the Brain Artery data, based on phylogenetic tree representations as data objects, can be found in Skwerer et al. (2014). Phylogenetic trees are a model for evolution of species (or other biological units) over time. In addition to the branching structure of the combinatorial trees in Section 10.1.1, the branches of phylogenetic trees have the attribute of length representing the time between species splits and/or the present. The motivation for addressing the artery data using phylogenetic tree methods is that the latter have been studied for a very long time. In particular, the ideas go back to Darwin (1859) with interesting early graphical representations already in Haeckel (1866), so much is known about them which should be useful for the study of trees as data objects. A fundamental point is that in a typical phylogenetic setting, one works with a common set of species (i.e. leaf set), and the goal is to explore (often to choose between) various ways in which the species could be reasonably organized into an ancestral tree. Hence, the main challenge to adapting this idea to the case of the brain artery trees is that the latter do not have a common leaf set. Instead, as noted above, arteries are collected only until they become too thin to show up reliably in the MRA (about 1 mm resolution). Hence each subject has a different number of arterial endpoints, none of which correspond across individuals in a meaningful way. To create a set of data objects appropriate for a phylogenetic type of analysis, common leaves were artificially generated as a set of corresponding landmarks, based upon the brain cortical surfaces of each subject (also collected in the original study), using an elegant algorithm of Oguz et al. (2008). An example of this for one subject appears in Figure 10.2, which is Figure 2.6 of the PhD dissertation Zhai (2016). In the left panel the blue curves are the original arteries and the red dots are landmarks on the cortical surface that correspond (with respect to shape of the surface) across all subjects in the sample. The red lines in the center panel show the projections of each landmark onto the arteries. The cyan part of the artery curves in the center panel show *orphans*, which are arterial portions that are beyond (with respect to arterial flow) the last such projection. The right panel shows the result of trimming those orphans which leaves the remaining parts as the blue tree representation. This representation is ideal for phylogenetic analysis, because all leaves now correspond across subjects.

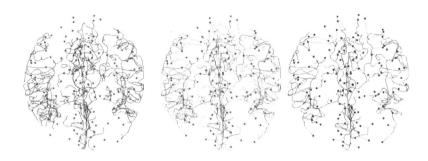

Figure 10.2 *Illustration of common leaf set construction for use of phylogenetic tree methods on the Brain Artery data. Dots are landmarks corresponding across all subjects. These are connected by projection to the artery tree shown as blue curves. Orphan curve segments, shown as cyan in the center panel, are trimmed to give the final tree in the right panel.*

Skwerer et al. (2014) go on to show that this approach gives some improved statistical inference (about age and sex) as summarized in Table 10.1 in Section 10.1.5.

Phylogenetic tree space is another example of stratified spaces, as discussed at the end of Section 2.3 and in Section 8.7 (both about covariance matrices as data objects). The particular geometry of the space creates a number of surprises, because it is locally flat (in fact exactly Euclidean on neighborhoods) almost everywhere, with a very notable exception being the origin. The singularity at that point results in the whole space having very strong non-positive curvature. This has some good consequences such as uniqueness of the Fréchet mean. However as discussed in Skwerer et al. (2018), even calculation of the Fréchet mean is quite challenging. For example the iterated pairwise weighted average algorithm of Sturm (2003) did not converge after 10,000 steps, and had a larger Fréchet variance than the origin at each of those steps. This motivated Skwerer et al. (2018) to develop a more sophisticated optimization approach to calculation of the Fréchet mean in phylogenetic tree space, which ultimately showed that the origin is the Fréchet mean of this data set. See Miller et al. (2015) for interesting related work.

Some interesting views of the strong curvature of phylogenetic tree space have been provided using MDS (recall Multi-Dimensional Scaling from Section 7.2), in the PhD dissertation Zhai (2016), as shown in Figure 10.3. Demonstration of curvature effects in these examples has been enabled by the ability to quickly compute geodesics using a clever algorithm invented by Owen and Provan (2011). The data objects are the $n = 67$ trees for which corresponding landmarks were found using the method of Oguz et al. (2008), represented as blue dots (with a few as blue squares) in each panel. In each case the Fréchet mean is shown as a red plus sign. The blue points in the left panel show the MDS distribution of the original data, suggesting an approximately Gaussian distribution with perhaps an outlier. The center panel shows a much different MDS distribution of the same data set. The green dots are additional data included in the MDS, which are chosen as lying along equally spaced grids connecting three of the data points. Note the addition of the green points gives a completely different MDS view, because the additional points result in more overall variation in their direction. A triangle is used here, because one characterization of non-positive curvature is that triangles (connecting three points with geodesics) "bend inwards" in the sense of having angle sums much less than the usual 180° of triangles in Euclidean space. This particular triangle is representative in the sense of having the median angle sum over all triples of data points. Those angles are between geodesics in the local Euclidean plane containing each vertex point.

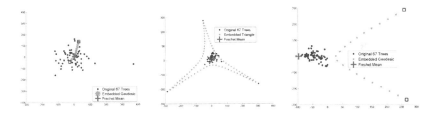

Figure 10.3 *MDS views of Brain Artery data (blue symbols), together with the Fréchet mean (red plus sign). The standard MDS is in the left panel. The MDS in the other two panels are based on all of the blue data, the red mean, and the green points, which are sampled along geodesics between a few data points. Inclusion of the green points shows how the strong curvature of the space introduces huge distortions in the MDS view. The left panel shows how the out of sample MDS can address this distortion by showing the same green points from the right panel embedded in the original MDS*

A related distortion is shown in the right panel of Figure 10.3. This time the MDS was based upon the union of the data together with only one geodesic (chosen to have the median pairwise distance), where the addition of the additional points again creates a very large distortion of the view. Another viewpoint on this is provided in Figure 8 of Skwerer et al. (2014), which shows a great deal of shrinkage in branch lengths as one follows such geodesics. Note that this time the distortion is so strong that the Fréchet mean actually lies outside the apparent convex hull (with respect to this embedding) of the data. This distortion can be effectively dealt with using the out of sample MDS approach (invented by Trosset and Priebe (2008) and refined in Chapter 3 of Zhai (2016)) that was discussed in Section 7.2. The result of this is shown in the left panel of Figure 10.3, where the blue dots have their original MDS configuration. The green points are the same as in the right panel, but do not influence the embedding because of the out of sample calculation. These points now approximately lie along a line and the Fréchet mean seems to be near the center.

See Nye (2011) for an early approach to PCA of phylogenetic tree data objects, and Nye et al. (2017) for updated ideas. Other approaches to PCA, based on *tree lines*, *sample limited geodesics*, and *principal rays* (making repeated use of the Owen and Provan (2011) fast geodesic algorithm) were proposed by Zhai (2016). These unfortunately did not give reasonable insights into the population structure of the Brain Artery data, apparently because of the strong curvature of the space, and the broad spread of the artery data around the space. An interesting open problem is to address these challenges by modifying MDS to target data representations in a negatively curved space, instead of the traditional Euclidean representations discussed in Section 7.2.

Asymptotic results have been established by Barden and Le (2018); Barden et al. (2018). An interesting way to quantify uncertainty in phylogenetic tree space can be found in Willis and Bell (2018), which was used to develop a notion of Confidence Sets in Willis (2018). An important point of the overview of Feragen and Nye (2020) is that phylogenetic tree space is another example of a stratified space as discussed in Section 8.7. This viewpoint is the key to the discovery of *stickiness* in Hotz et al. (2013) that seems to occur generally when studying limit behavior of sample means in phylogenetic tree space.

10.1.3 Dyck Path

A quite different choice of data objects was made in Shen et al. (2014). The key idea there was to use the Dyck Path idea of Harris (1952) (invented as a tool in the stochastic processes literature for the asymptotic analysis of branching processes) to represent each data tree as a curve, followed by the use of FDA techniques for the resulting statistical analysis.

The Dyck path analysis uses more tree information than the combinatorial approach of Section 10.1.1, including length of each branch as an edge attribute. As illustrated in the left panel of Figure 10.4, a challenge to that old approach was the correspondence issue caused by the arbitrariness of the embedding of the 3-d arterial trees into the plane. Different representations of the same tree can be obtained by *flipping vertices* as shown. The most compelling way to handle this ambiguity is through the concept of *invariance* discussed in Chapter 4 of Dryden and Mardia (2016). Proceeding as in Sections 1.2 and 8.3 (recall their context was shape analysis), the main idea is to call trees that can be reached from each other by vertex flips *equivalent*, and then to work with the equivalence classes (*orbits* with respect to the group of flipping operations) as the data objects. Note the strong parallel to the Fisher-Rao approach to curve registration taken in Section 9.1. Pursuing such an approach to this analysis is an interesting open problem. This was not done in Shen et al. (2014), who instead used the descendant correspondence idea of Section 10.1.1, where each equivalence class was represented by its member with most descendants on the left at each vertex.

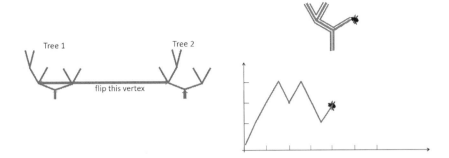

Figure 10.4 *Illustration of vertex flipping as an equivalence relation (left panel) and Dyck Path interpretation as an ant walking around a tree (right panel).*

The Dyck Path is intuitively expressed in terms of "an ant walking around a tree" in a two-dimensional embedding, as illustrated in the right panel of Figure 10.4. The tree is shown near the top in blue with the path of the ant shown in red. The lower plot shows the corresponding trace of the ant's distance from the root as a function of time (assuming uniform velocity). Such resulting red curves (hence amenable to classical Euclidean FDA) are the representations of the tree data objects. For better correspondence across data objects, Shen et al. (2014) also incorporate the idea of *support tree*, essentially the union of all the trees in the data set.

While the combinatorial approaches in Section 10.1.1 mostly focused on exploratory analyses, Shen et al. (2014) went on to do deeper confirmatory analyses. These motivated the new idea of *branch length representation*, which resulted in statistically significant correlations with age. This was not surprising as that connection was also found in the simple summary based analysis of Bullitt et al. (2005). However, a deeper analysis, based on *tree pruning* ideas found the first statistically significant connection of gender with tree structure. Actual p-values are summarized in Table 10.1.

10.1.4 Persistent Homology

A *Topological Data Analysis* (TDA) of the brain artery has been done by Bendich et al. (2016). That paper uses various *persistent homology* representations as data objects. In confirmatory analysis, these coordinate-free representations have given the strongest statistical significance found to date for both age and gender.

The basis of TDA is a *filtration* that can be viewed as extracting a topological representation of the data. An example of a filtration is shown in Figure 10.5. The input data points are the $n = 13$ plus signs in $\mathbb{R}^2$ shown in each panel. The filtration records how topological properties of the union of discs centered at the data points change as a function of disc radius (which is the *filtration index*). The process starts with 0 length radii, where there are 13 disconnected components represented as solid vertical bars in the right subplot of each panel (the dotted bars are explained below). In the top panel the radius is large enough that the eight points on the left now lie in a single connected region. Thus at this radius, represented by the horizontal line in the subplot, there are 6 bars that touch. There are 7 shorter bars, that ended at the radius where the circles became tangent, resulting in fewer connected components. The middle panel of Figure 10.5 shows a larger disc radius (represented as a higher horizontal line in the subplot), where only two connected components remain, since the 5 points on the left were connected up at the radius shown as the top of the 5 bars in the middle. In the bottom panel only one bar is still going, because the two circles of data have now been merged into a single connected component. As the radius continues to grow this single bar grows to infinite length.

The above discussed TDA representation of the data is based solely on connected components, which in topological terms are called H_0 (0-th order) *homological properties*. Also of use in TDA are higher order homologies, such as the first order H_1 (based on loops), the second order H_2 (focusing on cavities in 3 and higher dimensions), and also higher orders. In addition to demonstrating H_0 as solid bars, Figure 10.5 also shows an H_1 filtration, using dotted bars. Because there are no loops when the discs are disjoint the dotted bars do not start at the bottom. The first one starts when the right circle of data points first merges, just before the radius in the top panel. The middle row shows two dotted H_1 bars reflecting the two loops present at that radius. One bar has ended before the radius shown in the bottom panel as the right-hand loop disappeared. When a feature disappears, the convention is to end the shorter bar. Both loops disappear for large enough radii.

The classical use of TDA, as discussed by Bubenik and Kim (2007), Carlsson (2009), Zomorodian (2012), and Wasserman (2018), focuses on understanding the relationships between data objects.

However, TDA filtrations are also very useful for data object representations, as shown by Bendich et al. (2016), in an analysis of the Brain Artery data. The filtration proposed in that paper is quite different from the growing balls illustrated in Figure 10.5. It is instead based on the idea of filling a tank holding the arterial system, as demonstrated in Figure 10.6. An H_0 (connected component) filtration is indexed by the level of the water. Bars start when the water level touches new parts of the tree, and end when components merge together. Three such water levels for the same data tree are shown in Figure 10.6.

Bendich et al. (2016) used this and related object representations to explore age and gender for the artery data. A challenge was finding a Euclidean representation of the resulting bar graphs. A number of things were tried including various summaries of the bar lengths, together with the bar start times. The best of these gave results, using H_1, quite competitive with other analyses, as shown in Table 10.1 of Section 10.1.5.

More sophisticated Euclidean summaries of the bar graphs are the *persistence landscapes* studied in Bubenik (2015). A different approach to Euclidean representation of bar graphs are the *rank functions* of Robins and Turner (2016). An interesting type of filtration with some inversion properties (not usual in topological

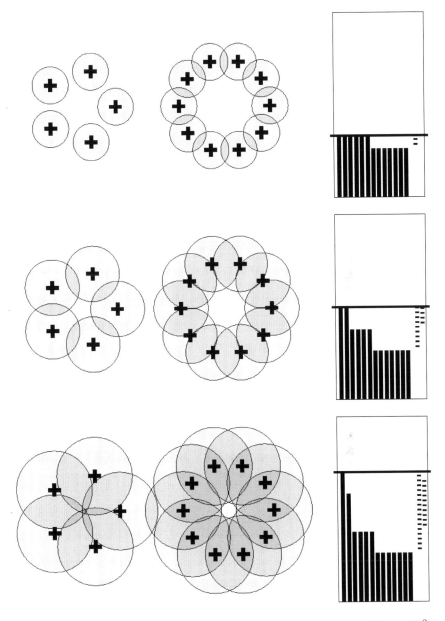

Figure 10.5 *Demonstration of a TDA filtration, based on plus signs as data objects in $\mathbb{R}^2$.* *Rows show different disc radii, i.e. filtration levels. Left panels show construction of bar code representation of data. Homological properties (before and at each filtration level) are displayed in the right panels: H_0 bars are solid and H_1 are dotted.*

Figure 10.6 *Three water levels (submerged parts shown in red) of the H_0 TDA filtration for one brain artery system.*

data analysis contexts) is the *persistent homology transformation* of Turner et al. (2014).

10.1.5 Comparison of Tree Analysis Methods

An interesting question is how to compare these quite diverse ways of under-standing the variation in the Brain Artery data set that were discussed in Sections 10.1.1–10.1.4. As noted in Section 2.2 it would be medically interesting to study effectiveness at finding early cancer or quantifying stroke tendency, but that infor-mation is not available here. So instead we work with the given covariates of age, sex, and handedness. Handedness has around 95% right-handed subjects, so that is not studied further. However, it is interesting to investigate any brain arterial structure associations with age and sex. To study age the first and third studies used slope of a regression line (not reported here), while the others reported cor-relations as shown in Table 10.1 (with blanks in cases where the correlation was not reported). Another approach to comparison is based on the p-values of hy-pothesis tests for a significant age effect, and also of significant sex effect. The sex data include two trans-sexuals that were not included in the sex-based calcu-lations, and no p-value was given in the first two studies (hence the blanks in the table). Note that there is a general trend toward stronger statistical inference for more sophisticated representations of the Brain Artery data objects.

A mathematically compelling approach to statistical analysis of tree-structured data objects can be found in Guo and Srivastava (2020). This uses equivalence classes as data objects ideas (discussed in Sections 1.2 and 8.3) to simultaneously summarize both the topological structure as well as the individual branch shapes. The latter is elegantly represented using the Fisher-Rao elastic registration ideas discussed in Section 9.2. Direct comparison with the other approaches summa-rized in Table 10.1 is unfortunately not possible, as that paper made a different choice of data objects (sub-trees as opposed to the full arterial system), but the results appear to be fairly similar.

	Section	Age ρ	Age p-val	Sex p-val
Comb.	10.1.1	-	0.0025	-
Phylo.	10.1.2	0.43	10^{-5}	-
Dyck	10.1.3	-	10^{-6}	0.033
TDA H_0	10.1.4	0.53	10^{-7}	0.11
TDA H_1	10.1.4	0.61	10^{-10}	0.031

Table 10.1 *Summary of results from various studies of the Brain Artery data. Shows improvement using more sophisticated methods.*

10.2 Networks as Data Objects

Networks are of wide interest and able to represent many different phenomena, for example social interactions between individuals and white matter fiber connections between regions in the brain, e.g. see Kolaczyk (2009) and Ginestet et al. (2017). If a sample of networks is available then it is natural to carry out OODA and, following the main theme of this book, important considerations are to decide what are the data objects and how they are represented.

10.2.1 Graph Laplacians

Consider a data set where each data object is an undirected graph (as introduced at the beginning of this chapter) comprising a set of m nodes (i.e. vertices), $V = \{v_1, v_2, \ldots, v_m\}$, and a set of edge attributes which are called *weights*, $\{w_{ij} : w_{ij} \geq 0, 1 \leq i, j \leq m\}$, indicating nodes v_i and v_j are either connected by an edge of weight $w_{ij} > 0$, or else unconnected (if $w_{ij} = 0$). We call this type of graph a *weighted network*. An *unweighted network* is the special case with $w_{ij} \in \{0, 1\}$. Following Severn et al. (2019) we restrict attention to networks that are undirected and without loops, so that $w_{ij} = w_{ji}$ and $w_{ii} = 0$.

There are several representations of weighted networks that are very useful. These include the weighted *adjacency matrix* $\mathbf{A} = \{w_{ij}\}$, which is the matrix of edge weights. In the unweighted case, $\mathbf{A}$ is the indicator matrix of edges. The *weighted degree* of a node is the sum of all edge weights associated with it. These are summarized on the diagonal of the *degree matrix*

$$\mathbf{D} = \text{diag}(\sum_{j=1}^{m} w_{1j}, \ldots, \sum_{j=1}^{m} w_{mj}) = \text{diag}(\mathbf{A}\mathbf{1}_{m,1})$$

where $\mathbf{1}_{m,1}$ is the m-vector of ones. Throughout this book, we will denote the matrix of ones as

$$
\mathbf{1}_{d,n} =
\begin{bmatrix}
1 & \cdots & 1 \\
\vdots & \ddots & \vdots \\
1 & \cdots & 1
\end{bmatrix}
\tag{10.1}
$$

The degree matrix contains useful information about the importance of each node, but it only contains partial information about the network. The weighted adjacency matrix does contain all the information about the network, but the degree information is only indirectly available. A representation that combines both adjacency and degree is the graph Laplacian matrix $\mathbf{L} = (l_{ij})$, defined as

$$
l_{ij} =
\begin{cases}
-w_{ij}, & \text{if } i \neq j \\
\sum_{k \neq i} w_{ik}, & \text{if } i = j
\end{cases}
$$

for $1 \leq i, j \leq m$. The graph Laplacian matrix can be written as

$$
\mathbf{L} = \mathbf{D} - \mathbf{A},
$$

in terms of the adjacency matrix, $\mathbf{A}$, and the degree matrix $\mathbf{D}$. An advantage of using the graph Laplacian matrix representation is that it explicitly includes both the degree information on the diagonal and the edge weight adjacency terms (as negative off-diagonal elements). Also, a graph Laplacian is a square symmetric positive semi-definite matrix, and so we have methods available to deal with such data objects as given in Section 7.3.5.

In more detail, the space of $m \times m$ graph Laplacian matrices is of dimension $m(m-1)/2$ and is

$$
\mathcal{L}_m = \{ \mathbf{L} = (l_{ij}) : \mathbf{L} = \mathbf{L}^t; \; l_{ij} \leq 0 \, \forall i \neq j; \; \mathbf{L}\mathbf{1}_{m,1} = \mathbf{0}_{m,1} \},
\tag{10.2}
$$

using the notation $\mathbf{0}_{d,n}$ for the matrix of zeroes from (4.2). In fact the space $\mathcal{L}_m$ is a closed convex subset of the cone of centered (meaning rows and columns sum to 0) symmetric positive semi-definite $m \times m$ matrices and $\mathcal{L}_m$ is a manifold with corners (Ginestet et al., 2017).

Figure 10.7 gives an indication of where the graph Laplacians, $\mathcal{L}_m$, lie within the space PSD_m (positive semi-definite matrices, from Section 3.3) in the case $m = 2$. The cone structure of PSD_m is highlighted by plotting $10{,}000$ realizations of 2×2 positive semi-definite matrices constructed as $\mathbf{X}\mathbf{X}^t$, where the entries x_{ij} of the 2×2 matrix $\mathbf{X}$ were generated as independent $N(0,1)$. The graphic is a 3-d plot of each realization as a red dot, of the form $\begin{pmatrix} y_{11} & y_{12} & y_{22} \end{pmatrix}^t$, where

$$
\mathbf{X}\mathbf{X}^t =
\begin{bmatrix}
y_{11} & y_{12} \\
y_{21} & y_{22}
\end{bmatrix},
$$

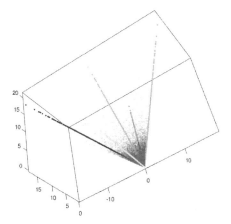

Figure 10.7 *Red: Illustration of the cone structure of the space* PSD_2, *using simulated members plotted as red dots. Various rank 1 subcones are shown using colors. Blue dots highlight the graph Laplacians* $\mathcal{L}_2$. *This shows this space (and hence that of the graph Laplacians) is naturally non-Euclidean.*

recalling that $\mathbf{XX}^t$ is symmetric. Important subcones of PSD_m are revealed by also plotting various rank 1 modifications of the coordinates of the red dots. Edges of the cone are explored using green points $\begin{pmatrix} 0 & 0 & y_{11} \end{pmatrix}^t$, yellow points $\begin{pmatrix} y_{11} & 0 & 0 \end{pmatrix}^t$, and cyan points $\begin{pmatrix} y_{11} & y_{11} & y_{11} \end{pmatrix}^t$. A more central subcone is multiples of the identity matrix shown as gray points $\begin{pmatrix} y_{11} & 0 & y_{11} \end{pmatrix}^t$. Some graph Laplacian matrices, i.e. members of $\mathcal{L}_2$, are shown as blue dots of the form $\begin{pmatrix} y_{11} & -y_{11} & y_{11} \end{pmatrix}^t$.

10.2.2 Example: A Tale of Two Cities

An example of a type of network is where the language structure of phrases of text or whole novels are studied using a network. In *corpus linguistics*, also called *natural language processing*, it is common to record word co-occurrences, where pairs of words occur within a scan of a number of neighboring words. Each distinct word is a node in the network, and the edges of the network between two nodes (words) have weights proportional to the number of co-occurrences of that pair of words. In Figure 10.8 we provide a network representation of the phrase *It was the best of times, it was the worst of times* that opens the Charles Dickens novel "A Tale of Two Cities". Here the thickness of the line denotes the weight w_{ij} which is the number of times the neighboring pair occurred. A neighbor pair is counted as occurring when they are within a scan window of just one word here.

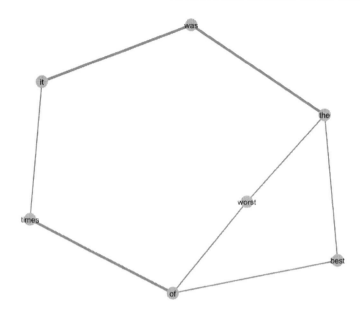

Figure 10.8 *A network representation of the phrase It was the best of times, it was the worst of times. An edge occurs when words are neighbors in the phrase, and the thickness of the edge is proportional to the weight.*

The weighted adjacency matrix is given by:

$$
\mathbf{A} = \begin{array}{c} \\ it \\ was \\ the \\ best \\ worst \\ of \\ times \end{array} \begin{bmatrix} it & was & the & best & worst & of & times \\ 0 & 2 & 0 & 0 & 0 & 0 & 1 \\ 2 & 0 & 2 & 0 & 0 & 0 & 0 \\ 0 & 2 & 0 & 1 & 1 & 0 & 0 \\ 0 & 0 & 1 & 0 & 0 & 1 & 0 \\ 0 & 0 & 1 & 0 & 0 & 1 & 0 \\ 0 & 0 & 0 & 1 & 1 & 0 & 2 \\ 1 & 0 & 0 & 0 & 0 & 2 & 0 \end{bmatrix}
$$

The degree matrix is:

$$
\mathbf{D} = \begin{array}{c} \\ it \\ was \\ the \\ best \\ worst \\ of \\ times \end{array} \begin{bmatrix} it & was & the & best & worst & of & times \\ 3 & 0 & 0 & 0 & 0 & 0 & 0 \\ 0 & 4 & 0 & 0 & 0 & 0 & 0 \\ 0 & 0 & 4 & 0 & 0 & 0 & 0 \\ 0 & 0 & 0 & 2 & 0 & 0 & 0 \\ 0 & 0 & 0 & 0 & 2 & 0 & 0 \\ 0 & 0 & 0 & 0 & 0 & 4 & 0 \\ 0 & 0 & 0 & 0 & 0 & 0 & 3 \end{bmatrix}
$$

Hence the graph Laplacian matrix is:

$$\mathbf{L} = \mathbf{D} - \mathbf{A} = \begin{bmatrix}
 & it & was & the & best & worst & of & times \\
it & 3 & -2 & 0 & 0 & 0 & 0 & -1 \\
was & -2 & 4 & -2 & 0 & 0 & 0 & 0 \\
the & 0 & -2 & 4 & -1 & -1 & 0 & 0 \\
best & 0 & 0 & -1 & 2 & 0 & -1 & 0 \\
worst & 0 & 0 & -1 & 0 & 2 & -1 & 0 \\
of & 0 & 0 & 0 & -1 & -1 & 4 & -2 \\
times & -1 & 0 & 0 & 0 & 0 & -2 & 3
\end{bmatrix}$$

Using such a representation with all the words in a novel provides a way of representing a novel as a graph Laplacian matrix, and we provide an example of analyzing the novels of Jane Austen and Charles Dickens in Section 10.2.4.

10.2.3 Extrinsic and Intrinsic Analysis

Since the sample space $\mathcal{L}_m$ for graph Laplacian data is non-Euclidean, as seen in Figure 10.7, standard approaches to statistical analysis cannot be applied directly. Severn et al. (2019) introduced a framework for extrinsic analysis of graph Laplacian data, in which "extrinsic" refers to the strategy of embedding the data into a Euclidean space, where analysis is performed, before mapping back to $\mathcal{L}_m$. An intrinsic distance for this manifold is the Euclidean distance restricted to the space of graph Laplacians. Discussion of extrinsic versus intrinsic distances was given in Section 7.3.3 (for shapes) and 8.1 (for directional data). Some advantages of an extrinsic approach on a manifold were given in Section 8.1. The choice of embedding enables freedom in the choice of metric used for the statistical analysis, and in various applications with manifold-valued data analysis there is evidence of the advantage in using non-Euclidean metrics, as discussed in Section 7.3.5.

10.2.4 Case Study: Corpus Linguistics

Severn et al. (2019) focussed on comparison of the novels of Jane Austen and Charles Dickens, as represented by networks with edges based on a scan of 5 words. In that paper the top 1000 words were chosen and compared using several of the distances that have been used for covariance matrices, that we described in Section 7.3.5. Interesting distinctions were made between the authors' styles. This approach takes advantage of the fact that the graph Laplacians are a convex subset of the space of covariance matrices. The data are from Mahlberg et al. (2016).

Here we consider some additional comparisons of the novels of Austen and Dickens as data objects, using all the $m = 48,385$ different words, as opposed to the top 1000 words used by Severn et al. (2019). We use MDS (Multi-Dimensional Scaling, Section 7.2) and hierarchical cluster analysis using Ward's linkage (described in Section 12.2) for different choices of metric, listed in Table 10.2. The graph Laplacians are very high-dimensional $m \times m$ matrices here, although they are very sparse. Hence these norms can all be efficiently computed using sparse

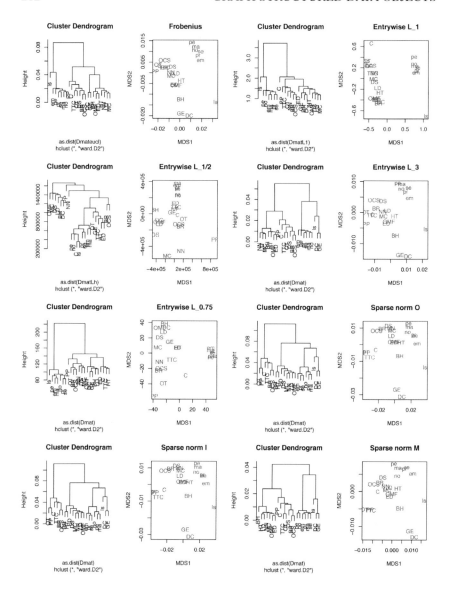

Figure 10.9 *MDS and cluster analysis for the Jane Austen (Blue, lower case) and Charles Dickens (Red, upper case) novels. The sparse norms that are used are: a) Frobenius of (7.9); b)–e) Entrywise norms of (7.10) for $p = 1, 0.5, 0.75, 3$; f) one norm (O) (maximum absolute column sum); g) the infinity norm (I) (maximum absolute row sum); h) the max norm (maximum modulus of all elements) (M). It can be seen from the various MDS plots that there is generally good separation between the Austen and Dickens novels. However, on inspection of the clusters for some metrics (entrywise $p = 3$, O, I, M) the Dickens novels Great Expectations (GE) and David Copperfield (DC) have been put in the same cluster as the Jane Austen novels.*

Panel Index	Symbol	Description
(a)	F	Frobenius norm, from (7.9)
(b)–(e)	E(p)	Entrywise norms, from (7.10), for $p = 1,\ 0.5,\ 0.75,\ 3$
(f)	O	One norm, maximum absolute column sum
(g)	I	Infinity norm, maximum absolute row sum
(h)	M	Maximum modulus of all elements

Table 10.2 *List of metrics, and indices, studied in Figure 10.9.*

matrix arithmetic in R. The results of clustering these very high-dimensional sparse graph Laplacians appear in the indicated panels of Figure 10.9. For each metric, the left panel is a dendrogram (also described in Section 12.2) showing exactly how the Austen (represented using lower case abbreviations of the titles) and the Dickens (upper case abbreviations) novels cluster. The left panel provides another view of the impact of each metric, as a scatterplot of the MDS scores with the Austen (Dickens) novels using the same abbreviated titles, colored blue (red, respectively). It is clear in Figure 10.9 that the MDS results vary widely using these choices of distances. Most choices of metric separate the Austen and Dickens novels (colors) in the two dimensional MDS plot. However, careful inspection of the dendrograms shows that the only methods that put all the Austen and Dickens novels in separate clusters are the Frobenius (F), and entrywise E(1), E(0.75) metrics, so these would be preferred here. For the sparse norms (O, I, and M) and entrywise E(3) the Dickens novels Great Expectations (GE) and David Copperfield (DC) have been put in the same cluster as Jane Austen's novels. This makes grammatical sense, because those Dickens novels were written in the first person, which is not inconsistent with some aspects of Austen's style. In the earlier analysis of Severn et al. (2019) based on the top 1000 words, the square root and Procrustes distances from Section 7.3.5 gave preferable clustering compared to the Frobenius distance (which again clustered GE and DC with Jane Austen's novels). These examples all show that the data object choice of metric is very important in such analyses.

Further extensions of regression models for network object data include the local Fréchet regression of Petersen and Müller (2019); functional models for time-varying networks of Dubey and Müller (2020); and non-parametric regression in Severn et al. (2021) investigating changes in writing style in the Austen and Dickens novels.

10.2.5 Labeled versus Unlabeled Nodes

In the above applications we have considered the nodes to be labeled, and the role of each node (e.g. a word in corpus linguistics) is the same for each data object. In some applications it may be appropriate to compare networks where the labeling of the nodes is also not of relevance. The network distances are then obtained by minimizing over all permutations of the labels. Such optimization is an NP-complete problem, although there are approximate algorithms available for such graph matching. As the set of permutations is a subset of the space of rotations the Procrustes distance provides another alternative approach. Statistical analysis of unlabeled graphs has been considered by Kolaczyk et al. (2020), Guo and Srivastava (2020), and Calissano et al. (2020).

CHAPTER 11

Classification–Supervised Learning

Statistical *classification*, also called *discrimination*, is easily understood in the context of data-driven automatic disease diagnosis. The process is based on a *training set* consisting of patients with known labels such as the presence or absence of the disease under consideration. The training data also includes a trait (i.e. feature) vector of relevant data for each patient. The goal of classification is a data-based rule for assigning new cases to either the disease present or absent classes on the basis of their data vector.

As noted in Section 4.3 and even this chapter title, classification is also called *supervised learning* in machine learning, see James et al. (2013) and Hastie et al. (2009), where the given class labels constitute the supervision. This provides a useful contrast with *unsupervised learning*, which is called by the more classical name of cluster analysis in Chapter 12. Both focus on subsets of the data, that are considered known in the former, and to be discovered in the latter.

Early work in this area, e.g. Fisher (1936), had strong connections with biology, where the term "classification" refers to development and refinement of the taxonomic organizational structure of kingdom-phylum-...-genus-species that has been at the heart of that field for many years. Hence as the term "classification" was already used, discrimination was once preferred terminology for the statistical task studied here. However, that term has since become so socially charged that classification has become more commonly used in statistical contexts. This may be connected to "discrimination" having been coined by advocates of the widely discredited area of eugenics. Another name, popular in the machine learning literature, is *pattern recognition*.

The general classification problem involves K classes, but for ease of notation typically the *binary classification* case of $K = 2$ will usually be considered here. There are many notational systems in use for discussing classification, but a particularly useful one (especially for linear classifiers) is to assume *training data*

$$(\widetilde{x}_1, \widetilde{y}_1), \cdots, (\widetilde{x}_n, \widetilde{y}_n), \tag{11.1}$$

where $\widetilde{x}_i \in \mathbb{R}^d$ and $\widetilde{y}_i \in \{-1, 1\}$ for $i = 1, \cdots, n$. As discussed around Table 7.1 the tilde on top of scalars and vectors are used to indicate that these are random quantities. Let $(\widetilde{x}_0, \widetilde{y}_0)$ denote a separate test case. The goal is to find a *classification rule* for using $\widetilde{x}_0$ to predict the corresponding class label $\widetilde{y}_0$, i.e. a function $c(x) : \mathbb{R}^d \to \{-1, 1\}$.

While not everyone in the machine learning literature does this, it is common to assume that the training data are drawn independently from an underlying joint probability distribution $f(x, y)$. It is useful to think in a Bayesian way, starting with a loss function of 0 for a correct classification and 1 for incorrect. Next define *prior probabilities* to be the marginals $p_+ = P\{\widetilde{y} = +1\}$ and $p_- = P\{\widetilde{y} = -1\} = 1 - p_+$. Performance of a classifier, c, is assessed in a natural way using the Bayes risk (expected loss), $P\{c(\widetilde{x}) = \widetilde{y}\}$. The likelihood ratio rule,

$$c_{LR}(\boldsymbol{x}) = \begin{cases} +1, & \text{when } \frac{f(\boldsymbol{x},+1)}{f(\boldsymbol{x},-1)} \geq 1 \\ -1, & \text{when } \frac{f(\boldsymbol{x},+1)}{f(\boldsymbol{x},-1)} < 1 \end{cases}$$

is Bayes risk optimal in the sense of minimizing the Bayes risk over all choices of c. Of course f is typically unknown, but the Bayes rule still provides an ideal to keep in mind and its Bayes risk is an upper bound on the classification performance of all possible methods.

An important general class of methods are the *linear classifiers*. Given a *direction vector* $\boldsymbol{w} \in \mathbb{R}^d$ (i. e. $\|\boldsymbol{w}\| = 1$) and an intercept $\beta \in \mathbb{R}$, define

$$c_{\boldsymbol{w},\beta}(\boldsymbol{x}) = 1 - 2 \cdot 1_{\{\boldsymbol{w}^t \boldsymbol{x} < \beta\}}, \tag{11.2}$$

where $1_{\{\cdot\}}$ denotes the *indicator function* (in contrast to the boldface $\mathbf{1}_{d,n}$ notation for the matrix of ones in (10.1)) that is 1 when the condition $\{\cdot\}$ is true, and is 0 otherwise. Note that $c_{\boldsymbol{w},\beta}$ assigns $\boldsymbol{x}$ to $+1$ exactly when it is on the positive (with respect to the direction of $\boldsymbol{w}$) side of the hyperplane whose normal vector is $\boldsymbol{w}$ and intercept is β. I.e. there is a *separating surface* between the two class regions that is a hyperplane.

An attractive aspect of linear classifiers, relative to some others discussed below is that they tend to be quite interpretable. This is because the entries of the vector $\boldsymbol{w}$ can be called *classification loadings*, in the spirit of PCA loadings as discussed around Figures 3.2 and 3.3. Thus the drivers of the classification can be understood using loadings plots of the type shown in Figure 4.11.

Fuzzy classification is a sometimes useful parallel to classification, where instead of trying to predict the class, the goal is to instead estimate an analog of the posterior probabilities $P\{\widetilde{y} = y|\boldsymbol{x}\}$. This gives not only the class prediction, but also a measure of confidence in that prediction. The terminology "fuzzy" in this context seems to go back to Zadeh (1965).

A comprehensive reference to many important ideas and methods in classification can be found in Duda et al. (2001). Another good overview can be found in McLachlan (2004).

Section 11.1 studies some classical statistical approaches to classification, but some non-standard ideas are revealed. Sections 11.2 and 11.3 provide a graphical introduction to the kernel trick and Support Vector Machines, which were at the heart of the early days of machine learning. An improvement of the Support Vector Machine called Distance Weighted Discrimination is introduced and discussed in Section 11.4.

Classification is an area where the "every dog has its day" quote in the preface is particularly applicable, as a large number of methods (each with relative strengths and weaknesses) are available.

11.1 Classical Methods

Perhaps the simplest classification method is called *one nearest neighbor*, where $c(\boldsymbol{x}_0)$ is the $\tilde{y}_i$ of the closest training point to $\boldsymbol{x}_0$. An extension that generally has better Bayes risk is the k nearest neighbors rule, where a vote is taken among the k nearest neighbors to $\boldsymbol{x}_0$. Besides the appealing property of simplicity a convenient aspect of these methods is that they are *distance-based* as discussed in Chapter 5. Thus they are easily constructed in any metric space. A serious drawback of the k nearest neighbor rule is that it requires choice of k, which is generally as hard as bandwidth selection as discussed in Section 15.3.

As discussed in Section 5.3, another very simple classifier is the *Mean Difference* (MD) or *Centroid* approach. One way of thinking of the MD is to simply take the class whose mean is closest. But MD can also be thought of as a linear classifier of the form (11.2). In particular the separating hyperplane has the line between the class means as the normal direction, and intercept as the midpoint between. A notational benefit of the ± 1 class labels is that $\widehat{\boldsymbol{w}} = \frac{\sum_{i=1}^{n} \tilde{y}_i \tilde{\boldsymbol{x}}_i}{\left\| \sum_{i=1}^{n} \tilde{y}_i \tilde{\boldsymbol{x}}_i \right\|}$ and $\widehat{\beta} = n^{-1} \sum_{i=1}^{n} \tilde{\boldsymbol{x}}_i$, where here and in the following the hat symbol is used to indicate an estimated quantity. When both classes are balanced and Gaussian with identity covariance, the MD is the likelihood ratio rule so it is Bayes risk optimal.

In situations where the variables may have arbitrarily different scalings (e.g. are measured in different units) it is sensible to first rescale each variable in the full training data set by subtracting its mean and dividing by its standard deviation (and applying the same standardization to $\tilde{\boldsymbol{x}}_0$). This variant of MD appears to have been named *Naive Bayes* by Domingos and Pazzani (1997). This is Bayes risk optimal in balanced Gaussian settings where the common covariance matrix is diagonal. The same issues (about the strengths and weaknesses of accentuating variables by standardization) raised during the analysis of the Two Scale Curves data in Figures 5.13 and 5.14 apply in this context as well.

The MD and Naive Bayes methods work well when there is no correlation among the variables, but otherwise they can often be substantially improved as demonstrated in using the Shifted Correlated Gaussians example in Figure 11.1. The left panel displays a $d = 2$ dimensional training data set with $n_+ = 20$ class +1 data points colored red and $n_- = 20$ labeled -1 colored blue, for a total of $n = n_+ + n_- = 40$ points. Green circled x signs denote the two sample means and the circled plus is the overall mean. The green direction vector from the overall mean shows the MD direction. The MD separating hyperplane (normal to the MD direction) is shown as the dashed green line. Projections of the training data onto the subspace (line) generated by the MD direction are shown in the right panel. Substantial overlap of the projected data indicates this is not an outstanding direction for linear classification. The left panel suggests that a direction that is more orthogonal to the point clouds could give much better performance.

The above ideas can be used to give a graphical derivation of the widely used and studied *Linear Discriminant Analysis* (LDA). In the case of binary classification, another common name for LDA is *Fisher Linear Discriminant*. Figure 11.1 suggests that MD failed because it only uses the class means and ignores the

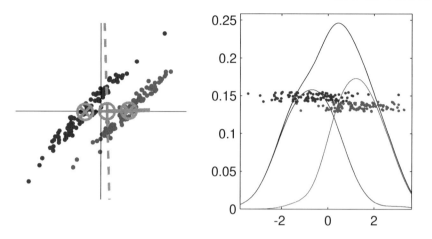

Figure 11.1 *Left panel shows the Shifted Correlated Gaussians training data set with color indicating classes, green direction vector (based at the circled plus sign) shows the MD direction, and separating hyperplane is the green dashed line. Right panel is projections in the MD direction. Shows poor class separation performance of MD when training variables are correlated.*

clearly important covariance information. LDA was proposed by Fisher (1936) who made the key observation that when both classes are assumed to be Gaussian with *common* covariance matrices the likelihood ratio classifier is linear, i.e. of the form (11.2). For this reason almost all (2 class) derivations of LDA are done from a Gaussian likelihood viewpoint. As in other situations, this can lead to the mistaken impression that LDA somehow requires Gaussianity to be effective. To counter this, a fully nonparametric visual introduction to LDA is given in Figure 11.2.

The same Shifted Correlated Gaussians data appears in the upper left panel. For simplicity of exposition, only the balanced case where the number of points in each class are the same is treated here. It is straightforward but notationally heavier to treat the general case by using appropriate class sample size weighting everywhere. Critical components of LDA are the class means $\overline{x}^+$ and $\overline{x}^-$, shown as circled x signs in both the left panel of Figure 11.1, and the lower left of Figure 11.2. Also important are the two within class sample covariance matrices, $\widehat{\Sigma}^+$ and $\widehat{\Sigma}^-$ (recall the definition (3.5)). Under the assumption that the underlying population covariance matrices are the same, it makes sense to estimate that common matrix by *pooling* the within-class sample covariance matrices to get $\widehat{\Sigma}^w$, by taking a weighted (according to class size) average of $\widehat{\Sigma}^+$ and $\widehat{\Sigma}^-$. Some algebra shows that $\widehat{\Sigma}^w$ can be written in a form that is quite similar to the overall sample covariance matrix $\widehat{\Sigma}$, with the important exception that the overall mean is replaced by the respective class means. For this reason $\widehat{\Sigma}^w$ is called the sample *within-class covariance*. Note that $\widehat{\Sigma}^w$ effectively captures the covariance structure (with strong positive correlation) that causes problems for MD in Figure 11.1. Hence it is natural to undo that by a *sphering* operation, i.e. transforming each data point to

$$\widetilde{z}_i = \left(\widehat{\Sigma}^w\right)^{-1/2} \widetilde{x}_i. \tag{11.3}$$

The success of this is seen in the plot of the $\widetilde{z}_i$ in the upper right panel of Figure 11.2, where each class indeed looks spherically distributed. That is the ideal situation for the MD classifier, which is applied in the lower right panel. The corresponding transformed class means $\overline{z}^+$ and $\overline{z}^-$ are shown as magenta circled x signs, with the overall mean (just the average of the two-class means) shown as the circled plus. The MD direction in this transformed space is shown as the magenta direction vector based at the overall mean, and the orthogonal separating plane appears as the dashed magenta line. Classification of the data in this space could be performed by applying the transformation to new data, but it is more insightful to transform the linear classification algorithm back to the original space. Note that the circled plus center point is transformed back to $\left(\widehat{\Sigma}^w\right)^{-1/2} \left(\frac{1}{2}\overline{z}^+ + \frac{1}{2}\overline{z}^-\right)$, while the back transformed normal vector is proportional to $\left(\widehat{\Sigma}^w\right)^{-1/2} \left(\overline{z}^+ - \overline{z}^-\right)$. Now inverting the sphering transformation (11.3) gives the LDA centerpoint $\widehat{\mu}_{LDA} = \frac{1}{2}\overline{x}^+ + \frac{1}{2}\overline{x}^-$ and the direction vector

$$\widehat{w}_{LDA} = \frac{\left(\widehat{\Sigma}^w\right)^{-1} \left(\overline{x}^+ - \overline{x}^-\right)}{\left\|\left(\widehat{\Sigma}^w\right)^{-1} \left(\overline{x}^+ - \overline{x}^-\right)\right\|}. \tag{11.4}$$

Thus this is a linear classifier (11.2) where the intercept is the inner product $\widehat{\beta}_{LDA} = \widehat{w}_{LDA}^t \widehat{\mu}_{LDA}$. The corresponding separating hyperplane is shown as the magenta dashed line in the lower left panel, which is clearly sensible for this data set.

This derivation of LDA is also interpretable in terms of the important idea of Mahalanobis distance, proposed by Mahalanobis (1936). Given independent random vectors $\widetilde{x}_1$, $\widetilde{x}_2 \sim N_d(\mu, \Sigma)$, a natural distance measure is

$$d_M = \left[\left(\widetilde{x}_1 - \widetilde{x}_2\right)^t \Sigma^{-1} \left(\widetilde{x}_1 - \widetilde{x}_2\right)\right]^{1/2}.$$

As discussed in Chapter 7, this distance is interpretable along the familiar lines of "standard deviations from the mean". Note that Mahalanobis distance is computable as the standard Euclidean distance applied to the sphered variables $\Sigma^{-1/2}\widetilde{x}_1$ and $\Sigma^{-1/2}\widetilde{x}_2$. Thus it is essentially Euclidean distance in the standardized space shown in the right column of Figure 11.2.

LDA has received far more attention than MD in the statistical literature, perhaps because it was perceived as doing much better in situations such as those studied in Figure 11.2, and being roughly similar otherwise. However that ignores the increasingly important high-dimensional case where $d > n$. This is a challenge because then the sample covariance matrix $\widehat{\Sigma}^w$ is not invertible. Various

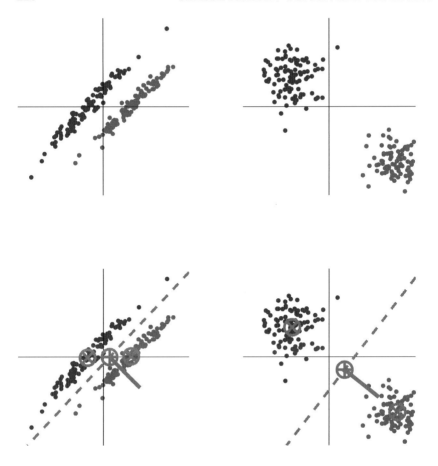

Figure 11.2 *Graphical introduction to LDA. Top left panel shows the same Shifted Corre-*
lated Gaussians data. Top right panel shows the result of transforming using the within-
class sphering operation. Bottom right panel shows the result of applying the MD classifier
to the transformed data. Bottom left panel shows the effectiveness of the MD classifier,
which is essentially an inversion of the sphering transformation.

domains of high-dimensional data are discussed in Chapter 14, where an important principle is that when matrix inverses are not available, often generalized
inverses contain the needed information. The Moore-Penrose generalized inverse
(also called pseudo-inverse) is a well-defined version of generalized inverse that
essentially consists of inverting nonzero eigenvalues (while leaving zero eigenvalues as is). See Section 17.1.2 for more discussion of eigenvalues. While use
of the Moore-Penrose generalized inverse gives a well-defined version of LDA,
performance in high dimensions is rather poor, as seen in Figure 11.3.

The panels in each column of Figure 11.3 are based on $n_+ = n_- = 20$
data vectors of dimension $d = 10, \ 40, \ 200$ dimensions in columns 1, 2, 3,

respectively. The $d = 200$ vectors in each column were simulated independently from Gaussian distributions having mean vectors $\boldsymbol{\mu}_+ = (+2.2, 0, \cdots, 0)^t$ and $\boldsymbol{\mu}_- = (-2.2, 0, \cdots, 0)^t$ for the $+1$ (red) and the -1 (blue) classes, respectively. To keep the realizations as similar as possible, the $d = 10$ and 40 versions are the subvectors of the first few entries. Thus each data set has the same signal, with noise level increasing with the dimension from left to right. Each panel shows the projection of the training data set onto a linear classification direction, i.e. a choice of $\widehat{\boldsymbol{w}}$ in (11.2). In the top row $\widehat{\boldsymbol{w}}$ is the MD direction. Note that for $d = 10$ in the top left panel, the projections seem close to what is expected from projection onto the optimal direction $\boldsymbol{w}_{OPT} = (1, 0, \cdots, 0)^t$. This is verified by displaying the angle between this estimated $\widehat{\boldsymbol{w}}_{MD}$ and $\boldsymbol{w}_{OPT}$ which is a very small $16°$ in this relatively easy case. Performance of MD remains good for $d = 40$, as seen in the center top panel, although the angle to $\boldsymbol{w}_{OPT}$ has grown somewhat. Even for the quite challenging case of $d = 200$ (top right panel) the MD direction continues to separate the classes, and perhaps surprisingly the classes seem to have actually moved farther apart, while the angle to $\boldsymbol{w}_{OPT}$ has also gotten larger. This phenomenon will be explained using high-dimensional concepts developed in Section 14.2. In the middle row, $\widehat{\boldsymbol{w}}$ is the LDA direction $\widehat{\boldsymbol{w}}_{LDA}$. Performance in the middle left panel ($d = 10$) is reasonably good, although as expected (because of the overhead of estimating $\widehat{\boldsymbol{\Sigma}}^w$) not as good as MD, neither visually nor in terms of the angle to $\boldsymbol{w}_{OPT}$. At first glance the $d = 40$ performance of LDA might be considered excellent, in the sense of separating the classes very well with respect to the projected within-class variances. But the angle to $\boldsymbol{w}_{OPT}$ of $80°$ gives substantial pause as that suggests that this classifier will have very poor generalizability properties when it is given new data. At $d = 200$, the LDA falls apart completely with a very poor angle to $\boldsymbol{w}_{OPT}$ as well as big overlap of the projected classes.

The bottom row shows a method (that is not commonly considered in discussions of classification methods) called the Maximal Data Piling (MDP) direction. MDP was first defined and studied in detail in Ahn and Marron (2010). In Figure 11.3, MDP has quite different properties from the above two methods in the higher dimensional cases. The $d = 40$ direction is quite spurious with an angle to $\boldsymbol{w}_{OPT}$ of $89°$, yet the projections have a very systematic structure of all the red points piled up at the *same value*, and similarly for the blue points. This piling of the two classes happens with probability one (for distributions absolutely continuous with respect to Lebesgue measure) for $d \geq n$, and is where the name MDP came from. In particular, as noted in Ahn and Marron (2010), out of all directions where both classes completely pile, MDP has the largest separation of the class piling points. For higher $d = 200$ the piling points spread substantially, and once again perhaps surprisingly the angle to $\boldsymbol{w}_{OPT}$ becomes quite comparable with that of MD. This suggests that in high dimensions the generalizability of MDP is fairly comparable to that of the optimal likelihood ratio rule, MD. Another high-dimensional quirk of MDP, shown by Ahn and Marron (2010), and explained in the PhD dissertation Miao (2015), is that for auto-correlated errors, MDP actually performs *better* than MD.

It is also seen in Ahn and Marron (2010) that the form of MDP is rather

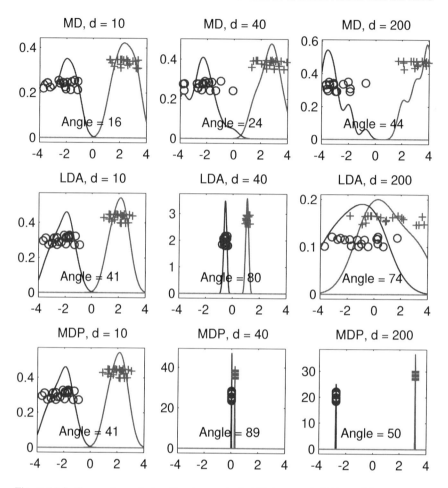

Figure 11.3 *Comparison of classification methods MD (top row), LDA (middle), and MDP (bottom), over varying dimensions. Angles to the optimal in degrees for each direction are included. Shows MD very robust, LDA has poor high-dimensional performance, and perhaps surprisingly good behavior of MDP.*

surprisingly similar to LDA. In particular the only difference is that the within class covariance matrix $\widehat{\Sigma}^w$ is replaced by the overall covariance matrix $\widehat{\Sigma}$, so that

$$\widehat{w}_{MDP} = \frac{\left(\widehat{\Sigma}\right)^{-1}(\overline{x}^+ - \overline{x}^-)}{\left\|\left(\widehat{\Sigma}\right)^{-1}(\overline{x}^+ - \overline{x}^-)\right\|}. \tag{11.5}$$

In addition, it is seen that for $d \le (n-2)$ the direction vectors $\widehat{w}_{LDA}$ and $\widehat{w}_{MDP}$ are the *same*, despite the different numerators in (11.4) and (11.5). This can be seen by a careful look at the $d = 10$ left center and bottom panels of Figure 11.3.

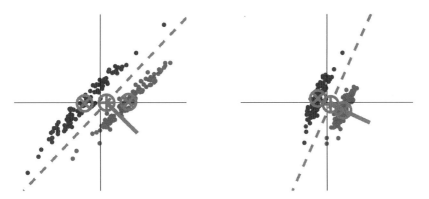

Figure 11.4 *Visual introduction to MDP, using the same Shifted Correlated Gaussians data and format as Figure 11.2. Shows how (for $d < n$) overall sphering results in the same classification rule as LDA.*

Note that both the horizontal and vertical coordinates of each symbol are exactly the same. This is an example of when using order in the data as heights in jitter plots conveys more information than using a random height, as discussed in Section 4.1. Daniel Peña has explained this phenomenon as illustrated in Figure 11.4 using the above viewpoint. This uses the same Shifted Correlated Gaussians data set as Figures 11.1 and 11.2 and this time only the corresponding bottom row is shown. The sphering that occurs in the right panel is done by $\left(\widehat{\Sigma} \right)^{-1/2}$ (as opposed to the $\left(\widehat{\Sigma}^{w} \right)^{-1/2}$ used to get LDA above). Instead of individually sphering each class separately (as done in the right column of Figure 11.2), now the full data set is sphered. The two-point clouds are still apparent, but they have been adjusted so that the overall sample covariance matrix is the identity. Note that this transformation also nicely handles the offset of the class means that was the original problem with MD highlighted in Figure 11.1. In particular, the sample means following this transformation are shown as magenta circled x signs, and the corresponding separating hyperplane is very sensible. The result of back transformation, following the same lines as in Figure 11.2 result in the linear classification rule in the lower left panel, which again is the *same* as in the LDA case.

A natural conclusion of the above is that the ubiquitous LDA should be *replaced* by MDP as the simple classifier of choice. This is because they are the *same* in the case $n < d$, and when they are different MDP is superior. Furthermore MDP is (very slightly) simpler to compute.

An unfortunately misleading presentation of these issues can be found in Section 3.8.2 of Duda et al. (2001), where it is stated that LDA optimizes the ratio of the difference between the means and the sum of the standard deviations of the projections onto the normal vector. While the standard LDA formula is given, it is in fact MDP that has the stated properties. That presentation is not mathematically incorrect, because it is assumed in Section 3.7.2 there that $d < n$, in which case all quantities are the same. But it is important to be aware that several statements in Section 3.8.2 are incorrect when $d \geq n$.

In situations where the within-class covariance matrices are not similar, one can still appeal to the Gaussian Likelihood Ratio to construct a classifier. While it is straightforward to compute the classification rule c_{GLR} the classification boundary (separating surface) is determined by quadratic equations and explicit representation is rather complicated and case-wise. This method is not treated further here, as it has been the subject of many classical texts, and because its high-dimensional performance is even worse than that of LDA shown in Figure 11.3.

Another branch of classification methods reviewed in Chapter 4 of Duda et al. (2001), assumes that each class has a smooth probability density, and uses smoothing methods as discussed in Chapter 15 to construct classifiers. Early results on optimal rates of convergence (of the type discussed in Section 15.2) were established in Marron (1983). For much more on optimality in related contexts see Devroye et al. (1996).

A point that will become relevant in Section 11.3 is that all of the methods developed in this section can be thought of as being based on probability distributions.

While the above discussion has been about binary classifiers for the sake of simplicity, an extension to K class methods of LDA is well worth describing. In the case of $K > 2$ the likelihood approach is no longer useful, but a direction-based approach is available. From Figure 11.2 and (11.4) it is clear that the LDA direction vector points in the direction of the difference of class means after appropriately adjusting by the pooled within-class covariance structure. For multiple means, the single direction between means can be usefully replaced by a PCA of the set of mean vectors. That would be computed as an eigen analysis of the covariance matrix of the set of class means, which is sometimes called the *between-class covariance* matrix $\widehat{\Sigma}^b$. As LDA got some benefit from adjusting the MD using the within-class covariance matrix $\widehat{\Sigma}^w$, the same approach makes sense here. This results in a set of direction vectors (analogs of the PCA loadings vectors detailed in Figures 3.2 & 3.3 of Section 3.1), called *Canonical Variate Analysis* (CVA) in Mardia et al. (1979) and *Multiple Discriminant Analysis* in Duda et al. (2001), as previously mentioned in Section 6.5. Potential differences between the Gaussian likelihood and algebraic derivation of these methods have been resolved by van Meegen et al. (2020)

The canonical variates can be calculated as an eigen analysis of $\left(\widehat{\Sigma}^w\right)^{-1}\widehat{\Sigma}^b$, by solving

$$\left(\widehat{\Sigma}^w\right)^{-1}\widehat{\Sigma}^b w = \lambda w$$

In practice we use the eigen decomposition of the matrix $\left(\widehat{\Sigma}^w\right)^{-1/2}\widehat{\Sigma}^b\left(\widehat{\Sigma}^w\right)^{-1/2}$ which is symmetric (hence more amenable to standard eigenanalysis algorithms) and has the same eigenvalues but with corresponding eigenvectors v related to the

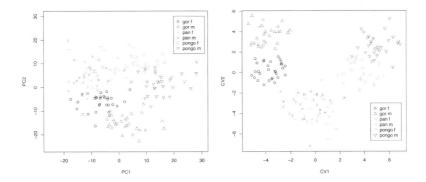

Figure 11.5 *PCA and CVA of a data set of six groups of great apes: males (m) and females (f) for each of Gorilla (gor), Chimpanzee (pan), and Orangutan (pongo). The left figure shows the first two PC scores and the right figure shows the canonical variate scores. There is much better separation of the groups using the canonical variates compared to PCA.*

canonical variate vectors w by $w = \left(\widehat{\Sigma}^w\right)^{-1/2} v$. This computation is a type of generalized eigen analysis, where eigenvectors of one matrix are computed with respect to another matrix (see Dryden and Mardia (2016), page 298). Both the high and low-dimensional issues of the relationship between LDA and MDP based on replacing $\widehat{\Sigma}^w$ with $\widehat{\Sigma}$ are likely to apply here, although this does not seem to have been explored as yet.

An example of using Canonical Variate Analysis is seen in Figure 11.5. In this data set there are six groups of great ape skull landmarks: males and females for each of Gorilla (gor), Chimpanzee (pan), and Orangutan (pongo). The data set is available in the `shapes` package in R, Dryden (2021). Generalized Procrustes analysis (as discussed in Section 7.3.4) is carried out to remove location, rotation and scale information to retain the shapes, which lie in the non-Euclidean shape space. Tangent space coordinates are obtained at the overall mean and both PCA and CVA are carried out. We plot the first two PC scores in the left-hand figure (which summarize 37.5% and 28.1% of the shape variability), and the first two canonical variate scores are shown on the right. The CVA gives much better visual separation of the groups than PCA. For CVA it is clear that there is good separation in shape for all six groups, except there is some overlap between the male and female chimpanzees. Careful inspection of the axis labels reveals how PCA directions maximize projected variation which is quite different from maximizing class differences as done by CVA. Note that Canonical Variate Analysis is equivalent to Canonical Correlation Analysis (Section 17.2.2) between the data matrix X and a new dummy indicator matrix Y of zeros and ones to indicate which group the observation belongs to (see Mardia et al. (1979), Exercise 11.5.4, page 330).

11.2 Kernel Methods

Kernel methods, as developed in the machine learning community are usefully viewed as a type of transformation, as discussed in Section 5.2, with the goal of making data more accessible to linear analysis. The fundamental idea goes back to Aïzerman et al. (1964). A canonical classification example called the Donut 2-d data is shown in Figure 11.6. Each panel is based on the same data set, with $n_+ = 100$ points in the positive class (donut hole) shown as red plus signs from 0.6 times a standard Gaussian distribution (i.e. $N\left(0, (0.6)^2\right)$, and $n_- = 100$ negative class points (forming the donut) appearing as blue circles generated using polar coordinates that are $2.5 + 0.5$ times a standard Gaussian (i.e. $N\left(5, (0.5)^2\right)$) in the radial component and uniformly over the angular component. Training data points $(x_{1,i}, x_{2,i})^t$ for $i = 1, \cdots, 200$ are plotted on the x_1 (horizontal) and x_2 (vertical) axes in each panel. Performance of the classifier based on each polynomial embedding is shown by treating each background pixel (recall this terminology from Section 6.1) as a new point and classifying it with yellow denoting a + (red) classification and cyan indicating − (blue). Both colors are scaled toward white for pixels near the boundary. In the upper left panel, LDA classification is done in the original $\mathbb{R}^2$, where LDA has very poor performance because it is constrained to be a linear classifier of the form (11.2), none of which can separate this data well.

The remaining panels of Figure 11.6 show the results of *polynomial kernel embedding*, where data are nonlinearly transformed, i.e. embedded, in a higher dimensional space where the linear method LDA is then applied. Again impact of the embedding process on the classification is shown by classifying the background pixels in the same way. The upper right panel shows the results of embedding into $\mathbb{R}^3$ by mapping the i-th data point to $(x_{1,i}, x_{2,i}, x_{1,i}^2)^t$. That mapping puts the data onto a sheet that is warped in the x_1 direction into a parabolic cylinder. LDA slices that surface in $\mathbb{R}^3$ in a horizontal way so that pixels with x_1 coordinate near 0 are assigned to the + (red) class and the rest are − (blue), which gives a marked improvement over the $\mathbb{R}^2$ version of LDA in the upper left panel. The lower left panel is similar to the upper right, except now the embedding is to $(x_{1,i}, x_{2,i}, x_{2,i}^2)^t$, so the parabolic cylinder in $\mathbb{R}^3$ is now folded in the x_2 direction, resulting in a rotation of the classification results. The lower right panel studies the embedding to $\mathbb{R}^4$, using $(x_{1,i}, x_{2,i}, x_{1,i}^2, x_{2,i}^2)^t$. It is hard to conceive of the 2-d sheet containing the data in $\mathbb{R}^4$, so interpretation is not so straightforward. However, the yellow-cyan coloring makes it clear that classification is happening mainly in terms of $x_{1,i}^2 + x_{2,i}^2$ (i.e. squared distance to the origin), giving a very good result in this case. In particular, this nonlinear transformation has taken a data set that is clearly intractable to linear methods and renders it to be quite amenable to the simple LDA.

While polynomial embedding is clearly ideal for the Donut 2-d toy data in Figure 11.6 it also has some drawbacks that include potential lack of flexibility as well as poor interpretability and non-obvious choice of polynomial degree. This is illustrated in Figure 11.7, which again starts with a data set in $\mathbb{R}^2$. But this time

Polynomials: x_1, x_2 only

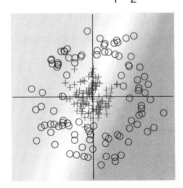

Polynomials: x_1, x_2, x_1^2

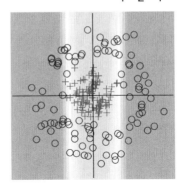

Polynomials: x_1, x_2, x_2^2

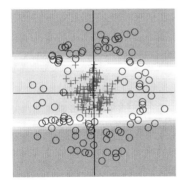

Polynomials: x_1, x_2, x_1^2, x_2^2

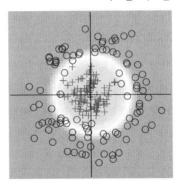

Figure 11.6 *Donut 2-d toy example, classified using LDA in various polynomial embedding spaces. Background reflects classification of each pixel location. Shows increasing benefits of higher order embedding.*

the data follow a checkerboard pattern that is even less amenable to linear classification. The format of Figure 11.7 is the same as for Figure 11.6. Each class consists of 8 standard Gaussian clusters of 25 points each, that are spaced 6 units apart in both directions. The polynomial embeddings shown in Figure 11.6 were not sufficiently flexible to give reasonable classification. Somewhat more flexible is the cubic embedding to $\left(x_{1,i}, x_{2,i}, x_{1,i}^2, x_{2,i}^2, x_{1,i}^3, x_{2,i}^3\right)^t$ shown in the left panel of Figure 11.7, which still gives very poor classification performance.

Much better performance appears in the right panel of Figure 11.7. This uses a much different embedding, using kernel density estimation ideas as discussed in Section 15.1. In particular the data are embedded in $\mathbb{R}^{49}$ by mapping $\left(x_{1,i}, x_{2,i}\right)^t$

Cubic Polynomials   Gaussian Kernels

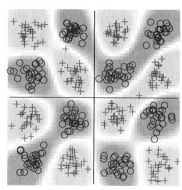

Figure 11.7 *Checkerboard 2-d toy example. Shows poor classification performance by a polynomial embedding, but a much better performance using LDA on a radial basis function embedding.*

to

$$\left(\phi\left((x_{1,i} - g_{1,1})/h\right) \cdot \phi\left((x_{2,i} - g_{2,1})/h\right), \cdots, \right.$$
$$\left. \phi\left((x_{1,i} - g_{1,49})/h\right) \cdot \phi\left((x_{2,i} - g_{2,49})/h\right)\right)^t,$$

where ϕ is the standard Gaussian density, where $(g_{1,1}, g_{2,1}), \cdots, (g_{1,49}, g_{2,49})$ are a 7×7 grid of kernel centerpoints, and where the spread of each Gaussian kernel is controlled by the bandwidth $h = 1$. This much more complicated embedding results in excellent classification where each training point is correctly classified. This shows why Gaussian kernels, a commonly used type of what are called *radial basis functions* in the kernel machine literature, have been very popular.

A clever variation of the kernel embedding idea is *kernel PCA*, proposed by Schölkopf et al. (1997). This is illustrated in Figure 11.8, based on the Four Cluster toy data set which is the basis of several examples in Chapter 12. This data set has two spherical clusters shown as blue circles and cyan squares, a stretched cluster of green plus signs, and a small two-point cluster (perhaps outliers?) shown as red x signs. Kernel PCA starts with conventional PCA performed in the kernel space, with visualization done in the original object space. The latter comes from projecting every point in the object space (essentially each pixel value) onto each direction vector, and coloring the image with the score, using gray for 0, and intensity of black (white) for magnitude of positive (negative, respectively) score.

The first component (i. e. mode of variation, shown in the left panel) highlights a coarse-scale clustering which provides a contrast between the stretched green cluster and the other data points, while treating the union of the blue and cyan as a single coarse-scale cluster. The second PC mode of variation refines the clustering view by separating the blue and cyan clusters. The third mode seems to split the long green cluster into two subclusters. The higher components (not shown here)

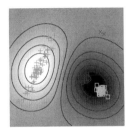  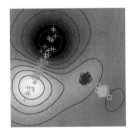

Figure 11.8 *Four Cluster toy example illustrating kernel PCA. Shows first three modes of variation. First (left) contrasts green cluster with union of blue and cyan. Second (center) contrasts blue and cyan clusters. Third (right) splits the stretched green cluster.*

are similarly interpretable as a further split, into 3, of the stretched green cluster, followed by a focus on the two-point red cluster. Gaussian kernel PCA and also other kernel methods depend on the choice of the window width. In Figure 11.8 this was chosen by trial and error to give good interpretation of the modes of variation. See Ahn (2010) for further useful ideas on this choice.

While examples like this are clearly compelling, this view of kernel PCA does not seem to have had much impact on real data analysis. This appears to be due to the fact that this visualization requires a low-dimensional (really $d = 2$) object space. A much more successful visualization based on kernel PCA is the t-SNE (t-distribution Stochastic Neighbor Embedding) approach of Maaten and Hinton (2008). This is usefully viewed as a type of inversion of the Gaussian kernel PCA, using a Cauchy kernel (i.e. very heavy tails) analog of kernel PCA. Inversion here is done much as in MDS (recall Multi-Dimensional Scaling from Section 7.2), where the key idea was to find a set of representers (of each data object) whose position best fits the set of distances between data objects. This approach is called *auto-encoding* in the machine learning literature. In t-SNE the Cauchy kernel embeddings of the representers are fit to the positions of the data objects in the kernel space. The heavy tails of the Cauchy kernel, relative to the Gaussian give a representation which keeps nearby points close to each other, while somehow pushing away points that are less close, which can visually accentuate clusters. This can result in impressive-looking graphics, although it does rely on choice of a tuning parameter, and does not resolve issues such as which clusters represent true underlying structure, as done e.g. by the SigClust method of Section 13.2. Finally note that the name t-SNE comes from the fact that the standard Cauchy distribution is also Student's t distribution with one degree of freedom.

Figure 11.9 shows three applications of t-SNE to the same Four Cluster toy data set. The 3 panels show different values of the *perplexity*, which is a tuning parameter. The default perplexity of 30 is shown in the center panel and gives a reasonable view of the data, except for the placement of the two red points in the small isolated cluster. The right panel, with perplexity 60, maybe somewhat better, although the blue and cyan clusters seem to be uncomfortably close. The perplexity 12 visualization in the left panel has the red points even farther apart

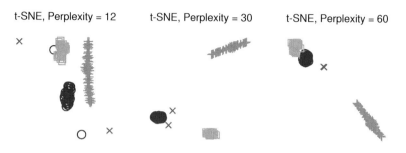

Figure 11.9 *Example of t-SNE visualization for the same Four Cluster data set, for different values of the perplexity parameter. Default of 30 is in the center panel, but the most reasonable choice maybe 60 as shown in the right panel.*

and two of the blue points separated from the rest. An important aspect is that the global structure is quite arbitrary, but the local clusters tend to be reasonably highlighted, which is the goal of t-SNE.

An application of t-SNE to the Osteo-Arthritis data set of Nelson et al. (2019) is shown in Figure 11.10, which is Figure 4 of that paper. This data set has four distinct subpopulations, which are first dichotomized as patients whose disease progressed over time (blue and cyan) versus those who did not (red and green), and second split on males (blue and red) versus females (cyan and green). The t-SNE visualization nicely highlights these as four clusters, suggesting these are useful partitions of this data set. However, the visual relationships between clusters appear to be symmetric. The black lines and numbers show the results of a quantitative investigation of the difference between these clusters, using the DiProPerm hypothesis test (discussed in Section 13.1) of the difference between each pair of subpopulations. The numbers are Z-scores which reflect statistical significance on the standard Gaussian scale, so larger than 2 is statistically significant. Larger Z-scores indicate stronger statistical significance. In Figure 11.10 this shows that the male versus female subpopulations are actually far more distinct than progression versus not. This highlights the fact that because of its stress on locality t-SNE cannot be counted on to visually indicate strength of relationships between clusters.

At the time of this writing, t-SNE visualization may be fading in popularity in the bioinformatics community, in favor of UMAP proposed by McInnes et al. (2018). Like t-SNE, UMAP seeks a set of visual representers of high-dimensional data, but deep ideas from differential geometry and fuzzy topology provide the motivation for this method.

A fundamental issue is how a kernel embedding relates to the inner products that lie at the heart of many data analytic methods such as PCA and LDA. In the examples shown in Figures 11.6 and 11.7 an *explicit kernel embedding* is used where the embedding to the kernel space is performed first, and then LDA is explicitly applied in that space. But there are major computational advantages to first computing the matrix of inner products of the data objects, and then mapping those to the kernel space, which is usefully labeled *implicit kernel embedding*.

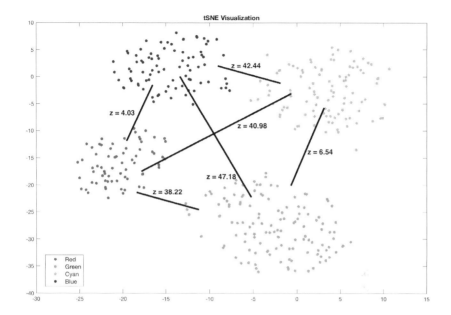

Figure 11.10 *Application of t-SNE to Osteo-Arthritis data. Highlights four subsets in the data, Red for male non-progressors, Green for female non-progressors, Cyan for female progressors, and Blue for male progressors. DiProPerm Z-scores show that, contrary to t-SNE visual impression, some subgroups are actually much more distinct than others.*

The details of the relationship between explicit and implicit embeddings have been carefully laid out in the kernel machines literature. A fundamental tool in that area is Mercer's theorem that relates embedding types. See Cristianini and Shawe-Taylor (2000); Shawe-Taylor and Cristianini (2004) for a good introduction, and Schölkopf and Smola (2002) for a more detailed investigation.

One implicit kernel embedding is the use of the matrix of centered inner products, which are used in the computation of distance matrices in MDS (Multi-Dimensional Scaling, Section 7.2). In particular in classical MDS the "centered inner product matrix" is formed to compute the matrix of pairwise Euclidean distances, see 14.2.1 of Mardia et al. (1979)). Although sometimes a horseshoe effect may be present in such representations as mentioned at the end of Section 7.2, the higher dimensional kernel embedding may be useful in classification tasks.

The great flexibility of kernel methods comes with great danger for overfitting. The Bidirectional Discrimination ideas of Huang et al. (2012) overcome this with a methodology that involves a product of two (or more) linear direction vectors for a more controllable enhanced flexibility.

While the above examples make clear the strong data analytical advantages available from kernel embedding, an interesting limitation has been pointed out by El Karoui (2010). The fundamental result presented there is that, in the case of very high dimension, kernel methods give the same classification performance as conventional linear methods. In particular, the analysis of El Karoui (2010) uses random matrix theory asymptotics, where both the sample size and the dimension $n, d \to \infty$, as studied in Section 14.1.

11.3 Support Vector Machines

There is an important contrast to notice between the classical methodologies reviewed in Section 11.1 and the kernel embedding context of Section 11.2. The methods in the former are strongly rooted in probability distributions (the foundations of the popular likelihood approach to statistics), which are very well developed and best understood in flat Euclidean spaces $\mathbb{R}^d$. However the key to the improvements demonstrated in the latter is to map the data onto a curved manifold (as studied in Chapter 8). For example the parabolic cylinders in the upper right and lower left panels of Figure 11.6, as well as the harder to understand 2-d curved surface in $\mathbb{R}^4$ that underlies the classification result shown in the lower right panel. Because probability distributions on manifolds are far less well developed (e.g. recall the discussion about Central Limit Theorems on manifolds in Section 8.5) and intuitively understood, it makes sense to consider non-probabilistic approaches to the invention of classification methods.

Such considerations appear to have motivated the invention of the *Support Vector Machine* (SVM) by Vapnik (1982, 1995). A very readable short introduction to the main ideas can be found in Burges (1998). While the SVM is aimed at data lying on curved manifolds, its characteristics are best understood in a simple Euclidean space such as the 2-d toy example shown in Figure 11.11. The data set shown in both panels is similar in spirit to the Shifted Correlated Gaussians data in Figures 11.1, 11.2, and 11.4 except there are now only $n_+ = 15$ red circles in the $+$ class, and $n_- = 15$ blue circles in the $-$ class. These data are *separable*, in the sense that there is a hyperplane between the classes with all of one class on one side, and the other class entirely on the other side. In fact there are many such separating planes, some of which are shown using colored dashed lines in the left panel of Figure 11.11. A useful view of the linear SVM is that it seeks to find the "best" among all such separating hyperplanes. Note this way of thinking is quite different from the mostly probability distribution based approaches in Section 11.1. The SVM notion of best is illustrated in the right panel of Figure 11.11, where given a candidate separating hyperplane the residuals $\tilde{r}_i$ from the projection of each data point x_i onto the plane are shown as thin magenta line segments. SVM chooses the hyperplane which is farthest from every data point in the sense of maximizing the smallest of these projected distances, as shown in the right panel. The minimizing distance is called the *margin*, and planes shown as black dotted lines parallel to the separating plane are shown. Note that there are 3 points on these lines, which are the data points that achieve the smallest distances. These are called *support vectors* (hence the name of the method) and are highlighted with black boxes.

To formulate the linear SVM as an optimization problem, let the $d \times n$ matrix

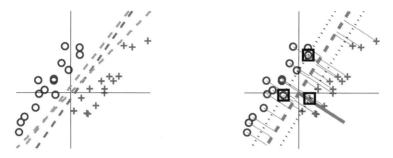

Figure 11.11 *Graphical introduction to the SVM. Left panel shows separable toy data, together with several separating hyperplanes, Right panel shows SVM direction vector (solid thick magenta line) and orthogonal SVM separating plane (dashed magenta line). Thin magenta lines show projected residuals. SVM margins are dotted lines with support vectors highlighted by black boxes.*

$X = [x_1, \cdots, x_n]$ and similarly combine the class labels into an $n \times 1$ vector $y = [y_1, \cdots, y_n]^t$. Note that the tilde notation used above to denote random quantities is deliberately not used here to reflect the common machine learning viewpoint that the input data are just numbers and not random variables. Also let Y denote the $n \times n$ diagonal matrix with $y_1, \cdots, y_n$ on its diagonal. Given a direction vector $w \in \mathbb{R}^d$ (thus $\|w\| = 1$) and an intercept $\beta \in \mathbb{R}$, the n vector of magenta residual lengths in the right panel of Figure 11.11 can be written in the form $\check{r} = YX^t w + \beta y$. The simplicity of this representation is due to the choice of ± 1 as class labels. In the separable case, the SVM optimization problem is then written as $\max_{w,\beta} \min_i \check{r}_i$. This can be solved using quadratic programming, by introducing a new variable τ, and maximizing τ subject to $\check{r}_i \geq \tau$ for $i = 1, \cdots, n$. Since $\check{r}$ scales with w and β, the maximization of τ subject to $\|w\| = 1$ is equivalent to minimizing $\|w\|$ subject to $\tau = 1$. When the data cannot be assumed to be separable, nonnegativity of the residuals is maintained by the use of slack variables $\xi = [\xi_1, \cdots, \xi_n]^t$ which are incorporated into the modified matrix of residuals $r = YX^t w + \beta y + \xi$. Given a tuning parameter λ, these modified residuals are then constrained to be nonnegative in the more general version of the SVM optimization:

$$\min_{w,\beta,\xi} \left(\|w\|^2 - \lambda \mathbf{1}_{1,n} \xi \right) \qquad (11.6)$$

subject to the constraints $r_i \geq 0, \xi_i \geq 0$, for $i = 1, \cdots, n$, using the notation from (10.1) of $\mathbf{1}_{d,n}$ for a $d \times n$ matrix of ones. Small values of λ give a solution called the *hard margin* SVM, which attempts to put as many data points as possible outside the margins (essentially the dotted black lines shown in the right panel of Figure 11.11). Larger values of λ relaxes that goal and allows more *violators*, i.e. points not outside of the margins.

Much more on SVMs from a kernel learning viewpoint can be found in the books by Cristianini and Shawe-Taylor (2000); Shawe-Taylor and Cristianini (2004) and Schölkopf and Smola (2002). A more statistically oriented overview can be found in Lin et al. (2002). For approaches to choice of the SVM tuning parameter, see Joachims (2000) and Wahba et al. (2000, 2003).

A detailed analysis of several important properties of the SVM can be found in Carmichael and Marron (2017). One is a study of the solution path, through the space of linear classifiers (11.2), as a function of the tuning parameter λ for a given data set. At one end of this path is the hard margin classifier studied in Figure 11.11. At the other end of the path is a trimmed version of the MD, which is exactly the MD when the classes are balanced. The path moves between these ends only over a rather narrow range and computationally useful bounds are given that guarantee the tuning parameter is in either ending state.

The SVM approach has been extended to image analysis data objects lying in a curved manifold (in the sense treated in Chapter 8) by Sen et al. (2008).

Another important issue is extension of the SVM to the multi-class case $K >$ 2. A simple extension involves repeated use of the binary SVM. There are two common approaches. First one trains a classifier on each class versus all the rest and chooses the class with the best result (in terms of projection onto the SVM direction vector). The second one runs all of the pairwise classifications, and then chooses a class by voting. As noted in Friedman (1996), the second option is generally better. Lee et al. (2004) proposed integrating the multi-class task into the SVM optimization problem.

While the SVM has become a workhorse method for many machine learning tasks, especially its hard margin form (as in Figure 11.11) has a deficiency in high dimensions that results in some loss of classification performance, as well as poor visualization properties. This is illustrated in Figure 11.12, whose format and data set are very similar to Figure 11.3. Again the same realization is used in all panels and the data are essentially Gaussian with shifted means, but this time the dimension is $d = 50$. Each panel is a projection of the data onto a direction in $\mathbb{R}^{50}$.

The top left panel is the optimal direction, in the sense that this is the direction in which the underlying population means have been shifted (recall the population means have been shifted to ± 2.2. Each panel displays an angle at the bottom, which is angle in degrees to this direction. Projections onto the MDP direction are shown in the top right panel. As expected from the discussion in Section 11.1, the data in each class pileup completely at a single point. To make this happen, MDP must clearly exploit small scale noise artifacts particular to this realization of the data, which should damage its generalizability as a classification rule. This is reflected by the large angle of $64°$ to the optimal direction.

The SVM direction shown in the lower left panel gives a better separation between the projected data, because it optimizes the spacing between classes (thus larger than MDP which must also make the data completely pile up). This will clearly result in much better classification performance because the angle to the optimal direction has now gone down to $37°$. However the SVM also reveals some data piling in the sense of both classes having a fair number of points that project to a common value at the SVM *margin,* shown as the thin black dotted lines in the right panel of Figure 11.11. These are the support vectors (basically ties for best

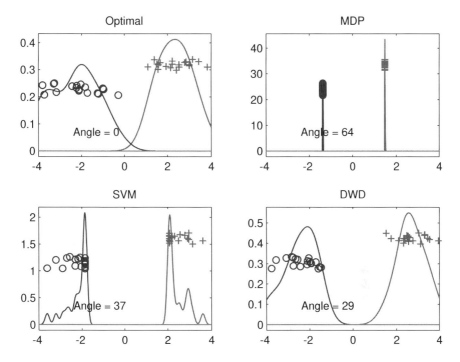

Figure 11.12 *Illustration of data piling for MDP (top right panel), SVM (lower left), DWD (lower right) on 50 dimensional Gaussian data set with means shifted along the optimal direction shown in the upper left. Angles to the optimal direction in degrees are shown for each direction. Reveals that SVM shares some of the data piling of MDP, which is avoided by DWD.*

in the SVM optimization) which tend to be quite numerous in higher dimensions. As discussed for the MDP this also indicates a lesser extent to which the SVM feels small scale noise artifacts which should also cause some loss in classification performance.

A related linear classification method, that has been motivated by this problem with SVM is *Distance Weighted Discrimination* (DWD) developed in Section 11.4. Note that projection onto the DWD direction (bottom right of Figure 11.12) has a smaller angle with the optimal of only 29° suggesting generally better classification performance (which has been verified using simulations in Marron et al. (2007)). Another useful property of DWD is the approximately Gaussian distribution of the projected classes, in sharp contrast to the apparently opposing triangle distributions for the SVM. As discussed in Section 11.4 this has serious implications for both data visualization and batch adjustment, as discussed in Benito et al. (2004) and Liu et al. (2009).

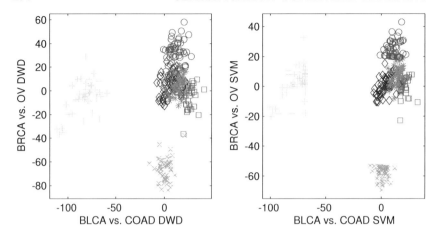

Figure 11.13 *Left panel is the scatterplot view (projections on DWD directions) of the Pan-Cancer data from the bottom middle panel of Figure 4.10. Corresponding SVM version appears in the right panel. Shows how data piling strongly impacts the use of SVM direction vectors in high-dimensional data visualization by creating spurious lines of data points.*

11.4 Distance Weighted Discrimination

The lower panels of Figure 11.12 demonstrate the high-dimensional data piling problem that is endemic to SVM, and that DWD can overcome it. The impact of this on data visualization is explored in Figure 11.13 using the Pan-Cancer gene expression data from Figure 4.10. The left panel of Figure 11.13 is a zoomed in view of the bottom middle panel of that figure, showing projections on the DWD directions trained on BLCA vs. COAD and on BRCA vs. OV. The right panel shows projections on the corresponding hard margin SVM directions. The big picture view as to how the clusters relate to each other is the same. But quite distracting to the human perception of these clusters in Figure 11.13 are the clear vertical lines of data points in the yellow (COAD) and magenta (BLCA) clusters, as well as the nearly horizontal lineups of points in the red (BRCA) and cyan (OV) clusters, which are all consequences of SVM partial data piling. Note also that in each of those clusters, the remaining data points all lie on the same side of the line as expected from the ideas illustrated in the lower left panel of Figure 11.12. This is why DWD is generally recommended for visualization tasks where the focus is on potential differences between pairs of given subgroups of data.

As noted in Marron et al. (2007), DWD achieves this improvement through a modification of the SVM optimization problem in (11.6). In the separable case, the idea is to change the way that the residuals $\check{r}_i$ impact the result, by letting all of them have some influence, instead of just the smallest ones. However, the smallest ones should play a more important role, which is achieved by studying the inverse of the residual, $\frac{1}{\check{r}_i}$. As any residual approaches 0, the inverse becomes large, so their sum works like a set of poles pushing the separating hyperplane toward the middle of the region between the data sets. Thus, the separable version of DWD uses the optimization problem $\min_{\boldsymbol{w},\beta} \sum_{i=1}^{n} \frac{1}{\check{r}_i}$. This is extended to the general case using the same slack variable approach used for the SVM in Section 11.3. Employing notation there, the DWD optimization problem is: given the tuning parameter λ

$$\min_{\boldsymbol{w},\beta,\boldsymbol{\xi}} \left(\sum_{i=1}^{n} \frac{1}{\check{r}_i} + \lambda \boldsymbol{J}_{1,n}\boldsymbol{\xi} \right) \qquad (11.7)$$

subject to the constraints $\check{r}_i \geq 0$, $\xi_i \geq 0$, for $i = 1, \cdots, n$. Note that this is the linear version of DWD. As for SVM, there are also appropriate kernelized versions.

The numerical solution to (11.7) proposed by Marron et al. (2007) is based on the *second-order cone* algorithm SDPT3, available at Toh et al. (2009). Like the nominally simpler *quadratic programming* that underpins SVM, second-order cone programming is a fast greedy search algorithm. The computational speed of this approach scales very well with respect to the dimension d, which was important for earlier bioinformatics data sets. However, scaling with respect to sample size n is much worse. This has motivated the development of FastDWD by Lam et al. (2018), which uses a different optimization approach based on a semiproximal *alternating direction method of multipliers*. The computational speed of Fast-DWD scales very well with respect to both d and n, and is sometimes even more efficient than the highly optimized LIBLINEAR and LIBSVM implementations of SVM.

Another issue is selection of the tuning parameter λ in (11.7). As noted in Section 11.3, the issue is quite important for SVM, because depending on the particular data set at hand, best performance can come from λ anywhere between the hard margin SVM and a trimmed version of the MD. The situation is much different for DWD, mostly because the analog of the hard margin in that situation gives much more broadly useful results. Thus Marron et al. (2007) recommend achieving this with the choice $\lambda = \frac{100}{d_t^2}$ (the number 100 is chosen as a "large number"), where d_t is a useful notion of scale of the data computed as $d_t = median \{\|\boldsymbol{x}_i - \boldsymbol{x}_{i'}\| : y_i = +1, y_{i'} = -1, \}$. An interesting open problem is analysis of DWD of the type done in Carmichael and Marron (2017).

As first shown in Benito et al. (2004) DWD provides a useful direction for the removal of batch effects from genetic data. At first this good performance was just an empirical observation, but the underlying reason was later revealed in Liu et al. (2009). The main idea is illustrated in Figure 11.14. The left panel shows a toy data set that models a typical situation in the area of batch adjustment. Symbols represent data objects with $n_+ = 200$ plus signs data collected by one lab, while $n_- = 200$ circles come from another. Colors highlight two subtypes whose difference is the focus of the experiment. Note that for each lab separately (both the plusses and the circles), there is a clear systematic difference between the colors, but there is a 4 : 1 imbalance between the subtypes. To boost statistical power it is natural to pool the data over labs, perhaps by subtracting the mean of the data from each lab. While this works well when the subtypes (colors) are balanced within each lab (symbols) it is challenging because of the imbalance here as shown by the

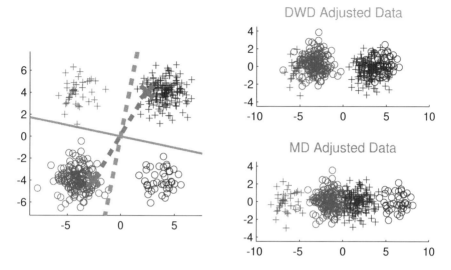

Figure 11.14 *Toy data set (left panel) showing value of DWD batch adjustment relative to an MD approach. Colors represent important subtypes to be studied, while symbols reflect irrelevant batch effects. MD shown in green, and poor results of simply subtracting the MD from the batches appear in the bottom right panel. DWD direction shown as a dashed pink line, with much better result of sliding the batches along the DWD direction in the upper right panel.*

sample means (green x signs). Subtracting the mean from each lab corresponds to bringing the labs together by shifting toward each other along the MD direction (dashed green line). The result of this manipulation is shown in the bottom right panel. Note that this MD adjustment is profoundly unsatisfactory because the difference between the colors (the goal of this study) has been seriously reduced to the extent that the main body of the subtypes overlap.

Better performance is available from use of DWD for batch adjustment. The DWD direction is shown in the left panel of Figure 11.14 as the dashed pink line, which is much closer to the ideal vertical direction than the green dashed MD direction. The result of sliding the labs along the DWD direction is shown in the upper right panel of Figure 11.14. This shows much better performance of DWD for pooling the labs (symbols) as now there is a substantial gap between the resulting colors. The DWD direction works better for batch adjustment because it is driven by the DWD separating hyperplane shown as the solid pink line. The $\frac{1}{r_i}$ poles at each data point push this plane away, which results in a much better direction for batch adjustment. The reason that DWD adjustment was seen to work well for cancer genetic data in Benito et al. (2004) seems to be due to the strong heterogeneity of cancer, which results in unknown unbalanced subtypes as illustrated in Figure 11.14.

Batch effects are becoming an increasingly important issue as there is a growing awareness that understanding very diverse diseases such as cancer need data sets larger than any single lab can collect. This has led to collaborative efforts such as The Cancer Genome Atlas, Weinstein et al. (2013), to integrate data across a large number of labs. While great care is taken about protocols and many other aspects of data collection, it is still impossible to completely eliminate unintentional batch differences. Handling these issues is the domain of the statistical area of *data heterogeneity*, discussed in Marron (2017a). Current approaches include *explicit* approaches such as DWD and other approaches to batch adjustment such as the empirical Bayes approach of Johnson et al. (2007), the Surrogate Variable Analysis method of Leek et al. (2012), as well as *implicit* approaches to statistical analysis that are robust against data heterogeneity as proposed by Meinshausen and Bühlmann (2015); Bühlmann and Meinshausen (2015).

A sparse version of DWD can be found in Wang and Zou (2016). Wang and Zou (2018) coined the term *generalized DWD* for a variation of DWD with important ramifications for batch adjustment and robustness against heterogeneity, where the impact of the poles $\frac{1}{\bar{r}_i}$ created by the residuals is explicitly controlled using a power q, i.e. $\frac{1}{\bar{r}_i^q}$. The case $q = 1$ is conventional DWD as defined in (11.7), and larger values of q magnify the effect illustrated in Figure 11.14 in the sense of pushing the solid pink line more toward horizontal. This results in a separating direction (dashed pink line) closer to the optimal vertical direction, i.e. improved robustness against heterogeneity. This also provides a connection between DWD and SVM which seems to be essentially the limit as $q \to \infty$.

As discussed in Section 11.3, extension of DWD to the multi-class case can be done by use of pairwise DWD in a one versus all scheme, or in the preferred multiple pairwise vote scheme. An intrinsically multi-class version of DWD (in the sense that the optimization problem is formulated in a multi-class way) has been proposed by Huang et al. (2013).

As noted in Section 18.3, Xiong et al. (2015) addressed a challenge in virology using data objects lying on the unit simplex. That challenge was a classification problem aimed at using DNA to identify the presence of viruses in blood samples, by aligning DNA in a given sample to the virus genome. The data objects were vectors of counts of DNA-seq reads, that were divided by the total counts to handle amplification effects. Such data objects are compositional data (i.e. lie on the unit simplex, as discussed in Section 18.3) as they sum to 1. The special challenge of that data was that while positive cases tended to lie fairly close to the center of the unit simplex, the negative cases tended to go off in many very different directions. The setting is essentially a very high-dimensional version of the donut example of Figure 11.6. However it is not clear that any of the kernel methods of Section 11.2 can handle the problem because test data tends to go off in directions that are not present in the training data. This concern is well justified in the left panel of Figure 11.15 which is from Figure 6.2 of the PhD dissertation Xiong (2015), reproduced as Figure 4 of Xiong et al. (2015), showing projections on the SVM separating direction in the radial basis function kernel space of the data. That classifier is trained on some positive cases shown as red plus signs and negative cases shown as blue circles. As in many other graphics, the vertical coordinates of these points are just jitter to give visual separation. The remaining symbols are independent test

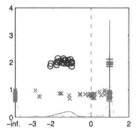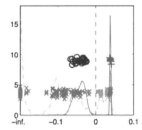

Figure 11.15 *Illustration of good performance of Radial DWD in virus detection in the right panel, contrasted with poor performance of SVM using a classical radial basis function kernel. Demonstrates ability of Radial DWD to correctly classify test cases that are very different from anything present in the training data.*

cases. The pink asterisks are positive human samples which all tend to be close to the center of the simplex and thus are correctly classified. The green asterisks are approximately positive samples representing viruses related to the positive class but from non-human mammals. Note that these are also correctly classified (i.e. to the right of the classification boundary shown as the vertical dashed line). More problematic are the cases shown as gray x signs, which are a large number of negative cases. While many of these are correctly classified (to the left of the boundary) a large number are not. These are cases that are far from the center of the simplex, but in directions that are not present in the training data, and hence are not reflected in the kernel space classification boundary. Xiong et al. (2015) address this problem by replacing the separating hyperplane used at many points above with a *separating hypersphere*. This is an option to keep in mind in other situations where one class is fairly centrally located and the other is spread very widely around, in a large number of other directions. Xiong et al. (2015) formulate a DWD-like approach to find the center and radius of the separating hypersphere (other linear classifiers could be adapted to spherical versions in a parallel way). This results in *Radial DWD* whose performance on the virus data is shown in the right panel of Figure 11.15. The horizontal axis is essentially the radius of a polar coordinate system, that has been shifted so that 0 is at the separating hypersphere. Cases shown at the left end (labeled "-inf") are those with all 0 counts, which are not on the simplex (and thus are naturally classified as negative). Colors and symbols are the same as in the left panels, with red plus signs for positive training data and blue circles for negative training cases. Once again, the positive test cases (pink asterisks) and related positives (green asterisks) are all correctly classified. But this time all of the negative test cases are correctly classified, verifying the above idea about using a separating sphere to correctly capture negative test cases that depart from the center in directions not present in the training data. An even more dramatic simulated example can be found in Figures 5 and 6 of Xiong et al. (2015).

11.5 Other Classification Approaches

There are other popular classification methods not treated in detail here. *Neural networks* were very popular in the 1980s and 1990s. A great success of that era was voice recognition software which went into widespread corporate use in the early 2000s. That success led to gross over-advertisement of the methodology, which resulted in many failed promises that in turn gave the whole field a bad name. However neural networks have more recently come back in a very strong way, under the new name of *Deep Learning*. While many of the basic ideas are the same, much of the current success (which is becoming almost completely dominant in fields such as computer vision) seems to be due to both size of modern data sets and also the far greater computational resources that are currently available. See Bengio et al. (2013) and Goodfellow et al. (2016) for good overview of this area. A current major challenge to deep learning methods is *interpretability*, i.e. the intuitive understanding of the drivers of differences between classes in classification contexts. In contrast, for simple linear methods, loadings plots as shown using the Pan-Cancer data in Figure 4.11, provide simple and direct insights into the main drivers of class differences.

Another approach to good interpretability is the tree-based approaches, such as the CART algorithm proposed in Breiman et al. (1984) The basic tree algorithms tended to get stuck in local optima, motivating many proposed solutions. The most widely accepted solution of this problem is the *random forest* approach of Breiman (2001). This is still widely used and is a frequent candidate in comparative classification studies, along with kernel SVM.

A quite different approach to combining a set of tree classifiers into a single more powerful classifier is one case of the idea of "combining weak learners" called *boosting* by Freund and Schapire (1995); Freund et al. (1999). See Friedman et al. (2000) for an interesting statistical perspective on boosting.

These many methods, all with their relative strengths and diverse properties make classification an area where the "every dog has its day" principle quite clear.

As noted above, all classification methods have potential for *overfitting*, i.e. resulting in a classification rule that is very effective on the training data but is not very generalizable, i.e. useful for new data. This potential typically gets worse for more complicated methods. This has led to a strong desire to measure performance of a classification method. Basic measurement is typically done using misclassification rates assessed on an independent *test data* set. This can be done either theoretically, or using a simulation study. But misclassification rates assume a particular relative weighting of classes. In situations where the desired weighting is not clear, an approach that essentially considers all weightings is the Area Under the Curve (AUC) summary of the Receiver Operating Characteristic curve introduced in Section 5.3.

For the analysis of a single real data set an objective approach is *cross-validation*, see e.g. Stone (1974), Kohavi et al. (1995), and Arlot and Celisse (2010). The idea is to divide the data into separate training and testing subsets, on which the method is, respectively trained and independently tested. Random choice is generally a good idea for this, and Monte Carlo variation should be damped by repeating the process and averaging. One approach generates repeats through independent random train-test divisions done at each step. Another common approach is k fold cross-validation where the data are partitioned into k subsets of approximate size n/k. Repeats consist of each subset being used in turn as the test set, with the others used for training. Choices for k can range from 2 to n. The choice $k = 2$ is generally not recommended as the size of the training data set is only $\frac{n}{2}$ which can be a substantially different classification problem than that being studied. The choice $k = n$ is essentially *leave one out* cross-validation. Common choices of k are 4 or 5. Generally recommended is to explicitly choose random subsets so that the classes are proportionally represented in the training and testing sets, as opposed to using randomization to make this only approximately true. Cross-validation is commonly used for two distinct tasks: for evaluation of the performance of a classifier and for choice of tuning parameters. When both are done together, it is important to use none of the test set for the latter, but instead to do a separate cross-validation completely within the training set. While cross-validation is broadly useful, it is important to keep in mind that it too can suffer strongly from sample variation, as illustrated in the context of smoothing in Section 15.3.

CHAPTER 12

Clustering–Unsupervised Learning

Clustering (unsupervised learning in the terminology of Chapter 11) is a broadly useful data analytic operation for many purposes. The main idea is to highlight cohesive subsets of the data that in some sense belong together. Major practical successes of clustering methods include the discovery of cancer subtypes, such as in Perou et al. (2000), which was a precursor to the currently very active research area of precision medicine. Clustering shares the idea of grouping data with classification as discussed in Chapter 11, with the big difference that here the goal is to determine the class labels. Hence clustering is also called unsupervised learning, as noted in Section 4.3 and this chapter title.

Also as noted in Section 4.3, good overviews of clustering are provided in Hartigan (1975) and Kaufman and Rousseeuw (2009). There are very many clustering methods and variations, some of which are discussed in the following sections. These include the k-means approach in Section 12.1, hierarchical clustering in Section 12.2 and simple visualization based methods in Section 12.3. An important method not further discussed here is called *model-based clustering*, which mainly uses Gaussian mixture models to identify clusters, see Bouveyron et al. (2019).

Key concepts are illustrated using the Four Cluster toy data set in the left panel of Figure 12.1. This data set has already been used in Figure 11.8. It features several important issues to keep in mind when choosing a clustering method. One of these is the elongated cluster of green plus signs, which could reasonably be considered one or more clusters. In particular its points are spread over a wider range than the distance between the round clusters of blue circles and the cyan squares, which can be reasonably treated as two separate clusters. The red x signs raise another relevant point. If there were only one it would usually be called an outlier. But can two such points be considered a cluster? What if there were more? It is useful to be aware of the conceptual fuzziness of the boundary that exists between sets of outliers and clusters.

12.1 K-Means Clustering

An intuitively appealing approach to clustering, named *k-means* by MacQueen (1967), was proposed by Steinhaus (1956). Given a set of (random as denoted by tildes) data $\tilde{x}_1, \cdots, \tilde{x}_n$ in $\mathbb{R}^d$, the main idea is to choose cluster index sets $C_1, \cdots, C_k$ that *partition* the full index set $\{1, \cdots, n\}$ (i.e. each index is contained in exactly one of the C_j), in a way that minimizes the *Within-Cluster Sum*

DOI: 10.1201/9781351189675-12

Figure 12.1 *Left Panel: Four Cluster data set for the illustration of the conceptual fuzziness and potential pitfalls of clustering. Right Panel: Another toy data set (shown twice: in the top and also in the bottom) providing insight into the Cluster Index introduced in (12.1). Sum of squared lengths of line segments in the top of the right panel are the Within, Cluster Sum of Squares, WCSS. The bottom of the right panel similarly demonstrates the corresponding Total Sum of Squares, TSS. Their ratio forms the Cluster Index (CI).*

of Squares

$$WCSS = \sum_{j=1}^{k} \sum_{i \in C_j} \|\widetilde{\boldsymbol{x}}_i - \overline{\boldsymbol{x}}_j\|^2 ,$$

where each within-cluster mean is denoted $\overline{\boldsymbol{x}}_j = \frac{1}{\#(C_j)} \sum_{i \in C_j} \widetilde{\boldsymbol{x}}_i$. The $WCSS$ is illustrated in the top of the right panel of Figure 12.1, as the sum of the squared lengths of the line segments, with the colored dots showing the within-cluster means. It can be conveniently rescaled by dividing by the total squared overall residuals from the mean,

$$TSS = \sum_{i=1}^{n} \|\widetilde{\boldsymbol{x}}_i - \overline{\boldsymbol{x}}\|^2 ,$$

where $\overline{\boldsymbol{x}}$ is the conventional overall mean of the full data set. The TSS is usefully contrasted with the $WCSS$ in the bottom of the right panel of Figure 12.1 (again the black dot shows the overall mean). That rescaling results in the *Cluster Index*

$$CI = \frac{WCSS}{TSS}. \tag{12.1}$$

Careful consideration of the right side of Figure 12.1 reveals that $CI = 0$ when each data point lies at its cluster mean, and $CI = 1$ (largest possible value) when all cluster means are the same as the overall mean. In between smaller values of CI tend to indicate tighter clusters.

Some properties of CI are illustrated in Figure 12.2, using just a toy data set in one dimension. The toy data consist of four Gaussian clusters, two of which have 5 points centered at ± 20, respectively, and two of which have 1000 points centered at ± 2 shown as black asterisks. The black curve is a kernel density estimate as discussed in Section 15.1. The behavior of CI is shown by the red curve, whose height is the value of CI for the clustering of the data set determined by the horizontal coordinate. For example, at the coordinate shown by the vertical dashed

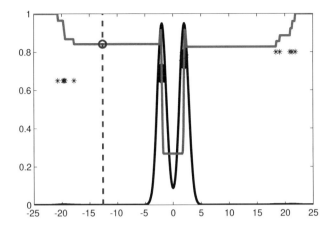

Figure 12.2 *One dimensional example, highlighting local minima challenges faced by CI. Many random starts can be stuck in a local minimum whose CI is far from the global minimum CI.*

blue line, the first cluster is the set of points to the left, while second is those to the right. The CI for that clustering is about 0.84, indicated by the height of the blue circle. As the blue line is horizontally shifted, the circle traces out the red curve. Note that $CI = 1$ at either end, because $WCSS = TSS$ when one of the two clusters is empty. As the blue line moves through the leftmost small cluster, the red line comes down. It stays flat in regions where there are no data points. There are two peaks near ± 2 as the CI goes up in the middle of a cluster (because the points in the cluster contribute to both terms of $WCSS$ which has a major impact on both cluster means). In the center region, both cluster means are essentially in the middle of the respective clusters resulting in both terms of the $WCSS$ being much smaller. This shows both how CI can easily have multiple local minima, and also the potential for many search methods to get stuck in a quite poor (in terms of the resulting CI being far from the optimum value) local minimum. Note this happens even in the simplest case of one dimension. It is natural to expect far more treacherous behavior in higher dimensions.

The standard k-means algorithm starts with some (perhaps random) set of candidate cluster means. The algorithm iterates through assignment of each data point to the closest mean followed by re-computation of the cluster means. This iterative process tends to converge to local optima, so sometimes several random restarts are used. There is a large literature on choosing initial values and restart methods. A faster algorithm has also been proposed in Pelleg and Moore (1999).

The Four Cluster toy data set from Figure 12.1 is used to illustrate some of the properties of k-means clustering in Figure 12.3. These graphics were created using the Matlab function kmeans with default parameter choices. The upper left panel shows the case $k = 2$. The optimal 2-means partition puts the green plus signs into one cluster (still colored with the same green), and all the other points

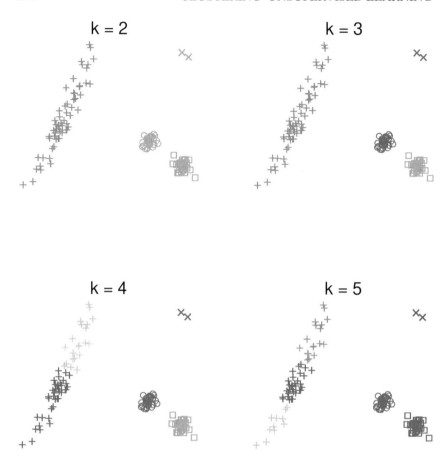

Figure 12.3 *Demonstration of k-means clustering, using the Four Cluster data, for $k =$ 2, 3, 4, 5. Shows care needs to be taken in interpretation.*

into the other cluster (colored a common pink chosen as between the blue, cyan and red colors). It is useful to imagine analogs of the right side of Figure 12.1 while interpreting these plots.

The $k = 3$ panel in the upper right shows a reasonable 3 cluster labeling of this data set, which nicely separates the 3 biggest clusters that were shown as green plusses, cyan squares, and blue circles in the left panel of Figure 12.1. The latter cluster is now colored magenta to reflect the union of the blue circles and the red x signs.

One might hope that the lower left $k = 4$ clustering would yield the original clusters as colored in Figure 12.1. However the stretch of the green plus signs is such that the CI is made smaller by splitting that (note the darker and lighter shades of green) while keeping the blue circles and red x signs in the same cluster.

The red x signs finally become their own small cluster in the case $k = 5$ shown in the lower right panel, however the blue circles and cyan squares have now been labeled as a single cluster colored an intermediate shade of blue, while the green plusses are now split into three clusters indicated with different shades of green. Deeper investigation showed that this is a local minimum of CI. Different starting values gave a smaller CI, which gives the 4 colored clusters shown in the left of Figure 12.1, with the green cluster split into two. This reveals an important aspect of k-means clustering: care is needed about local minima, and the results can be intuitively slippery, not always giving expected results. Hence it is best used in situations where visual confirmation (e.g. with PCA or CVA, as illustrated in Figure 11.5, scores distribution plots) can be done.

Cabanski et al. (2010) proposed the *SWISS score* (a renaming of the CI) for comparison of various preprocessing methods in bio-informatics settings. It essentially rates different approaches on how distinctly they allow a data set to be clustered. A perhaps surprising finding is that when $k > 2$ clusters are of interest, instead of using the k-means version of CI, it is better to use the average of the pairwise CIs. This seems be related to the observation of Friedman (1996), discussed in Section 11.3, that when extending a binary classification approach to a multi-class context, a vote among pairwise classifications is better than the one class versus the union of others approach. Permutation based inference for SWISS, in the spirit of that described in Section 13.1, for assessing the statistical significance of differences between preprocessing methods was also developed in Cabanski et al. (2010).

The SigClust approach to confirming the presence of a cluster, described in Section 13.2, uses k-means clustering in a direct and natural way. SigClust also provides useful information when facing the challenging problem of choosing the number of clusters k. A natural approach is to try several values of k. A drawback of k-means clustering is that the clusters are not *nested*, i.e. there need not be any relationship (such as inclusion or overlap) between clusters for a given k and $k + 1$. An approach which does give a nested sequence of clusters is hierarchical clustering discussed in Section 12.2.

12.2 Hierarchical Clustering

Hierarchical clustering gives a full *dendrogram* (i.e. nested set) of clusterings. An example, again using the Four Cluster toy data set is shown in Figure 12.4. This can be thought of in two equivalent ways.

One way of thinking, called *divisive* or *top-down*, is to start with the entire data set in one cluster, followed by a nested series of binary *splits*, until only singletons (clusters with just one member) are left. Such splits are represented as horizontal line segments in Figure 12.4, where the vertical line segments represent clusters. Colored lines are used for those clusters which contain only points with that color in the left of Figure 12.1. The length of each vertical segment indicates strength of the cluster. The first split near the top separates the two red points from the rest. The remainder is then split between the green cluster and the union of the blue and cyan. Next, the blues and cyans are split. Vertical line segments above

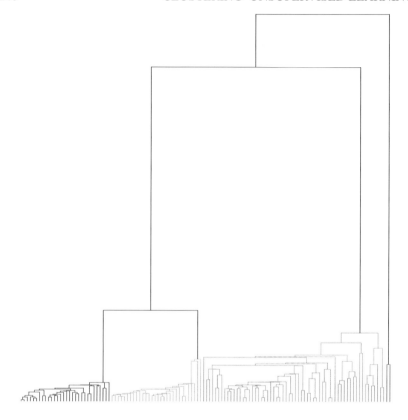

Figure 12.4 *A dendrogram for the Four Cluster data, based on Euclidean distance and single linkage. Reveals full nested series of clusters from one large cluster at the top to singleton clusters at the bottom.*

this split are relatively long because the blues and cyans are closer together than the other colored clusters. Remaining splits all happen within the colored clusters. The corresponding vertical segments are relatively shorter because these smaller clusters are much less distinct.

The other way to think of the dendrogram, called *agglomerative* or *bottom-up*, is to start with all singletons (shown as the bottom of the dendrogram), and then sequentially combine pairs of clusters until only the full data set in one big cluster is left. As one moves up the dendrogram, first the points with common colors are combined, and then finally those clusters get merged, with the blue-cyan and green merge happening late which reflects the big distances between those.

As both ways of thinking result in the same dendrogram, either is fine for understanding the nested set of clusters. However, the difference becomes important when considering computational issues.

There are many ways to construct hierarchical dendrograms, most of which are indexed by a *distance* (between data objects) and by a *linkage* (which basically gives a notion of cluster separation). Several linkage functions are studied for the Euclidean distance in Figure 12.5, for the same Four Cluster data set from the left of Figure 12.1. In each case, the dendrogram has been cut at a level that mostly results in four clusters (five in the lower right).

The upper left panel shows the results from *Ward's* linkage (Ward (1963)), which tends to target roughly equal splits. Hence the relatively large green cluster is split (quite similar to the $k = 4$ results of k-means in Figure 12.3), and the red and blue clusters are treated as just one. This is accomplished by minimizing the within-cluster $WCSS$ as done by k-means. The results are generally different (despite being similar in the bottom left of Figure 12.3 and the top left of Figure 12.5) because hierarchical clustering imposes nesting on the overall set of clusters, minimizing $WCSS$ only at each split.

The upper right panel shows *average* linkage, where the round but nearby cyan and blue clusters are combined. The red points become their own small cluster, because average linkage does not prioritize similar cluster size. Finally because of its length the long green cluster is split.

The lower left panel shows the result of *single* linkage. Single linkage essentially takes the distance between clusters to be the minimum pairwise distance between points, which gives the perhaps intuitively expected result in this two-dimensional case.

Single linkage is explored a little further in the lower right panel, where the five cluster version is shown (corresponding to a lower cut in Figure 12.4). Note that the fifth cluster comes from splitting the long green cluster. It is worth noting that split is not near the center of the cluster as for some of the other linkages, but instead happens at the biggest gap between green points as expected.

The distance measure also has a large impact on the clustering result, as shown in Figure 12.6, for the same toy data set. All panels are again based on dendrograms (all using single linkage) cut at a level that results in four clusters. The lower right panel shows the result using the Spearman distance (one minus the Pearson correlation between the vectors of the ranks), which is particularly unappealing, but makes the point that choice of distance measure can have a very large impact on the result. The remaining panels are variations on cosine distance, defined at (7.3). This is useful in situations where it is desirable to ignore scaling. This distance is essentially based on angles (ignoring radii) in a polar coordinate representation. The three panels make the point that choice of *center* (shown in each case as the large black dot) of the analysis is critical to the result when using this cosine distance. In the upper left panel the center is taken as the overall sample mean, which results in a fairly good clustering, that is very close to the average linkage result in the upper right of Figure 12.5. A deliberately poor choice of center appears in the upper right panel of Figure 12.6, which results in combining the blue and cyan clusters with most of the green, while pulling off small bits of the greens and the reds as separate clusters. A better choice of center point appears in the lower left panel, resulting in the perhaps expected result. This behavior is intuitively understood as dividing the data set into sectors whose boundaries are determined by rays coming out of the center point. The upper left sample mean

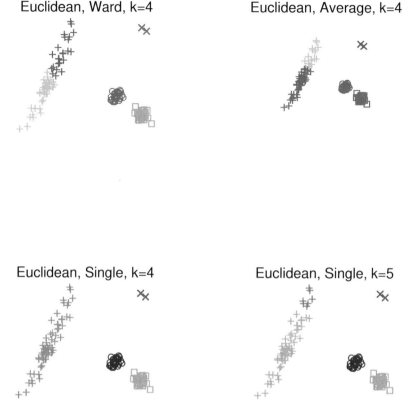

Figure 12.5 *Exploration of several common linkages in hierarchical clustering for the Four Cluster data, all based on Euclidean distance. Shows single linkage gives perhaps the most intuitive clustering in this 2-d case.*

combined the blue and cyan clusters, as they must lie within the same sector. The good clustering in the lower left used a center that exploits the large gaps in angle (as targeted by single linkage) between the desired clusters.

In summary, for this two-dimensional toy data set, single linkage seemed to give the most intuitively expected result, while Ward's tendency to give even splits was less useful. Other linkages such as average and complete (not shown here, but is fairly similar to Ward's) tend to lie between these extremes.

But this relative relationship between linkages changes markedly in higher dimensions, as shown using the High-Dimensional Gaussian data in Figure 12.7. That data set is $n = 30$ vectors of dimension $d = 500$ simulated from the standard normal distribution. The clustering in both panels is based on Euclidean distance, with single linkage used in the left panel, and Ward's linkage in the right. Both dendrograms are most easily interpreted from the viewpoint of starting at the top and considering consecutive splits.

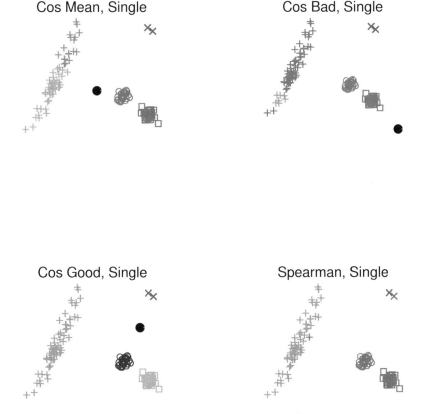

Figure 12.6 *Study based on the Four Cluster data, of cosine distance with different choices of center point, shown as a black dot. Mostly shows choice of center point has a major impact on the quality of the result. In addition the lower right panel demonstrates very poor performance of the Spearman distance.*

The single linkage (left) analysis starts by splitting off a singleton (one point cluster), with all other points forming the remaining cluster. This pattern continues through most of the dendrogram. Because many such singletons is not a typical goal of a cluster analysis, single linkage tends to be less appealing in higher dimensions (despite the intuitively sensible 2-d performance in Figure 12.5).

On the other hand Ward's linkage (right panel of Figure 12.7) shows a far different behavior. This analysis favors splits which are relatively balanced, which may be more useful, depending on the application. As for the 2-d examples considered in Figure 12.5, other linkages (such as average and complete) tend to give results that lie in between single and Ward's linkage. It usually makes sense to try several linkages, and ultimately use the one giving the most useful visual results for each given situation.

Further insights into this clustering behavior comes from the scatterplot matrix

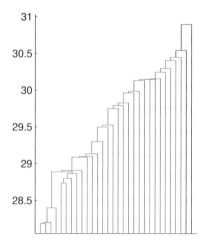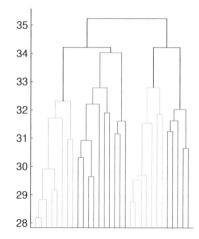

Figure 12.7 *Dendrograms for the High-Dimensional Gaussian data set ($d = 500$), using Euclidean distance. Contrasts single linkage (left), which sequentially separates individuals, with Ward's linkage (right), which tends to give more balanced splits. Colors highlight important aspects in other plots.*

view of the High-Dimensional Gaussian data set shown in Figure 12.8. Instead of being based on PC directions as has been done in many other scatterplot matrices, now the directions are chosen to highlight the first two singleton clusters corresponding to the black vertical line segments in the left panel of Figure 12.7. This is done using axes which are simply unit vectors pointing in the direction of each of the two data points (shown as black circles in all panels), i.e. vectors of the form $\frac{x_i}{\|x_i\|}$. These points may seem surprisingly far from the rest (shown as red circles in all panels) for Gaussian data. Also perhaps surprising is the orthogonality of these outlier directions. However as shown in Section 14.2 this is actually quite natural behavior for high-dimensional Gaussian data. For contrast the third axis shows the PC1 scores, i.e. the projections onto the PC1 eigenvector. Note that the sample standard deviation of these scores is clearly larger than in the singleton directions, which is why this is the PC1 direction. That sample standard deviation of these PC1 scores may also appear to be surprisingly large for projections of standard normal data. This also turns out to be a natural aspect of high-dimensional data, as discussed in Section 14.1.

A quite different scatterplot matrix view of this same data set is shown in Figure 12.9, this time aimed at revealing the performance of the Ward's linkage clustering in the right panel of Figure 12.7. Colors come from the dendrogram in the right panel of Figure 12.7, which was cut at a level resulting in four clusters. These clusters are usefully visualized using Mean Difference directions as axes. The first contrasts the Cyan vs. Blue clusters, and the second the Green vs. Red. Again it is useful to compare with the PC1 scores shown as the third axis. As above, the standard deviation in the PC1 direction is larger than for either of these two cluster

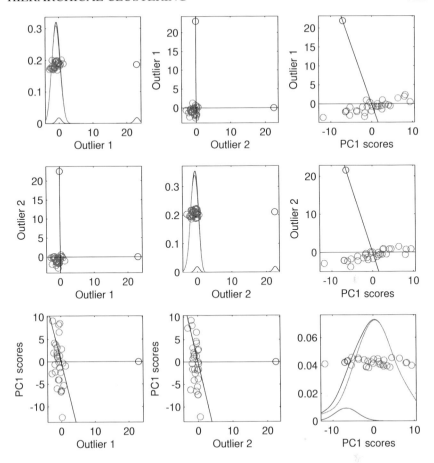

Figure 12.8 *Scatterplot matrix view of the* ($d = 500$) *High-Dimensional Gaussian data in Figure 12.7, using single linkage colors from the left panel. The first two axes show how well the first two singleton clusters are separated from the rest of the data, which is why they were chosen by single linkage clustering. The third axis is in the PC1 direction, for contrast.*

differences. In parallel to the above discussion, the gap between these clusters may appear to be quite large for standard normal data, but this turns out to be quite natural using calculations as in Section 14.2.

In summary, hierarchical clustering is a very flexible set of methods with great potential for discovering interesting and important clusters in data. However, because there is no uniformly best clustering method (once again the "every dog has its day" concept applies here), substantial prior knowledge is essential to use it effectively. Also because there are so many choices and options available, confirmatory analysis becomes very important. The SigClust approach discussed in Section 13.2 is useful for this.

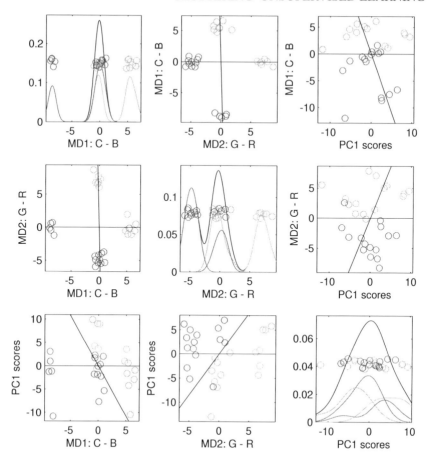

Figure 12.9 *Another scatterplot matrix view of the High-Dimensional Gaussian data as in Figure 12.8, this time illustrating the Ward's linkage analysis, with colors from the right panel of Figure 12.7. First two axes are based on colored cluster Mean Differences, Cyan - Blue in the first direction, and Green - Red in the second. Third axis again shows the PC1 scores.*

12.3 Visualization Based Methods

While formal clustering methods such as those described in Sections 12.1 and 12.2, are useful and direct ways to target clusters, the latter often appear naturally in various data visualizations. Examples of this include the toy data set in Figures 4.3 and 4.4, as well as the Pan-Can RNAseq data in Figure 4.7. Another example of this is the *Mass Flux* data shown in Figure 12.10. This functional data set (curves as data objects) was provided by Enrica Bellone of the National Center for Atmospheric Research. Each curve is a characterization of an atmospheric cloud. The top row shows the initial object centering operation, with the input curves shown in the top left panel, the mean curve in the top center, and the recentered

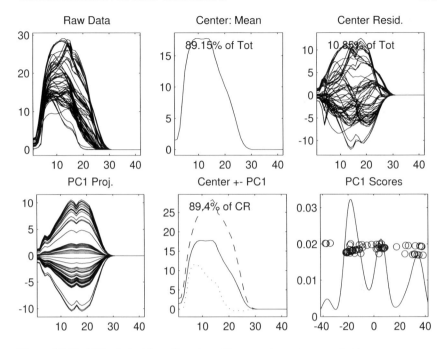

Figure 12.10 *PCA of the Mass Flux data. Top row shows impact of object mean center-ing (raw data, mean, and residuals). Bottom left shows the first mode of variation, center shows relationship with mean, and bottom right shows the scores. Note smooth histogram indicates 3 clusters in the PC1 scores.*

data (the mean residuals) in the top right. The first mode of variation (best rank-one approximation of the centered data matrix) appears in the bottom left panel, showing strong vertical variation mostly on the left side. The bottom center plot shows the mean plus the largest (dashed) and smallest (dotted) of the curves in the bottom left. Because of the sloping nature of the mean, this view reveals that this mode of variation is actually about both the curve heights and also the location of the peaks. The bottom right focuses on the scores (the coefficients of the curves shown in the bottom left). The smooth histogram (kernel density estimate, Section 15.1) provides a strong indication of 3 bumps, suggesting clustering in this data set. An important issue is whether or not these clusters represent important and re-producible structure in the data, or are spurious sampling artifacts. This question is addressed for this data set using the SiZer method in Section 15.4.3, where it is seen that these three peaks are statistically significant in some sense. This turned out to be consistent with three known important cloud types.

Other visualization methods are also often used to discover clusters. In bioin-formatics this is commonly done with t-SNE as illustrated in Figure 11.9, or with UMAP as discussed in Section 11.2.

With any visual method for finding clusters, there are limits to how many clus-ters can be perceived. In complicated situations, it can be useful to *drill down*.

That means to take each cluster found in an earlier step, and to separately visualize that (e.g. using a new PCA on just the data in that cluster). Vidal et al. (2016) have proposed an automated version of this called *Generalized PCA*.

A quite different approach to finding clusters in high dimensions is based on the Maximal Data Piling (recall the acronym MDP) idea discussed around Figure 11.4 in Section 11.1. Note that (with probability one for data from a distribution that is absolutely continuous with respect to Lebesgue measure) every binary partition of the data set admits the MDP direction $\widehat{w}_{MDP}$ (as defined at (11.5)). The relative clustering strength of each candidate partition is directly measured as the difference between the two projection points of each class onto the $\widehat{w}_{MDP}$ vector, and the partition that maximizes this can be regarded as the best partition in high dimensions. A computational benefit to such an approach is that each $\widehat{w}_{MDP}$ can be computed relatively quickly, because the inverse of the sample covariance matrix, $\left(\widehat{\Sigma}\right)^{-1}$, only needs to be computed once. A major hurdle is that there are 2^{n-1} such binary partitions. An interesting open problem is to develop a heuristic to make this approach computationally tractable.

12.3.1 Hybrid Clustering Methods

Clustering, i.e. unsupervised learning, has also been combined with other methods in several interesting ways including

- Semisupervised Learning, see Chapelle et al. (2006). The idea here is that some but not all of the class labels are known, which gives a methodology which is somewhat between classification and clustering.

- Combining clustering with curve registration (as studied in Chapter 9), see Bernardi et al. (2014b,a).

- Combining clustering with data integration (as studied in Section 17.2.3), see Hellton and Thoresen (2014).

CHAPTER 13

High-Dimensional Inference

Confirmatory analysis is a nearly completely dominant part of classical statistics, as typically taught in modern courses. Much of this is based on fitting parametric probability distributions to data, and basing inference such as hypothesis tests and confidence intervals, on such distributions. As noted in Marron (2017a) and Carmichael and Marron (2018), this approach lies at the roots of the *scientific method*, which has provided great benefits to science over the years.

So far this approach has not been well developed in OODA contexts, often because in many OODA data settings such as the manifold data objects of Chapter 8 and the tree-structured data objects in Chapter 10, suitable probability models are not yet in common use. However as shown using the Overlapping Classes 1000 dimensional Gaussian data set in Figures 4.12 and 4.13, statistical confirmation is a critically important task, to ensure that non-spurious and reproducible structure has been discovered. Two main approaches to this are discussed in this chapter. The DiProPerm hypothesis test, for testing the difference between two *previously labeled* subgroups is discussed in Section 13.1. While it is tempting to use DiProPerm to assess the significance of subgroups found by clustering methods, it is seen in Section 13.2 that this can be quite inappropriate. A test for significance of clustering, that is appropriate in the challenging high-dimensional context, called SigClust, is described in Section 13.2.

13.1 DiProPerm–Two Sample Testing

A sensible way of understanding statistical significance of features observed in a visualization is to use a hypothesis test directly related to that visualization. This goal is achieved by the DiProPerm (for DIrection-PROjection-PERMutation) hypothesis test proposed by Wei et al. (2016), discussed in Section 4.2. For example the PCA scatterplot analysis of the $d = 20,000$ dimensional Two Class Gaussian data set in Figure 6.6 shows two distinct classes, which are not at all visually apparent in the heat map view of the same data shown in Figure 6.5. Statistical significance of this visual difference is investigated using DiProPerm in Figure 13.1. The upper left panel shows the first step of DiProPerm: visualization based on the MD (recall the acronym for Mean Difference from Section 11.1) direction and projection of the data onto it, which clearly highlights this class difference. The latter is quantified by subtracting the means of the projected classes, which gives about 23.1, indicated using a vertical green line in the top right panel. Statistical significance of this class difference is assessed using a permutation test.

The permuted null distribution comes from randomly relabeling the data, and

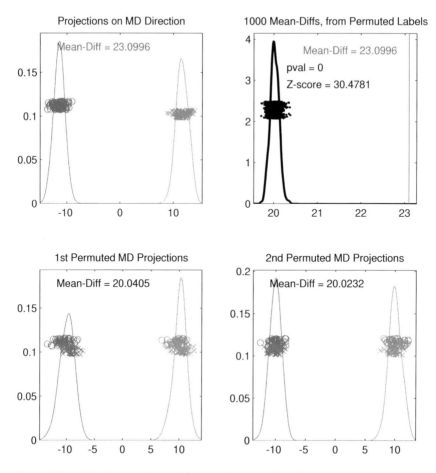

Figure 13.1 *DiProPerm analysis of the significance of the difference between the means of the magenta and green classes (same symbols and colors) of the Two-Class Gaussian data shown in Figure 6.6. Upper left panel shows projections onto the MD direction, with class contrast summarized by the difference of the class means (green line in the permutation plot, top right panel). Bottom panels show two realizations of permuted class labels, whose class mean differences are summarized in the top right panel. Shows class difference is strongly significant, despite not being visible in the heat map view of Figure 6.5.*

for each of those recomputing both MD and the projections followed by recomputation of the projected class mean differences. Two realizations of this process are shown in the bottom panels of Figure 13.1. The colors and symbols indicate how the permutation works. In the top left panel the magenta data are all indicated with circles, with x signs for the green cases. In the bottom panels the same symbols are used for each data point, but about half of the magenta circles have had their label (i.e. class color) switched to green and similarly for the x signs. Note that the horizontal scales in the bottom panels are different from the top left, and in fact the two permuted group means, 20.04 and 20.02 are substantially smaller than the 23.1 observed for the original data. Those are two of the black dots shown in the *permutation plot* in the upper right panel of Figure 13.1, which are projected class differences from 1000 independent permutations. Note the black dots reflect permutation differences that are all much smaller than the green line showing this difference is strongly statistically significant. A quantification of this is the *empirical p-value*, which is the proportion of the black dots to the right of the green line, which in this case is 0. In situations where it is desired to compare the significance of several DiProPerm results, empirical p-values are not useful when both are 0, e.g. as for the Pan-Cancer data shown in Figure 13.3. This issue is usefully addressed with the *Z-score*, an alternate measure of significance. The Z-score summarizes the distribution of the black dots using its mean and standard deviation, and measures significance of the DiProPerm test as the number of standard deviations that the green line is above the mean, in this case about 30.5.

The choice of MD direction made in Figure 13.1 results in a hypothesis test which is equivalent to a standard permutation mean test, i.e. projection is not needed except for the visualization. However, as noted in Section 11.4 DWD provides a direction which can be better in a number of senses, including data visualization, in which case the projection step of DiProPerm is essential beyond just visualization. The potential benefit of using the DWD direction in DiProPerm is illustrated in Figure 13.3.

Figure 13.2 shows a DWD DiProPerm analysis of the Overlapping Classes data from Figures 4.12 and 4.13. Recall the classes shown there were both sampled from the standard normal distribution, so there should be no significant difference, yet the DWD projection suggested an apparent strong difference in the upper left panel of Figure 4.13 (repeated in the left panel of Figure 13.2). The realizations of the permutations (analogs of the bottom panels of Figure 13.1) are quite similar to the left panel and are thus not shown here (nor in other DiProPerm examples in this section) to save space. The right panel shows the permutation plot with the mean differences of the projections onto the DWD direction for 1000 permutations. Note that this time the green line lies in the middle of the null distribution of black dots, as quantified by an empirical p-value of 0.82, showing this projected difference is clearly not significant (again despite the contrary visual impression in the left panel). As discussed in Section 4.2, this reveals how important it is for exploratory data analysis to be complemented by confirmatory analysis.

Given the goal of connecting inference with visualization by DiProPerm, there are several reasonable choices of direction vector. E.g. the MD was used in Figure 13.1, while Figure 13.2 was based on DWD. An advantage of the MD is that it is faster to compute. However the sharper class distinctions available from DWD may be expected to give stronger inference, when there actually is a difference between the groups. An example of this property is shown in Figure 13.3, using the same log-transformed Pan-Cancer data (same colors and symbols) as in Figure 5.16. The top left panel shows the MD projection, which does separate the classes, but the bottom left panel shows that projecting on the DWD direction gives a cleaner separation. This can be seen from carefully studying the kernel density

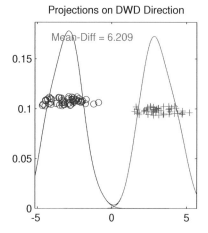 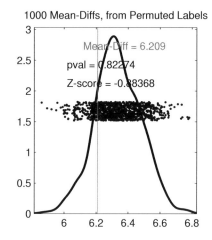

Figure 13.2 *DiProPerm analysis of the Overlapping Classes data shown in Figures 4.12 and 4.13, using the DWD projection. Reveals that the apparent large gap between the red and blue classes is not statistically significant. i.e. is an artifact of natural sampling variation.*

estimates in the valley between the cancer types. This example also demonstrates the value of the Z-score over the empirical p-value. In particular both of the latter are 0, yet the much larger Z-score (about 28.8 for DWD versus 15.9 for MD) demonstrates that DWD gives much more powerful inference in this case. An even more dramatic example of DWD giving stronger DiProPerm results appears in the analysis of the Lung Cancer data discussed around Figure 13.12.

Also important is the statistic used to summarize the difference between the projected means, shown in green text in the projection plots and as a green vertical line in the permutation plots. A perhaps natural choice is the 2 sample t statistic. This is classical in conventional hypothesis testing because it incorporates sample variability into the statistic. However, such incorporation is not needed here, as the permutation already deals with this variation. Wei et al. (2016) showed that using the t statistic actually gives some instability causing a potential loss of power, so the simple difference of the means is recommended instead. There are other reasonable choices for quantifying differences in 1-d distributions, such as the AUC from the ROC curve, discussed in Section 5.3.

Several DiProPerm analyses of the Drug Discovery data studied in Chapter 5 are summarized in Table 13.1. Recall from the full raw data PCA plot in Figure 5.3 a major goal for this data set was to contrast actives (red) from in-actives (blue), so these are contrasted using DiProPerm hypothesis tests. All empirical p-values are 0, so again meaningful comparisons come from the Z-scores shown in the right column of the table, with a relevant figure number shown for quick reference in the middle. The second row shows the initial result of removing all variables with any missings, which were coded as -999 (recall the finding of these was highlighted in Figure 5.6). While the PCA for that data set has not been shown here, as it

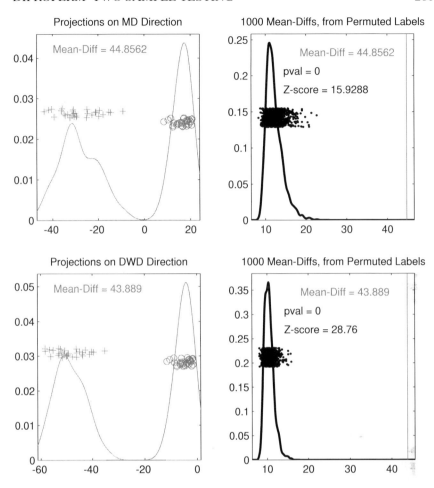

Figure 13.3 *Comparison of MD and DWD directions in DiProPerm, using log Pan-Cancer data from Figure 5.16. The larger Z-score in the permutation plot (lower right panel) shows that sharper class separation from DWD results in a more powerful hypothesis test.*

is essentially the same as Figure 5.3, this shows that removal of these variables actually has a meaningful effect.

The third row of Table 13.1 summarizes the application of DiProPerm to only the binary data, whose PCA appears in Figure 5.12. Perhaps surprisingly this filtering of the data contains more active–inactive information than the raw data, as evidenced by the larger Z-score. The benefits of standardizing the non-binary part of the data are clear from the larger Z-score in the fourth row (PCA in Figure 5.15). Some (fairly marginal) improvement comes from application of the automatic shifted log transformation discussed in Section 5.3.

DiProPerm was motivated by the desire to quantify the reproducibility of lessons learned from data visualizations. That goal is not standard in the area of

Data	Figure Number	Z-score
Raw	5.3	10.4
No Missings	5.6	11.6
Binary Only	5.12	14.6
Standardized	5.15	17.3
Auto Log Trans'd	-	17.8

Table 13.1 *Summaries of DiProPerm Z-scores for various versions of the Drug Discovery data studied in Sections 5.1–5.3. Shows improved separation of classes following improved preprocessing.*

high-dimensional hypothesis testing, where the focus is usually on optimization of test power. Important references in that area include Bai and Saranadasa (1996), Srivastava and Du (2008), and Chen and Qin (2010). A method quite similar to DiProPerm was proposed by Ghosh and Biswas (2016). The statistical power of DiProPerm has been substantially improved in strong signal cases by using *balanced permutations* in Yang et al. (2021).

13.2 Statistical Significance in Clustering

Another important type of confirmatory analysis is determining when clusters discovered using say a method from Chapter 12 are "really there", i.e. represent important and reproducible underlying population structure as opposed to being spurious sampling artifacts. One might attempt to use a mean based hypothesis test, such as the DiProPerm method of Section 13.1 to the discovered classes for this purpose. However, as pointed out by Andrew Nobel, this is inappropriate as illustrated in Figure 13.4. That analysis is based on the $n = 100$ two-dimensional standard normal data points displayed as a scatterplot in the lower right panel. The colors and symbols used there show the result of an application of 2-means clustering, which essentially splits the data in half. These can be considered to be two classes whose mean difference can be tested with DiProPerm. Projection onto the MD direction is shown in the upper left panel. One realization of the permutation test is shown in the lower left panel, revealing a difference of projected class means that is *much* smaller than in the upper left panel. This is born out in the 1000 replication summary (upper right panel) where the Z-score of about 11.5 shows the red and blue class means are clearly different. However, because the data are a single standard normal cluster, it does not make sense to conclude that there are two distinct clusters here. The lesson is that DiProPerm is about the difference in means of *given classes*, which is *different* from determination of clusters, so another approach is needed for clustering applications.

The first major hurdle to construction of a test for statistically significant

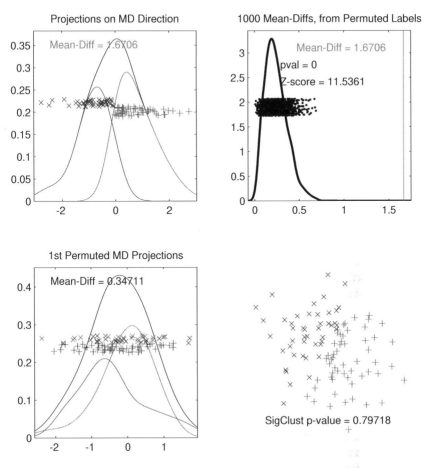

Figure 13.4 *Simple 2-means clustering example to illustrate how DiProPerm applied in clustering contexts may inappropriately indicate clusters. Raw data in the bottom right are clearly a single (in fact Gaussian) cluster, yet the DiProPerm (applied to the two colored groups in the lower right) result shown in the permutation plot (top right) is clearly strongly significant.*

clusters is to define the notion of *cluster*. Tibshirani et al. (2001) suggest thinking about circular uniform distributions, as a conservative approach, but there are not natural probability mechanisms which generate that. A reasonable null hypothesis for naturally generated clusters is the Gaussian distribution, proposed by Sarle and Kuo (1993). That idea underlies the *SigClust* hypothesis test proposed by Liu et al. (2008) and substantially improved by Huang et al. (2015).

SigClust assesses how well a single Gaussian distribution fits the data set, by measuring goodness of fit using the 2-means CI (Cluster Index, defined at (12.1)). A p-value (or Z-score) can be estimated by comparing the CI from the data with a null population generated by simulating data sets of the same size n from a fitted

Gaussian distribution. It is worth noting there are two useful SigClust tasks, which are handled in a similar way.

1. *Confirmatory SigClust.* Testing whether a *given* clustering (i.e. set of labels) is significantly different from a fitted Gaussian distribution. Such labels could come from some extrinsic information, or just from a visual impression, when there is interest in whether the clusters are "really there". Here the focus is on whether or not the given CI is smaller than those of the null Gaussian distribution.

2. *Exploratory SigClust.* Investigating whether a data set should be considered to be more than one cluster. In this case comparison is done between the 2-means CI (recall the minimizer over all potential clusterings) for the data and the null Gaussian CI distribution.

Such a p-value from this second approach, for the data in Figure 13.4, appears in the lower right panel. This is clearly insignificant as expected from underlying Gaussian data, so SigClust gives an appropriate answer of one cluster in this case.

The next major hurdle is estimation of an appropriate Gaussian null distribution. In most situations it will be inappropriate to use a spherically symmetric Gaussian distribution. Thinking about the Four Cluster data set in the left panel of Figure 12.1, the blue circles and the cyan squares each seem to constitute separate clusters. But what about the collection of green plus signs? Using the definition in terms of coming from a single Gaussian this is taken here to be a single cluster, although others might prefer a different choice. Also thinking about the Gaussian based definition of cluster, the two red points are regarded as their own small cluster, as single sufficiently extreme outliers are also regarded as singleton clusters.

Various subsets of the two-dimensional Four Cluster data in the left panel of Figure 12.1 are used to build intuition for SigClust. The straightforward case of just the blue circles and cyan squares is considered in Figure 13.5. Note that just those two clusters are in full color in the left panel, while the other points are gray because they are not used in this example. Also shown in the left panel is an elliptical contour of the fit Gaussian density together with a sample of the same size drawn from that density (black stars). Finally the CI using the blue and cyan labeling (hence using SigClust in confirmatory mode) is shown near the top ($CI = 0.085$), and also is the horizontal coordinate of the green line in the right panel. As expected, that is much smaller than the CI of the black points shown near the bottom ($CI = 0.462$). The cluster labeling used for the black points comes from applying two means clustering, which aims to minimize the CI over all possible class labelings, as discussed in Section 12.1. The comparison of the data CI with the null distribution is shown in the right panel. The black dots are 1000 realizations of the CI from data drawn from the Gaussian fit. Generally in SigClust, a real data CI (shown in green) that is *smaller* than the simulated null CI population provides strong evidence of more than one cluster (in contrast to DiProPerm where larger values are more significant). In this example, the green line is much smaller than the simulated null distribution, indicating strong significance. These SigClust results are reported in a similar fashion as in Section 13.1,

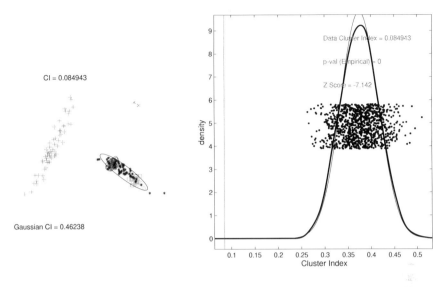

Figure 13.5 *SigClust analysis, using the given cluster labels of the blue and cyan clusters in the Four Cluster data. Shows a strongly significant result.*

where the empirical p-value is the proportion of black dots to the left of the green line, and the Z-score represents the number of standard deviations the green line is to the left of the black dot null population (again being useful for comparing results with 0 p-values).

Potential splitting of the long green cluster is studied in Figure 13.6, using a similar graphic. This time only the green points are not grayed out, and are used in the calculation. Here SigClust is used in the exploratory mode described above, so no input clustering is used, and instead SigClust only aims to assess whether or not the long green cluster should be split. Because there is no input clustering, no CI is shown in the left panel. However, everything else is similar to before, with an overlay of the fit density contour, and a corresponding simulated Gaussian data set of the same size. This time the distributional summary plot in the right panel shows the data CI (green line) is well in the middle of the null population of black dots. Hence there is no evidence for more than one cluster in this case, also as expected.

A much more challenging case is the union of the long green cluster and the red outliers, shown in Figure 13.7. The format is the same as the two figures above, but now the surprising feature is that the data CI is actually larger than the simulated null population. This is caused by the severe imbalance between the given cluster sizes. The 2-means clustering-based null distribution of CIs essentially starts with the black points in the left panel, which are more spread than the green points because of the influence of the red points on the sample covariance matrix estimate. These black points are split in half (as shown in the right panel of Figure 12.1) and the corresponding $WCSS$ is compared with the overall TSS (recall

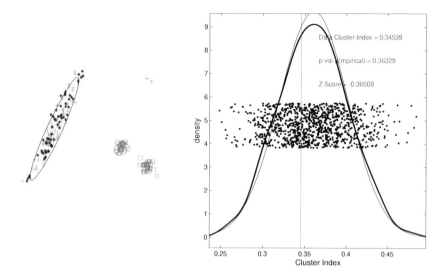

Figure 13.6 *SigClust investigation of whether to split the long green cluster in the Four Cluster data. Large p-value indicates a statistically significant split has not been found, i.e. the green plus signs are a single cluster.*

this Sums of Squares notation from (12.1)). On the other hand the given clustering CI (the vertical green line in the right panel) is relatively closer to 1 because the very few red outliers have a relatively small impact on either the $WCSS$ or the TSS, which are thus close to each other. Note that the resulting hypothesis test in this case is still valid, but it is important to realize that it can be quite conservative in the case of strongly unbalanced cluster sizes. An interesting open problem is the development of a modified SigClust that gives better performance in the unbalanced case.

13.2.1 High Dimensional SigClust

The above two dimensional examples are relatively simple, since good estimates of the fit Gaussian distribution, $N_2(\boldsymbol{\mu}, \boldsymbol{\Sigma})$ on $\mathbb{R}^2$, are available. But construction of decent estimates is much more challenging in higher dimensional cases. Some serious reduction of this problem can be made by exploiting invariance properties of the Cluster Index, CI. The first one is *shift-invariance*, which is the fact that if an entire data set is shifted in any direction by a constant amount the CI remains the same. For this reason, there is no need to estimate the population expected value $\boldsymbol{\mu}$ of the $N_d(\boldsymbol{\mu}, \boldsymbol{\Sigma})$ distribution, instead the simulations can equally well be drawn from the $N_d(\mathbf{0}_{d,1}, \boldsymbol{\Sigma})$ (using the zero matrix notation from (4.2)) distribution. The second key property of the CI is *rotation invariance*, where a rigid rotation of the full data set similarly leaves the CI unchanged (because all the line segments in the right panel of Figure 12.1 will keep the same lengths in a rigid

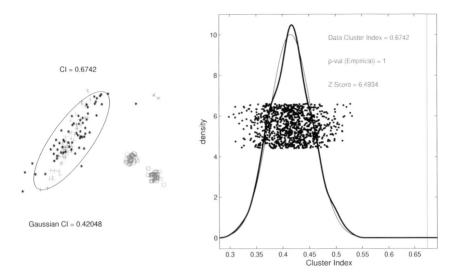

Figure 13.7 *Challenging (subset of the Four Cluster data) case for SigClust, with the long green and small red clusters input. Given CI (based on green (plus) vs. red (x) labels) is actually larger than the null distribution, because of the severely unbalanced class sizes.*

rotation). The value of rotation invariance comes from the eigen (spectral) analysis of the underlying population covariance matrix $\boldsymbol{\Sigma} = \boldsymbol{U}\boldsymbol{\Lambda}\boldsymbol{U}^t$ where $\boldsymbol{U} \in O(d)$ (the orthogonal group of matrix transformations defined in Section 7.3.5), and $\boldsymbol{\Lambda}$ is the diagonal matrix of eigenvalues. See Section 17.1.2 for more discussion of eigenanalysis. Hence by a suitable rotation of the data, it is enough to draw the null hypothesis data from the $N_d\left(\boldsymbol{0}_{d,1}, \boldsymbol{\Lambda}\right)$ distribution. Given a set of eigenvalues $\lambda_1, \cdots, \lambda_d$ along the diagonal of $\boldsymbol{\Lambda}$, this simulation is straightforward by drawing standard normal vectors and rescaling the jth entry by λ_j for $j = 1, \cdots, d$. Note that for high d this is computationally much more efficient than using packaged multivariate Gaussian data generation functions, because those tend to entail working through a $d \times d$ covariance matrix, as discussed in Section 17.1.4.

It remains to estimate $\boldsymbol{\Lambda}$, which is challenging in the case of $d > n$. There are two major problems in this case. The first is that only the first n eigenvalues (or maybe $n - 1$ when mean centering has been done) of the sample covariance matrix are nonzero. The second is that the nonzero sample eigenvalues tend to overestimate their corresponding underlying theoretical eigenvalues, because the additional energy (quantified as sums of squares about the mean as discussed in Section 3.1) generated by the theoretical eigenvalues that are not represented gets added in. This seems to result in a conservative null distribution, first pointed out in Liu et al. (2008), because simulating data in a lower-dimensional subspace results in a smaller CI despite the larger eigenvalues. An open problem is theoretical verification that use of the sample covariance matrix $\hat{\boldsymbol{\Sigma}}$ (defined in (3.5)) always leads to conservative inference in SigClust.

Some assumptions need to be made to address this. A common assumption, which makes sense in many (but it is important to note not all) real data situations is that there is essentially a strong low-rank signal together with some relatively small scale noise that is *isotropic*, i.e. has the same variance in all directions. One model for isotropic noise is a Gaussian distribution with all eigenvalues the same (i.e. diagonal covariance matrix). However this holds for any multivariate distribution generated by independent marginals with the same variance (not even the same distribution is needed). To see why, note that for independent random variables X_1, X_2 each with variance σ^2, the variance of the projection of (X_1, X_2) onto the arbitrary unit direction $(\cos \theta, \sin \theta)$ is

$$var\left(X_1 \cos \theta + X_2 \sin \theta\right) = var\left(X_1\right) \cos^2 \theta + var\left(X_2\right) \sin^2 \theta = \sigma^2. \quad (13.1)$$

Similar calculations are straightforward for directions in higher dimensions. In the isotropic errors case, the problem of estimating Λ has been effectively solved by Huang et al. (2015).

An example where the assumption of a strong low rank signal plus additive isotropic noise makes sense is the Pan-Cancer gene expression data studied in Section 4.1.4. Here only the $n = 50$ Kidney Cancer cases (blue diamonds in Figure 4.9) are studied. The data objects are vectors where each of the $d = 12478$ entries represents activity of a gene. These entries are essentially counts measured with random variation. Those counts are large enough that on the log scale the random variation is roughly Gaussian. This (appropriate for high dimension eigenvalue estimation by SigClust) behavior is verified using the diagnostic plot in the left panel of Figure 13.8. That graphic studies the distribution of all $50 \times 12478 = 623,900$ entries of the data matrix (called pixels in titles in Figures 13.8 and 13.11). Because the full jitter plot of this data set would result in massive overplotting, just a random sample of 5000 green dots is shown. Those points show a distribution which is highly concentrated near 0 with some substantial outliers. The latter is driven by the low-rank signal structure (much of which is appropriately captured by the first principal components), while the smaller scale noise appears in the middle. This behavior is nicely reflected by the blue curve which is a kernel density estimate (discussed in more detail in Section 15.1). That suggests the random noise part of this distribution is approximately Gaussian (i.e. suggests strong potential for an isotropic background noise distribution). The variance of this background noise distribution is very useful for essentially interpolating those zero sample eigenvalues.

The full sample standard deviation (about 1.15 for this data set) provides an inappropriate estimate of the background standard deviation, because it strongly feels the relatively few, but quite large, signal values. Scale measures that are much less affected by large values have been developed in the field of *robust statistics*, which is overviewed in Chapter 16. For the estimation of the background noise standard deviation, it is desirable to use a method which nearly ignores values outside the quartiles such as the (Gaussian rescaled) Median Absolute Deviation (MAD) from the median, $\widehat{\sigma}_{MAD}$. For a data set of size m (in Figure 13.8

Robust Fit Gaussian Q-Q, All Pixel values

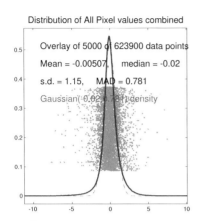

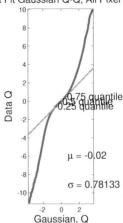

Figure 13.8 *SigClust diagnostics for the Kidney Cancer type from the Pan-Cancer data. Left panel studies the distribution of all entries of the data matrix, suggesting an isotropic Gaussian background noise distribution. Right panel is fit Gaussian Q-Q plot indicating good fit of the suggested distribution within the quartiles.*

$m = d \times n$), $\widetilde{x}_1, \cdots, \widetilde{x}_m$, this is defined as

$$\widehat{\sigma}_{MAD} = \operatorname*{median}_{i=1,\cdots,m} \left(\left| \widetilde{x}_i - \operatorname*{median}_{j=1,\cdots,m} (\widetilde{x}_j) \right| \right) / \Phi^{-1}(0.75), \qquad (13.2)$$

where Φ denotes the standard Gaussian cumulative distribution function. The purpose of that rescaling by $\Phi^{-1}(0.75)$ (which is the theoretical MAD of the standard Gaussian distribution) is to put the $\widehat{\sigma}_{MAD}$ on a Gaussian scale in the sense that when applied to a large Gaussian data set it appropriately estimates the population standard deviation. The Gaussian density with standard deviation taken to be $\widehat{\sigma}_{MAD} = 0.781$ (and centered at the sample median $\widehat{\mu}_{MED} = -0.02$ which similarly ignores the outlying values) is shown using the red dashed curve in the left panel of Figure 13.8. Note that it appears to fit the central "background noise" part of the distribution quite well. A more precise view of this goodness of fit appears in the QQ plots shown in the right panel of Figure 13.8. This version of the QQ plot shows the sorted data $\widetilde{x}_{(1)}, \cdots, \widetilde{x}_{(m)}$ on the vertical axis versus the corresponding estimated Gaussian quantiles $\widehat{\mu}_{MED} + \widehat{\sigma}_{MAD}\Phi^{-1}\left(\frac{1}{m+1}\right), \cdots, \widehat{\mu}_{MED} + \widehat{\sigma}_{MAD}\Phi^{-1}\left(\frac{m}{m+1}\right)$ on the horizontal axis, as the red curve. When the given distribution fits the data, the QQ curve should roughly follow the green curve, which is the 45-degree line. In this case the fit is good for the part of the data between the quartiles, but quite bad for the outlying data points (which again represent important signal in the data) suggesting that the background noise standard deviation has been effectively estimated.

Use of this isotropic background noise estimate to provide a reasonable full set of estimated eigenvalues is shown in Figure 13.9. Each panel shows various eigenvalues on different scales. The top two panels study the full set of $d = 12478$ eigenvalues, while the bottom panels are the same as the top, but are zoomed into just the first 100 eigenvalues (as the structure of those important eigenvalues is essentially squished into the vertical axis because $100 \ll 12478$). Both panels on the left use the original variance scale on the vertical axis, while the $\log_{10}$ scale is used in both right-hand panels. In all panels the sample eigenvalues (this is the scree plot introduced in Figure 3.5) are shown using black circles. Note that there a few extremely large eigenvalues and many much smaller ones. Only $49 = n - 1$ of the log scale sample eigenvalues appear (because the $\log_{10} 0$ is undefined and thus not plotted).

The magenta horizontal line shows the background variance that is estimating the variance of the noise part of the model. An important goal of the eigenvalue estimation is to keep the total energy of the estimated eigenvalues (i.e. their sum) equal to the total energy of the data (i.e. the sum of squares about the mean, which is the sum of the eigenvalues). Replacing the 0 sample eigenvalues with the estimated background variance adds a lot to the total, which thus motivates some shrinkage of the nonzero sample eigenvalues. Appropriate shrinkage formulas have been developed in Huang et al. (2015), and the resulting estimated eigenvalues are shown as the red dashed curve in each panel of Figure 13.9. Note that on the right side (indeed most) of each panel, this dashed red curve is the same as the magenta line. It rises for the first few eigenvalues, but represents substantial shrinkage from the black circle sample eigenvalues, again with the goal of making the total energy the same. For some data sets, the total energy in the sample eigenvalues (their sum) may be less than d times the background variance. In this case, the background noise is clearly not isotropic so this estimation scheme cannot be used. Hence one must resort to simply using the sample eigenvalues. As noted in Liu et al. (2008), this results in a conservative SigClust, so the result is valid but the test may be less powerful in those situations.

The results of using SigClust to investigate splitting the Kidney Cancer data from Section 4.1.4 (thus working in the second mode described above) are summarized in the p-value plots of Figure 13.10. The left panel uses Gaussian data simulated from the carefully estimated eigenvalues shown as the red dashed curve in Figure 13.9. Note that none of the Gaussian simulated black dots lies to the left of the vertical green line, giving an empirical p-value of 0. This is strong evidence that there are at least two clusters in the Kidney Cancer data indicating the existence of cancer subtypes, which is consistent with the results of Ricketts et al. (2018) and with the analysis in Section 15.4.4.

The right panel of Figure 13.10 shows the result of replacing the careful eigenvalue estimates (the red dashed curve in Figure 13.9) with the sample eigenvalues (the black circles). With that estimate the p-value is only about 0.14, no longer statistically significant in the classical 0.05 sense. So this is an example where the method of eigenvalue estimation clearly matters and the sophisticated approach of Huang et al. (2015) is worthwhile. It also is an example of the conservative nature

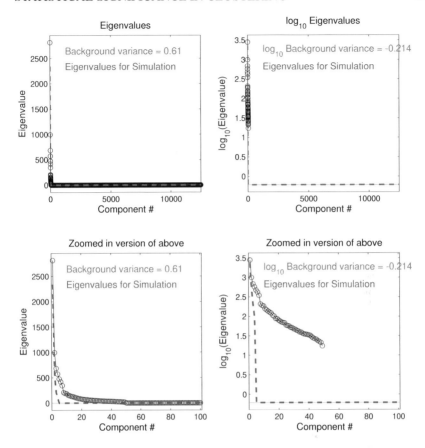

Figure 13.9 *Various estimated eigenvalues for the Kidney Cancer data. Sample eigenvalues are shown as black circles. Improved estimates, that both incorporate the estimated isotropic background noise variance (shown as the horizontal magenta line) and also keep the total energy constant, are shown as the red dashed curve.*

(especially in high dimensions) of the use of naive sample eigenvalues in SigClust discussed above.

For effective use of SigClust it is also important to realize there are cases where the sophisticated shrinkage eigenvalue estimates demonstrated in Figure 13.9 are inappropriate. An example of this is the Lung Cancer data studied in Section 4.1.3 where clusters were seen to be of keen interest. Figure 13.11 investigates whether the red and blue clusters visually brushed in Figure 4.7 are a statistically significant separation using the confirmatory version of SigClust (i.e. based on given, in this case manually selected, cluster labels). The left panel is the first SigClust diagnostic plot showing the distribution of all entries in the data matrix, which is far different from that shown in Figure 13.8. While both distributions are based on logs of counts (plus 1), here there are far more 0s and many other very small

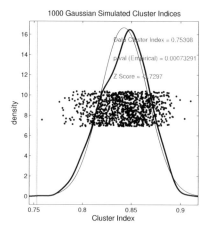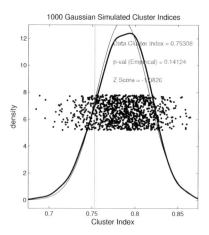

Figure 13.10 *SigClust p-values for the Kidney Cancer data. Left panel shows that careful eigenvalue estimation (red dashed curve in Figure 13.9) gives a statistically significant result, indicating this cancer type has potentially important subtypes. Right panel reveals that naive sample eigenvalue estimation (black circles) gives a too conservative (not even significant) result in this same case.*

discrete values (appearing at $\log_{10}$ of small positive integers), so no Gaussian background distribution is apparent. In this case it is recommended to use the conservative sample covariance eigenvalues in the application of confirmatory Sig-Clust. One can plunge ahead with the formulas developed above for the isotropic background variance case, but there are serious consequences for this data set. In particular d times the estimated background noise is now larger than the total sum of squares about the mean (energy in the data), so the estimated eigenvalues are all simply the background variance, i.e. the dashed red curve in the analog of Figure 13.9 is completely flat. This puts the simulated Gaussian data at the heart of the SigClust p-value computation into the HDLSS isotropic domain that will be studied in Section 14.2, where all simulated data are essentially equidistant from each other, as characterized in (14.5). This results in simulated CIs being very similar to each other, so the Z-score (whose denominator is the standard deviation of the black dots) is unrealistically negative (in the thousands in this case, since the standard deviation of the black dots is so small). Hence this violation of the isotropic background noise distribution can result in a seriously anti-conservative SigClust result. Thus the diagnostics are important in high-dimensional situations.

The result using the sample eigenvalues, recommended when the isotropic background diagnostics fail, is shown in the right panel of Figure 13.11. This shows that the red and blue clusters in Figure 4.7 are clearly statistically significant. This is consistent with the explanation of those clusters from the discussion of Figure 4.8.

Figure 13.12 shows another analysis of the Lung Cancer data studied in Figures 4.5–4.8. An application of SigClust, to separate the blue cluster from the union of the red and gold (shown as an intermediate orange color in Figure 13.12) gave a p-value of 1. The p-value plot looks much like the right panel of Figure 13.7,

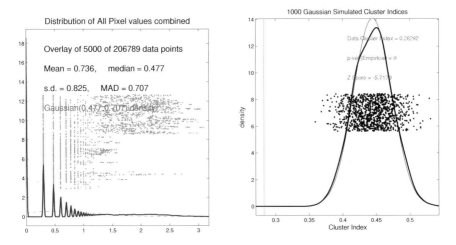

Figure 13.11 *SigClust for the red vs. blue clusters in the Lung Cancer data of Figures 4.5–4.8. Diagnostic plot in the left panel reveals no underlying Gaussian background noise. SigClust using sample covariance eigenvalues shows these clusters are statistically significant in the right panel.*

and hence is not shown here. The scatterplots of Figure 13.12 show that this was caused by the orange cluster really consisting of two clusters so the WCSS is much bigger than the WCSS that results from just splitting all of the data. This shows that SigClust is best used together with data visualization, not as a blind hypothesis test. This data set also provides an example where the DWD version of DiProPerm (which had a Z-score of 55.3) gave a much stronger result than the MD version (Z-score of 9.2). The left panel of Figure 13.12 shows the projections

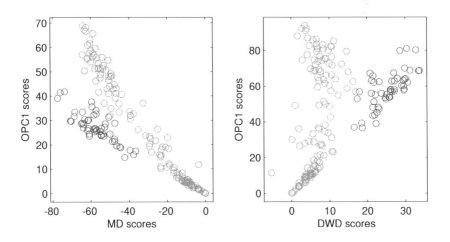

Figure 13.12 *Scatterplots of projections of Lung Cancer data from Figures 4.5–4.8.*

of the data onto the orthogonal MD direction. Note that MD gives a rather poor separation of the two classes. On the other hand, the projection onto the DWD projection in the right panel (again together with the orthogonal PC1 direction) shows much better class separation which is why the DWD Z-score is so much better. Of course the use of DiProPerm is not really appropriate here as these clusters were visually chosen.

Other approaches to testing significance of clustering can be found in McShane et al. (2002), Tibshirani et al. (2001); Tibshirani and Walther (2005) and Suzuki and Shimodaira (2006). A different approach is Bayesian inference based on normal mixtures, as in Richardson and Green (1997) and Bouveyron et al. (2019). Yet another interesting way to think of importance of clusters is via their stability using *consensus clustering* proposed by Monti et al. (2003), with a Bayesian variation by Lock and Dunson (2013).

CHAPTER 14

High Dimensional Asymptotics

A key insight of classical mathematical statistics is that asymptotic analysis provides a useful tool for understanding the fundamentals that underlie complicated statistical contexts. Limiting operations as the sample size $n \to \infty$ have been a workhorse approach that has revealed very many foundational statistical discoveries, e.g. those related to laws of large numbers and the central limit theorem. These have been so useful that some have developed the mistaken idea that asymptotics should only be about growing sample sizes and their impact on statistical analyses. A central theme of this chapter is that in the modern era of very large and complex data sets, many less familiar but very useful insights are available from the use of more general limiting operations involving *both* the sample size n and the dimension d.

As noted in Shen et al. (2016b,a), there are several *domains* of quite different and important asymptotic insights available, that are usefully indexed by the relationship between the sample size n and the dimension d. Important early work on the asymptotics of high-dimensional statistical methods was done by Portnoy (1984, 1985, 1988), who studied various robust methodologies using the limit as $n \to \infty$ together with the dimension d growing at the rate $d \sim n^{1/2}$. A different domain is the large literature on random matrix theory, where n and d grow at the same rate, some basics of which are briefly reviewed in Section 14.1. Perhaps surprising is the idea that anything useful at all can be shown in yet another domain where $d \to \infty$ for fixed n (i.e. growing and hence increasing noise, but a fixed number of data objects), as discussed in Section 14.2. A few ideas about the case where both grow, but d grows more quickly than n are studied in Section 14.3.

It is worth noting that at the time of this writing, a number of the ideas and views expressed here are directly contrary to the currently prevailing opinion of the theoretical statistics community. The latter is perhaps summarized in Section 8.3 of Wainwright (2019): 'does PCA still perform well in the high-dimensional regime $n < d$? The answer to this question turns out to be a dramatic "no".' That view notably clashes with the experience of practitioners who actually analyze data: PCA frequently gives useful visual insights for very high-dimensional data. Clear examples in this book include the Lung Cancer data in Section 4.1.3, the Pan-Cancer data in Section 4.1.4, and the GWAS data in Section 16.2.2. The resolution of this apparent paradox, as well as some others also based on such misinterpretations of the asymptotic theory, can be found in Section 14.2.

DOI: 10.1201/9781351189675-14

14.1 Random Matrix Theory

Random matrix theory is centered on the eigenvalues of the sample covariance of a $d \times n$ matrix of independent standard Gaussian random variables. Motivated by a model for the nuclei of heavy atoms in nuclear physics, Wigner (1955) first studied square $d \times d$ matrices of more general independent random variables (with common variance) as $d \to \infty$, including providing a semi-circle law for the distribution of the eigenvalues of the random matrix. Marčenko and Pastur (1967) explored the asymptotic distribution of the eigenvalues (studied around (17.21)) of sample covariance matrices (defined in (3.5)). Those matrices are based on i.i.d. $N_d \left(\mathbf{0}_{d,1}, \sigma^2 \mathbf{I}_d \right)$ (using the notation of (4.2), $\mathbf{0}_{d,1}$ is the $d \times 1$ vector of zeros) random vectors $\widetilde{z}_1, \cdots, \widetilde{z}_n \in \mathbb{R}^d$. The limiting operation lets both $n \to \infty$ and $d \to \infty$ with $\frac{d}{n} \to c$. The density of the asymptotic eigenvalue distribution is given in (14.1). That distribution tends to Wigner's semi-circle law as $c \to 0$ and also (ignoring zero eigenvalues) as $c \to \infty$.

Figure 14.1 provides an illustration of the basic ideas. For simplicity, all panels are based on a $d \times n$ matrix $\widetilde{X}$ of independent standard Gaussian random variables where the common variance σ^2 is taken to be 1. An approximation of the sample covariance matrix, which is particularly simple to work with, is the outer product $\widetilde{\Sigma} = \frac{1}{n} \widetilde{X} \widetilde{X}^t$. This is not quite the $d \times d$ sample covariance matrix (when thinking of the columns of $\widetilde{X}$ as data objects), because the sample mean has not been subtracted. However, because the expected value of each entry is 0, this is still a reasonable estimate of the underlying theoretical covariance matrix, $\Sigma = \mathbf{I}_d$. In fact it is the maximum likelihood estimate of Σ under the assumption that the sample mean is $\mathbf{0}_{d,1}$. The left panel of Figure 14.1 shows scree plots (recall from Figure 3.5 this is a plot of the sorted eigenvalues of $\widetilde{\Sigma}$ as a function of the PC index $j = 1, \cdots, d$) in the case $d = 100$, for several values of n. Since $\widetilde{\Sigma}$ estimates $\mathbf{I}_d$, each eigenvalue can be viewed as an estimate of 1, shown as the horizontal dashed line. For the very large $n = 10,000$ (shown in green) the eigenvalues are rather close to 1, although there is some natural variation as seen. That variation grows as n is reduced to 1000 (shown in red) and to 300 (blue). There is a growing asymmetry which becomes especially strong at $n = 100$ (black). The asymmetry can be thought of as being caused by the constraint that eigenvalues must be non-negative (actually positive in this case).

While the left panel of Figure 14.1 explores differing matrix shapes, the case of constant matrix shape is considered in the center panel. The shape is constant in the sense that $\frac{d}{n} = c$, with $c = \frac{1}{5}$ there. While sampling variation is visible for the magenta eigenvalues for $n = 500$ and $d = 100$, it diminishes for the cyan case where $n = 2000$ and $d = 400$, and looks essentially like a smooth curve for the yellow $n = 8000$ and $d = 1600$. To make this apparent convergence occur, the horizontal axis has been rescaled to $\frac{j}{d}$ for each case. The right panel shows a useful conceptual viewpoint of the limit of this process, by studying the distribution of the cyan ($n = 2000$ and $d = 400$) eigenvalues. The vertical axis shows the binning of these into a 50 bin histogram. That histogram approximates the $c = \frac{1}{5}$ version of the Marčenko-Pastur distribution, which essentially reflects the limit of the apparent curves in the center panel as $n, d \to \infty$ (with $\frac{d}{n} = \frac{1}{5}$). Good overview of many aspects of Marčenko-Pastur theory can be found in the monographs Bai and Silverstein (2010) and Yao et al. (2015).

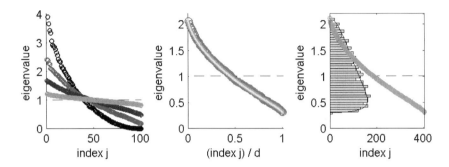

Figure 14.1 *Scree plots motivating Marčenko-Pastur distributions, using i.i.d. standard Gaussian samples. Left panel illustrates differing distributional shape over $n = 100$ (black), $n = 300$ (red), $n = 1000$ (blue), and $n = 10,000$ (green), all for $d = 100$. Center panel shows stability of distribution for common $c = \frac{d}{n} = \frac{1}{5}$, in the cases $n = 500$ (magenta), $n = 2000$ (cyan), and $n = 8000$ (yellow), using decreasing symbol size for clear visibility. Right panel shows how binning of eigenvalues generates a histogram estimate of the Marčenko-Pastur distribution (thin curve) for the cyan $n = 2000$, $d = 400$ case.*

Generally in that limit as $n, d \to \infty$ with $\frac{d}{n} = c$ the vertical histograms, as illustrated in the right panel of Figure 14.1 converge to the *Limiting Spectral Density* (LSD) shown as the thin curve, which is the density of the Marčenko-Pastur distribution indexed by c. First consider the case $c < 1$ (i.e. $d < n$), where the sample covariance matrix $\widehat{\Sigma}$ has no zero eigenvalues (with probability one). Figure 14.2 shows the LSD for several values of c. In the more general case where the Gaussian variance is σ^2, the LSD has the form

$$LSD_{c,\sigma^2}(x) = \frac{1}{2\pi c \sigma^2 x} \sqrt{(b-x)(x-a)} 1_{[a,b]}(x), \qquad (14.1)$$

where $1_{[a,b]}$ denotes the indicator function (defined at (11.2)) on the interval $[a, b]$, $a = \sigma^2 \left(1 - \sqrt{c}\right)^2$ and $b = \sigma^2 \left(1 + \sqrt{c}\right)^2$. This probability distribution has the perhaps surprisingly simple expected value σ^2 and variance $c\sigma^4$. The $LSD_{c,1}(x)$ curves (note the case $\sigma^2 = 1$) for several values of c are plotted in Figure 14.2.

The red curve shows the case $c = \frac{1}{3} = \frac{100}{300}$, which gives the LSD corresponding to the red scree plot in the left panel of Figure 14.1. The lower density height on the right corresponds to the increased spacing between red circles on the left side of the scree plot. The upper end of the support of the LSD, $b = \left(1 + \sqrt{\frac{1}{3}}\right)^2 \approx 2.49$, also fits well with the top red circle. The higher density to the left of 1.0 is essentially caused by the eigenvalues being constrained to be positive. Note that there is a distinct lower support point $a = \left(1 - \sqrt{\frac{1}{3}}\right)^2 \approx 0.18$ to the LSD as well, i.e. even in the limit as $n, d \to \infty$ (with $c = \frac{d}{n} = \frac{1}{3}$) these eigenvalues are bounded above 0. Another aspect of the LSD that is visually clear is that it is vertical at the support points, in the sense that its slope is infinite there (recall the unit semi-circle has the same property where it intersects that horizontal axis). This can be mathematically checked by calculating $\lim_{x \downarrow a}$ and $\lim_{x \uparrow b}$ of the derivative of the LSD.

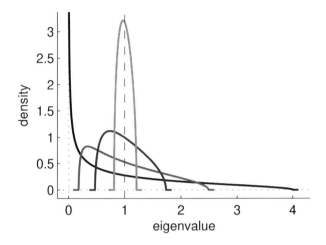

Figure 14.2 *Marčenko-Pastur distribution LSDs for examples in the left panel of Figure 14.1 using same colors; black for $c = 1$, red for $c = \frac{1}{3}$, blue for $c = 0.1$, and green for $c = 0.01$. Shows wide array of shapes of this distribution.*

The black curve in Figure 14.2 is the LSD for $c = 1$, which explains the structure of the black scree plot in the left panel of Figure 14.1. Note that the highest black circle is around the upper support point $b = \left(1 + \sqrt{1}\right)^2 = 4$. This LSD is very strongly skewed and in fact has a pole at its lower support point $a = \left(1 - \sqrt{1}\right)^2 = 0$. That again can be checked by calculating the $\lim_{x \downarrow 0} LSD_{1,1}(x)$.

The case of $c = 0.1 = \frac{100}{1000}$ is the blue curve, which models the blue scree plot in the left of Figure 14.1. As the larger sample size $n = 1000$ results in the blue circles having much less variation, the support points have moved toward 1 (indicated by the vertical dashed line in Figure 14.2). In particular, they are now $a = \left(1 - \sqrt{0.1}\right)^2 \approx 0.47$ and $b = \left(1 + \sqrt{0.1}\right)^2 \approx 1.73$, which fits with the top and bottom blue circles in the left panel of Figure 14.1. Also note that the shape of the LSD has now changed substantially, in particular it is concave everywhere, versus having the convex region that was present for the smaller values of c.

The support points come even closer to 1 in the case $c = 0.01$ shown as the green curve, again predicting the structure of the green scree plot in the left of Figure 14.1. Note that now the shape is becoming rather symmetric, as it converges to a rescaled version of Wigner's semi-circle distribution as discussed above. Also the support points $a = \left(1 - \sqrt{0.01}\right)^2 \approx 0.81$ and $b = \left(1 + \sqrt{0.01}\right)^2 \approx 1.21$ are now becoming closer to equidistant to 1.

A parallel Marčenko-Pastur theory for the alternate matrix shape case of $c > 1$, i.e. $d > n$, can also be derived. In this case the sample covariance matrix $\widehat{\boldsymbol{\Sigma}}$ (defined at (3.5)) has (with probability one) $d - n + 1$ zero eigenvalues (the $+1$ is due to object mean centering). Hence the limiting spectral distribution of the eigenvalues is a mixture of a discrete point mass at 0 (with weight $1 - \frac{1}{c}$) together with a continuous density which is a rescaling of (14.1) (with weight $\frac{1}{c}$). Using Singular Value Decomposition ideas as in Section 17.1.2, the nonzero eigenvalues of $\widetilde{\boldsymbol{X}}\widetilde{\boldsymbol{X}}^t$ are the same as the eigenvalues of the corresponding *Gram Matrix* $\widetilde{\boldsymbol{X}}^t\widetilde{\boldsymbol{X}}$, i.e, the matrix of inner products of the columns of $\widetilde{\boldsymbol{X}}$. The normalization factor of $\frac{1}{n}$ in the sample covariance matrix results in diagonal entries which are sample variance estimates (of the underlying theoretical variances which are all 1) computed as sums of squares of n Gaussians. In the Gram Matrix case, the n diagonal entries are now sums of squares of d Gaussians, so a normalization of $\frac{1}{d}$ is needed for these n diagonal entries to estimate 1, i.e. the sample variances essentially estimate $c = \frac{d}{n}$. This calculation reflects the fact that for $d \gg n$ there is far more variation being summarized by each of the n nonzero eigenvalues in the $d \times n$ matrix. Hence, the nonzero eigenvalues are roughly $c = \frac{d}{n}$ times those following $LSD_{\frac{1}{c}, \sigma^2}$. Several elegantly symmetric facts follow from this. First considering the mixture structure, the expected value of the full mixture LSD is still σ^2 (this comes from $\left(1 - \frac{1}{c}\right)0 + \frac{1}{c}(c\sigma^2)$) and similarly the variance remains at $c\sigma^4$. Second the above formulas for support points $a = \sigma^2\left(1 - \sqrt{c}\right)^2$ and $b = \sigma^2\left(1 + \sqrt{c}\right)^2$ require no modification in this new case of $c > 1$. Finally the continuous density part of the LSD has exactly the same form (14.1) scaled by $\frac{1}{c}$.

This theory can be used in the analysis of hierarchical clustering in Section 12.2 to understand the standard deviation of PC1 of the Overlapping Classes data in Figure 4.12. As the largest principal component, the PC1 variance should be at the upper support point of the Marčenko-Pastur LSD. In particular, that is $b = \sigma^2\left(1 + \sqrt{c}\right)^2 = 1^2\left(1 + \sqrt{\frac{1000}{100}}\right)^2 \approx 17.3$. Taking the square root (to get back to the standard deviation) gives 4.16 which fits well with the spread of points in the upper left panel of Figure 4.12 and the center panel of Figure 4.13.

A novel data-based approach to choosing the number of non-noise principal components, which is based on Marčenko-Pastur mathematics, has been proposed by Choi and Marron (2018).

Random matrix theory in general was brought to the statistics literature by Johnstone (2001). A particularly useful tool has been the null distribution of the largest eigenvalue, derived by Tracy and Widom (1994). See Zhou et al. (2018a,b) for a deep application of the Tracy-Widom distribution in genetics.

An interesting issue is the extent to which the assumption of Gaussianity can be weakened and still obtain the Marčenko-Pastur distribution for the PCA eigenvalues. Figure 14.3 explores this using scree plots based on three data sets with $d = 100$ and $n = 1000$. Each data set consists of a matrix of independent random variables all with expected value 0 and variance 1. But the data sets differ in terms of the marginal, with a Gaussian marginal scree plot shown as black circles, an exponentially distributed plot shown as magenta plus signs, and the Bernoulli scree plot in green x signs. Note that all three are very similar, except for some natural sampling variation. In particular the Marčenko-Pastur LSD is not affected by either the strong skewness of the exponential distribution or the strong discreteness of the Bernoulli. Except in the case of heavy tails, marginal distributions matter very little, but essentially independence of all entries of the matrix (thus in particular independence of the variables forming the data vectors) is crucial.

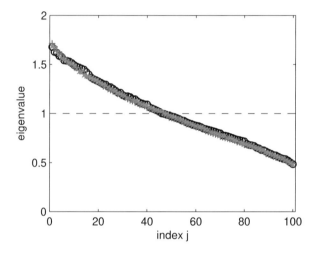

Figure 14.3 *Scree plots indicating LDS is the same for different marginal distributions. Gaussian is black circles, Exponential shown as magenta plus signs, Bernoulli using smaller green x signs.*

However there are important situations where classical Marčenko-Pastur theory does not hold, such as scale mixtures of Gaussians. The simplest of these are of the form

$$f\left(\boldsymbol{x}\right) = \omega\phi_d\left(\boldsymbol{x}, \sigma_1^2\right) + (1 - \omega)\,\phi_d\left(\boldsymbol{x}, \sigma_2^2\right) \qquad (14.2)$$

where the weight $\omega \in (0, 1)$ and where $\phi_d\left(\boldsymbol{x}, \sigma^2\right)$ denotes the multivariate Gaussian density with expected value $\boldsymbol{0}_d$ and covariance matrix $\sigma^2\boldsymbol{I}_d$. Recall from the discussion around Figure 11.14 that mixtures of Gaussians are quite natural in applications where data heterogeneity is an important issue. Note that scale mixtures of normals also provide good illustration of the important point that 0 covariance does not imply independence. In particular the covariance matrix of (14.2) can be shown to be a multiple of the identity matrix (so all pairwise covariances are 0), yet entries of the vector can be strongly dependent. E.g. when $\sigma_1^2 \gg \sigma_2^2$ a large entry suggests the vector came from the first component, so all other entries tend to be relatively large as well. Li and Yao (2018) have discovered an analog of the Marčenko-Pastur distribution for more general scale mixtures of normals.

In simple cases, these can be computed using the numerical solution of integral equations of Dobriban (2015) (see Choi and Marron (2018) for a gentle introduction).

14.2 High Dimension Low Sample Size

As noted in the survey paper Aoshima et al. (2018), the study of *High Dimension Low Sample Size* (HDLSS) asymptotics, where $d \to \infty$ for fixed n, was popularized by Hall et al. (2005). However, earlier work on that type of asymptotics was done in Casella and Hwang (1982), in the context of Stein estimation.

Interesting HDLSS insights arise even in the apparently straightforward case of a single standard Gaussian random vector

$$\widetilde{z} = \begin{pmatrix} \widetilde{z}_1 \\ \vdots \\ \widetilde{z}_d \end{pmatrix} \sim N_d \left(\mathbf{0}_{d,1}, \mathbf{I}_d \right).$$

It is natural to think that observations drawn from this distribution lie relatively near the origin, because the probability density

$$f_Z \left(\widetilde{z} \right) = (2\pi)^{-d/2} \, e^{-\widetilde{z}^t \widetilde{z}/2} \tag{14.3}$$

has a peak there and decreases very rapidly in all directions as $\left\| \underset{\to}{z} \right\| \to \infty$. However, using the notation (7.2), the Euclidean distance from $\widetilde{z}$ to the origin is

$$\delta_2 \left(\widetilde{z}, \mathbf{0}_{d,1} \right) = \left(\sum_{j=1}^d \widetilde{z}_j^2 \right)^{1/2},$$

whose distribution is easy to understand, since $\sum_{j=1}^d \widetilde{z}_j^2$ has a Chi-Squared distribution with d degrees of freedom. That distribution has expected value d and variance $2d$ (i.e. standard deviation $\sqrt{2d}$). A simple Taylor series (sometimes called the *delta method*) calculation shows that in the limit as $d \to \infty$ (for fixed n)

$$\delta_2 \left(\widetilde{z}, \mathbf{0}_{d,1} \right) = \sqrt{d} + O_p \left(1 \right). \tag{14.4}$$

This shows that most of the distribution of $\widetilde{z}$ lies near the surface of a growing sphere of radius $\sqrt{d}$ in $\mathbb{R}^d$, and perhaps paradoxically there is essentially no chance of finding $\widetilde{z}$ anywhere near the origin, $\mathbf{0}_{d,1}$. This apparent contradiction to the density (14.3) being much larger near the origin is resolved by recalling that this is density *with respect to Lebesgue measure*. To see how Lebesgue measure can give perhaps non-intuitive results such as (14.4) consider the calculation of the volume of the (solid) unit ball in $\mathbb{R}^d$. This is most easily calculated as an integral using polar coordinates. Writing the integral with respect to radius last, that integrand goes up extremely rapidly near the end of the interval $[0, 1]$ for large d. This shows that high-dimensional Lebesgue measure tends to strongly push probability mass outwards. The balance point between this effect and the rapidly diminishing tails of the Gaussian density (14.3) to keep mass near the origin is seen to be at

$\sqrt{d}$ in (14.4). This has interesting implications for individuals in that when enough measurements are made (i.e. d is sufficiently large) it is essentially impossible for any individual to be "average", i.e. near the population expected value vector. The result (14.4) also explains the perhaps surprising observation in the discussion of Figure 12.8 that two standard Gaussian data points are so far from the rest of the data. Recall that $d = 500$ and $n = 30$ for that High-Dimensional Gaussian data set. Thus $\sqrt{500} \approx 22.4$ well explains the distances from the two "outlying" points to the origin in the first diagonal panels of Figure 12.8.

A result in the same spirit as (14.4) holds for independent $\tilde{z}_1, \tilde{z}_2 \sim N_d(\mathbf{0}_{d,1}, \mathbf{I}_d)$,

$$\delta_2(\tilde{z}_1, \tilde{z}_2) = \sqrt{2d} + O_p(1). \tag{14.5}$$

This is not surprising since the difference of independent standard normals is again normal with standard deviation $\sqrt{2}$. Applying this idea to a sample $\tilde{z}_1, \cdots, \tilde{z}_n$ results in the idea that all of these points are equidistant from each other. Thus for $n = 3$, they lie at vertices of an equilateral triangle (a similar phenomenon was discovered by Kendall (1988) in the context of shape diffusion), and for $n = 4$ at vertices of a regular tetrahedron. Even conceptualizing this for $n > 4$ is challenging because the point configuration lies in a space with dimension higher than 3. Humans tend to be good at perception in three dimensions, but quite poor at higher dimensional perception. This may be because our perceptual systems come from our ancestors for whom a major task was finding food. Food exists in three dimensions which seems to be why we are so good at understanding three-dimensional space.

One more related result in this spirit comes from studying the angle (with vertex at the origin) between independent vectors $\tilde{z}_1, \tilde{z}_2 \sim N_d(\mathbf{0}_{d,1}, \mathbf{I}_d)$,

$$\angle(\tilde{z}_1, \tilde{z}_2) = 90° + O_p\left(d^{-1/2}\right). \tag{14.6}$$

Note that this right angle completes the Pythagorean relationship in terms of the right triangle with both leg lengths $\sqrt{d}$ from (14.4) and hypotenuse $\sqrt{2d}$ from (14.5). The orthogonality of data directions observed in the High-Dimensional Gaussian data in the top center panel of Figure 12.8 is also explained by (14.6).

Equations (14.4), (14.5), and (14.6) have some interesting implications for the very nature of high-dimensional variation. While letting $d \to \infty$ (for a fixed set of n data objects) clearly adds more and more noise to a Gaussian data set, that noise manifests itself in a rather special way. In particular, given a sample $\tilde{z}_1, \cdots, \tilde{z}_n$ consider the subspace of dimension n generated by the data. This is a hyperplane of dimension n passing through the origin. This plane can be rotated in $\mathbb{R}^d$ to the subspace of the first n coordinates. Within that subspace, all data points are essentially distance $\sqrt{d}$ from the origin, and $\sqrt{2d}$ from each other and can be again rotated so that each data point is close to a coordinate axis. This results in the data lying near the vertices of the $\sqrt{d}$ rescaling of the n-dimensional *unit simplex* in $\mathbb{R}^d$. Hence while increasing d creates increasing variation, that goes mostly into very random rotation. Modulo that rotation and scaling, the data actually converge to a rigid deterministic structure. This HDLSS phenomenon has been called *geometric representation* by Hall et al. (2005).

A simple illustration of geometric representation is shown in Figure 14.4. The top left panel shows a conceptual model of $n = 2$ points in $\mathbb{R}^3$ as red circles. The top right panel shows the subspace generated by these points (i.e. the 2-d plane containing them and the origin) in cyan. Red solid lines are also used to depict the concept from (14.4) that each point is approximately distance $\sqrt{d}$ from the origin. Similarly a dashed line illustrates the roughly $\sqrt{2d}$ distance between points from (14.5). The first step of geometric representation is to rotate this cyan subspace into the coordinate system of the first $n = 2$ axes in $\mathbb{R}^3$ as shown in the lower left panel of Figure 14.4. The second rotation happens within this $\mathbb{R}^2$, to move the points close to the axes, resulting in the unit simplex (in this case just a line segment) approximation.

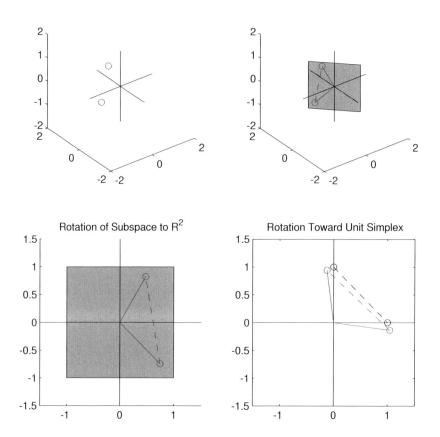

Figure 14.4 *Illustration of HDLSS geometric representation using just two data points. Shows how added noise from increasing dimensionality goes mostly into random rotation, while modulo rotation data points (red circles) converge to vertices of a rescaled unit simplex (black circles).*

Another illustration of geometric representation is shown using a small simulation study in Figure 14.5. Each panel shows seven (the number of colors in Matlab's default color palette) $N_d\left(\mathbf{0}_{d,1}, \boldsymbol{I}_d\right)$ data sets of size $n = 3$, in the dimensions d shown above each panel. Each data set determines a hyperplane of dimension 2, which is rotated into the $\mathbb{R}^2$ shown in each panel. Next within that $\mathbb{R}^2$ each data set is rotated so the first two data points are equidistant from the vertical axis, with the minimum of those two below the origin. These two operations result in modding out the increasing amount of random rotation, so the points coalesce to vertices of the 3-dimensional unit simplex, which in this view are the vertices of the shown equilateral triangle in $\mathbb{R}^2$ with side length $\sqrt{2d}$. For $d = 2$ shown in the upper left panel, the points appear to be rather random. Yet at only

$d = 20$ in the upper right panel the points are already beginning to coalesce. In the lower left panel for $d = 200$ the points are now quite close to the vertices of the triangle. The much larger $d = 20,000$ is used in the lower right panel to show how the points are now so close to the vertices that the seven different data sets are barely distinguishable due to over-plotting.

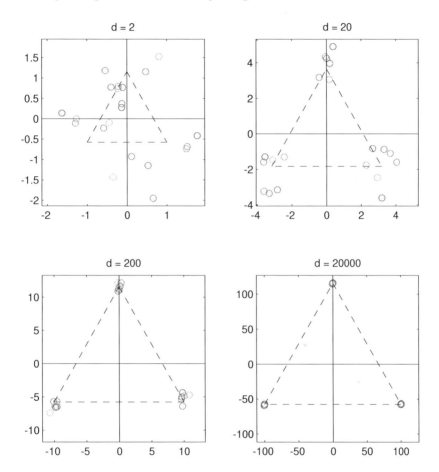

Figure 14.5 *Simulations demonstrating geometric representation for several dimensions d, each with seven data sets (shown in different colors) of size $n = 3$. Shows for increasing dimension that modulo rotation and rescaling data points converge to vertices of a rescaled unit simplex.*

Geometric representation calculations can explain a number of the apparently strange phenomena observed earlier. For example in the discussion of the Overlapping Classes data analysis in Figure 4.13, it was stated (and verified by the DiProPerm test) that the perhaps surprisingly distinct DWD difference observed between the two classes (which were generated from the same probability distribution) in the top left panel was simply the result of natural variation in Gaussian data. The first observation is that the n_1 red points determine a hyperplane of dimension $(n_1 - 1)$, and similarly for the n_2 blue points. The pairwise distance relationship (14.5) indicates that all pairs consisting of one red and one blue point are approximately $\sqrt{2d}$ apart. This shows that the red and blue hyperplanes are essentially parallel. Applying DWD to such data will result in a direction vector that is essentially in the direction of the common normal vectors of the planes, explaining the two very distinct clusters in the top left panel of Figure 4.13. Furthermore the distance between the clusters can be approximated by the distance between the hyperplanes. The first step of that computation is to find the distance from the origin to each plane. It follows from (14.4) that each point is distance $\sqrt{d}$ from the origin. To understand how this can be used to calculate the distance from the origin to the plane, rotate say the red plane so that the data points essentially lie in the first n_1 positive coordinate directions (which is possible because of the approximate orthogonality result (14.6)). Thus the red data points can be thought of as lying near the vertices of the $\sqrt{d}$ rescaled, n_1 dimensional unit simplex

$$\left\{ \left(\begin{array}{cccccc} x_1 & \cdots & x_{n_1} & 0 & \cdots & 0 \end{array} \right)^t \in \mathbb{R}^d : \sum_{j=1}^{n_1} x_j = \sqrt{d},\ x_j \geq 0 \right\}.$$

A good approximation of the mean of the red points can be found from the value of c for which the vector $\left(\begin{array}{ccccc} c & \cdots & c & 0 & \cdots & 0 \end{array} \right)^t$ lies on the rescaled simplex, which entails $\sum_{j=1}^{n_1} c = n_1 c = \sqrt{d}$, i.e. $c = \frac{\sqrt{d}}{n_1}$. Note that the distance from this approximate mean to the origin is $\sqrt{\sum_{j=1}^{n_1} \left(\frac{\sqrt{d}}{n_1} \right)^2} = \sqrt{\frac{d}{n_1}}$. A similar idea applies to the approximate simplex of n_2 blue points in Figure 4.13. In particular the distance from the mean of the blue points to the origin is approximately $\sqrt{\frac{d}{n_2}}$. Note that this is very consistent with one of the key concepts from the Central Limit Theorem that averaging n independent random variables reduces the standard deviation by a factor of $n^{-1/2}$. Now the direction separating the red and blue clusters in the upper right panel of Figure 4.13 is essentially driven by the difference between these two means. Since the points in the two clusters are all orthogonal, by (14.6), their respective subspaces are orthogonal as well. It follows that the two mean vectors are orthogonal to each other, and thus the distance between them can be computed using the Pythagorean theorem, $\sqrt{\sqrt{\frac{d}{n_1}}^2 + \sqrt{\frac{d}{n_2}}^2} = \sqrt{\frac{d}{n_1} + \frac{d}{n_2}}$. Plugging in the values $d = 1000$ and $n_1 = n_2 = 50$ results in approximate distance between the means of $\sqrt{\frac{d}{n_1} + \frac{d}{n_2}} \approx 6.3$. This fits well with the observed cluster means shown in the upper left panel of Figure 4.13. A related HDLSS analysis of clustering, as defined in Chapter 12, can be found in Borysov et al. (2014).

A parallel calculation similarly predicts the apparently very large distances between the colored clusters, for the $d = 500$, $n = 30$ High-Dimensional Gaussian data shown in Figure 12.9. For the cyan - magenta clusters the mean difference (left column) prediction is $\sqrt{\frac{500}{8} + \frac{500}{5}} \approx 12.7$, while for the green - red (center column) $\sqrt{\frac{500}{7} + \frac{500}{10}} \approx 11.0$.

A similar explanation of the behavior observed (for MD using a simple Gaussian toy data set) in the top row of Figure 11.3 is a little harder because the underlying theoretical means are not 0, but instead are 4.4 units apart in the first coordinate direction. Again reflecting the approximate orthogonality of the first coordinate with both means the Pythagorean relationship is $\sqrt{\frac{d}{20} + 4.4^2 + \frac{d}{20}}$. Plugging in $d = 10$ (top left panel of Figure 11.3) gives about 4.5, while $d = 40$ (top center) yields around 4.8 and $d = 200$ (top right) gives about 6.3. An interesting open problem is similar explanation of the differences (for the more complicated classifiers LDA and MDP) in the other panels of Figure 11.3.

The above calculations all assumed a standard Gaussian $N_d\left(\mathbf{0}_{d,1}, \mathbf{I}_d\right)$ distribution for simplicity of illustration. Hall et al. (2005) extended this to non-Gaussian distributions by allowing dependence in the form of a mixing condition along the data vector. Mixing conditions are a natural way to allow some dependence in the laws of large numbers that lie at the heart of geometric representation. See Bradley (2005) and White (2014) for good overviews of the large literature on mixing conditions. The first submission of Hall et al. (2005) was rejected by a top statistical journal on the grounds that the reviewers were unaware of applications where such a dependence assumption was sensible. That later turned out to be an ironic decision, as that same year of 2005 was the date of the first publication on Genome-Wide Association Studies (GWAS) by Klein et al. (2005). As discussed in Zhou et al. (2018a,b) and in Section 16.2.2 GWAS is an excellent example of a data analysis context where geometric representation ideas have given clear insights, and the mixing condition assumption of Hall et al. (2005) is a very natural consequence of the splitting and merging of chromosomes that constitutes the reproductive process.

A number of other assumptions beyond normality for geometric representation have been developed by Ahn et al. (2007) and Yata and Aoshima (2012). It was pointed out by John Kent that when exploring sufficient conditions for geometric representation a useful example to keep in mind is the scale mixture of Gaussians in (14.2). In that case the sharp $\sqrt{d}$ in (14.4) becomes just $O_p\left(\sqrt{d}\right)$ because with probability ω the observation is $\sigma_1\sqrt{d}$ from the origin, and at distance $\sigma_2\sqrt{d}$ otherwise. However it is important to note that geometric representation is more a property of the *dependence structure* than of the marginal distribution (consistent with the ideas of Figure 14.3). As noted in the discussion of (13.1), an important consequence of assuming independent variables is the joint distribution has common variance of projections in all directions. Another consequence is that in the case of GWAS discussed above, the marginal distributions are discrete (often binary), yet geometric representation ideas still give useful insights.

Because visualization of PCA scores has been so useful at many points above, the HDLSS mathematical underpinnings of these are considered here. In view of the tendency of angles to converge to 90° (as seen in (14.6)) a natural starting point is the behavior of the direction vectors used for projection in PCA, i.e. the eigenvectors of the sample covariance matrix as detailed in (17.21). For this assume the random sample $\tilde{x}_1, \cdots, \tilde{x}_n$ is drawn (independently) from the Gaussian distribution $N_d(\mathbf{0}_{d,1}, \Sigma_d)$, where the d dimensional theoretical covariance matrix Σ_d has eigenvalues $\lambda_{1,d}, \cdots, \lambda_{d,d}$ and first eigenvector $u_{1,d} \in \mathbb{R}^d$. Further assume a *spike covariance model*, in the spirit of Paul (2007), where most eigenvalues are fixed, but the first eigenvalue grows with the dimension. In particular assume

$$\lambda_{1,d} = d^\alpha, \ \lambda_{2,d} = \cdots = \lambda_{d,d} = 1, \tag{14.7}$$

where $\alpha > 0$ controls the growth rate of the first eigenvalue. This distribution has a hot dog shape, with a single long axis in the direction $u_{1,d}$, and smaller spherically symmetric variation in all orthogonal directions. Next let $\widehat{\Sigma}_d$ denote the sample covariance matrix defined at (3.5), which is a natural estimate of Σ_d and similarly define estimators $\widehat{u}_{1,d}, \hat{\lambda}_{1,d}, \cdots, \hat{\lambda}_{d,d}$ using the corresponding sample eigenanalysis of $\widehat{\Sigma}_d$ (again see (17.21)). Because $u_{1,d}$ and $\widehat{u}_{1,d}$ are unit vectors, it is natural to study their relationship using angles. Jung and Marron (2009) showed that the power α is critical to this angle. In particular, in the limit as $d \to \infty$ with fixed n,

$$\angle(\widehat{u}_{1,d}, u_{1,d}) \to \begin{cases} 0° & \alpha > 1 \\ 90° & \alpha < 1 \end{cases}. \tag{14.8}$$

The first part of (14.8) has the form of standard *consistency*, in the sense that when the first eigenvalue is large enough the empirical eigenvector $\widehat{u}_{1,d}$ appropriately converges to the underlying true eigenvector $u_{1,d}$. The second part of (14.8) is much less standard in mathematical statistics. It says when the eigenvalue is too small, not only do these directions not converge, but in fact they diverge to the greatest extent possible (recall eigenvectors are determined only up to $\pm$ sign flips, so angles $> 90°$ should be replaced by their supplements). Hence this condition is called *strong inconsistency*. To understand why $\alpha = 1$ is critical here, recall from (14.4) that Standard Gaussian data tend to lie near the surface of the unit sphere with radius $\sqrt{d}$. When $\alpha > 1$ the stretched Gaussian distribution has standard deviation $d^{\alpha/2} \gg \sqrt{d}$, where the exponent is $\alpha/2$ because eigenvalues are on the scale of variance. Thus most of the probability mass of the stretched Gaussian is outside the sphere of radius $\sqrt{d}$ pulling the empirical eigenvector $\widehat{u}_{1,d}$ in the direction of $u_{1,d}$. Furthermore when $\alpha < 1$, it follows that $d^{\alpha/2} \ll \sqrt{d}$ so essentially all of the mass due to the spike lies within the standard sphere. In this case $\widehat{u}_{1,d}$ will take some random direction, but according to (14.6) random directions tend to be orthogonal, resulting in the 90° part of (14.8). Important extensions of these results, including to more than one spike eigenvalue and also to a broader range of asymptotics can be found in Shen et al. (2016a,b).

Note that (14.8) leaves open the boundary case of $\alpha = 1$. That was resolved in Jung et al. (2012b) with the establishment of appropriate limit distributions that appropriately meld the two results in (14.8). That set of mathematics provides the basis for an important proposal for selection of the number of eigenvalues in PCA by Jung et al. (2018).

Jung and Marron (2009) also studied the behavior of the first eigenvalue and showed that under the eigenvalue assumption (14.7), again in the limit as $d \to \infty$,

$$\frac{\hat{\lambda}_{1,d}}{\lambda_{1,d}} \xrightarrow{d} \frac{\chi_n^2}{n}. \tag{14.9}$$

Thus the first spike sample eigenvalue is inconsistent, although in a rather controlled and interpretable way. For example $E\chi_n^2 = n$, so $\hat{\lambda}_{1,d}$ is unbiased in a crude sense. This issue, together with a broad set of different conditions for consistency and strong inconsistency along the lines of (14.8) has been explored in a series of paper by Yata and Aoshima (2009, 2010a,b, 2012, 2013). Some landmarks of those results are proposed variations of PCA that features a consistent first eigenvalue, thus overcoming the perhaps disappointing result (14.9), as well as obtaining consistency for some values of $\alpha < 1$ in (14.7). A less elegant way to overcome the inconsistency of (14.9) is to follow the $\lim\limits_{d \to \infty}$ with a $\lim\limits_{n \to \infty}$ as discussed in Section 14.3.

Given the popularity of *sparse* methods (those that approach high-dimensional challenges by assuming large numbers of 0s) in the world of mathematical statistics, it is natural to investigate parallels of (14.8) in sparse contexts. Shen et al. (2013) developed a complementary theory based on a sparsity index β (in addition to the signal strength parameter α), and laid out interesting regions for consistency and strong inconsistency in the α, β parameter space.

Results of the form (14.8) and (14.9) have generated some considerable skepticism. An interesting objection is that (14.8) holds even for sample size $n = 1$. This may be surprising, however in that case the direction $\hat{u}_{1,d}$ points in the direction of the single data vector $\tilde{x}_1$. The stretch of the spiked Gaussian density described above in the direction $u_{1,d}$ then explains why (14.8) holds. Discussing consistency in the case $n = 1$ is certainly not typical, and one might interpret this as an indication that the spike assumption (14.7) is too strong to have any practical relevance. However the two sides of (14.8) reveal that when no assumption in the spirit of $\alpha > 1$ holds, PCA cannot give any useful results. Yet at many points above, e.g. Figures 4.5–4.8, PCA in very high dimensions ($d = 1709$ in that case) clearly gives nonrandom, useful results. Hence while (14.7) may appear strong to those theoreticians whose thoughts are unencumbered by real data, in fact it evidently is reasonable in interesting practical contexts.

Another potential objection to this theory is that (14.9) hints at potential inconsistency of PC scores in high dimensions. To investigate this, for $i = 1, \cdots, n$ let $\hat{s}_{i,j}$ denote the ith PC score for the jth component, i.e. the projection of $\tilde{x}_i$ onto the eigenvector $\hat{u}_{1,d}$. Now $\hat{s}_{i,j}$ can be viewed as an estimate of the theoretical score $s_{i,j}$, which is the projection of $\tilde{x}_i$ onto the theoretical eigenvector $u_{1,d}$. Using results from Shen et al. (2016b), it can be shown that under the appropriate analogs of the eigenvalue assumption (14.7)

$$\frac{\hat{s}_{i,j}}{s_{i,j}} \xrightarrow{d} R_j, \tag{14.10}$$

as $d \to \infty$, where R_j is a non-degenerate random variable. Consistency of PC scores would correspond to a degenerate $R_j = 1$, thus revealing general inconsistency of PC scores in HDLSS settings. This seems like an apparent contradiction to the clear usefulness of PCA in finding interesting structure in data, e.g. as in Figures 4.6 and 4.7. That is resolved by noting that while the limiting random variable R_j depends on the component index j, it does not depend on the data index $i = 1, \cdots, n$. Thus for each data point, it has the same realization, so while the scores are indeed off by a random factor of R_j, it is the *same factor* for each $\tilde{x}_i$. Hence in PCA scatterplot matrices in high-dimensional situations, such as Figures 4.6 and 4.7, the numbers on the axes do not reflect the underlying probability distribution. However the *relative positioning* of the data points is correct which is why PCA is very useful in high-dimensional data analysis. In addition to the results mentioned above, Yata and Aoshima (2010a, 2012) have proposed modifications of PCA which *are* consistent in the sense that $R_j = 1$.

Another point of skepticism that has been raised is that it is unclear how to check assumptions such as (14.7) "in practice". But this loses sight of the fact that these results are merely mathematical statistics, and such issues exist almost everywhere in that area. For example, the Central Limit Theorem also relies on assumptions such as independence and second moments which similarly cannot be proven from any data set, yet its value in explaining naturally observed phenomena is unquestioned. This same principle applies as well to the asymptotics described here, as indeed it applies to all of mathematical statistics.

A particularly interesting HDLSS characterization of the usefulness of PCA scores can be found in Hellton and Thoresen (2017), which explores the impact of measurement error on HDLSS scores. Interesting work generalizing rank based nonparametric statistics to HDLSS contexts can be found in Biswas et al. (2014, 2015, 2016) and Ghosh and Biswas (2016).

As noted in Aoshima et al. (2018), the mathematics behind many of these results can be established using the theory called *concentration of measure*, which has a large literature started by Talagrand (1991, 1995). See Ledoux (2001) for good overview, and Koltchinskii and Lounici (2017b, 2016, 2017a) for useful directly related results. This is the theory that seems to have generated the common misunderstanding, typified by the quote from Wainwright (2019) at the beginning of this chapter.

Closely related asymptotics also based on fixed sample size n, but following the tangential path of hypothesis testing for repeated measures, have been explored by Pesarin and Salmaso (2010), who propose a notion called *finite sample consistency*.

14.3 High Dimension Medium Sample Size

An asymptotic domain that lives between the random matrix theory discussed in Section 14.1 and the HDLSS context of Section 14.2 is where both $d, n \to \infty$ with $d \gg n$. This has been called *High Dimension Medium Sample Size* (HDMSS) by Yata and Aoshima (2013) and *ultrahigh dimensionality* by Fan and Lv (2008). The added $\lim_{n \to \infty}$ results in a theory that corresponds somewhat more with conventional mathematical statistics. In particular some of the inconsistency results such as (14.9) and (14.10) turn into more conventional consistency statements. There is still a sharp boundary between consistency and strong inconsistency of the sample eigenvectors of the type in (14.8), but as clearly expressed in Figure 1 of Shen et al. (2016a), the $\alpha = 1$ threshold that was critical to the boundary is now replaced by one driven by the relative rates at which d and n tend to ∞.

A conceptual issue is the parameterization of the asymptotics in the HDMSS domain. Fan and Lv (2008) took a conventional approach working with the limit as $n \to \infty$ and taking for example $d \sim n^\theta$. On the other hand, Yata and Aoshima (2013) base their analysis on the limit as $d \to \infty$ and working with $n = d^\rho$. Of course they are essentially the same domain but note that the Yata-Aoshima formulation gives a much more natural interface between the HDMSS and HDLSS worlds.

CHAPTER 15

Smoothing and SiZer

Smoothing methods are useful approaches to obtaining information from low dimensional (often just 1-d) sets of data. They are often called *nonparametric* because they do not rely on typical parametric models, instead aiming to produce a useful visual interface for data analysis. The two main smoothing contexts are *density estimation* and *nonparametric regression.*

Density estimation starts with a random sample (recall independent and identically distributed from Section 7.1) of scalars $\widetilde{x}_1, \cdots, \widetilde{x}_n$ (using notation from Table 7.1) from an unknown probability density curve $f(x)$. The goal is to recover f which as noted in Section 3.3.1 often provides useful population structure insights. The first thing most people would try is a histogram, but as noted in Section 15.1 that approach has problems that are not all that widely understood.

Regression starts with a random sample of pairs $(\widetilde{x}_1, \widetilde{y}_1), \cdots, (\widetilde{x}_n, \widetilde{y}_n)$ and the goal is to recover the conditional expected value (i.e. regression function) $m(x) = E[\widetilde{y}|\widetilde{x} = x]$. Again the smoothing approach is nonparametric in the sense that instead of imposing an assumed parametric structure by something in the spirit of a classical linear model, the data are allowed to speak for themselves. Smoothing methods provide a much more flexible view, which can lead to unexpected discoveries that can be totally obscured by an inappropriate rigid parametric structure. An elegant example is the discovery of pre-pubertal growth spurts in human growth data, in a series of papers summarized in Müller (1988).

In both cases, for the target to be identifiable, some assumptions must be made. Such assumptions are almost always some variant of the theme of *smoothness.* This can be defined in terms of continuity and differentiability, or in terms of various function spaces such as Sobolev and Besov spaces.

Perhaps because of the intuitive appeal and elegant simplicity of smoothing, there is a very large literature. Important monographs on density estimation include Tapia and Thompson (1978) who point to a biblical reference for early density estimation, Devroye and Györfi (1985) who present compelling reasons why theoretical analysis should be based on the L^1 norm, Silverman (1986) who gave a very accessible early account, Scott (2015) whose treatment is very comprehensive and Klemelä (2009) who proposed some novel high-dimensional visualizations. Monographs on regression are more diverse and tend to advocate specific methods, such as Müller (1988) and Härdle (1990) for two different kernel smoothers, Fan and Gijbels (1996) who made a strong case for local polynomial methods, Wahba (1990), Eubank (1999), Green and Silverman (1994), Wang (2011), and Gu (2013) who advocated smoothing splines, Stone et al. (1997) who make a compelling case for regression splines, and Györfi et al. (2002) who

DOI: 10.1201/9781351189675-15

take a distribution free approach. Additional accounts of smoothing generally can be found in Prakasa Rao (1983), Thompson and Tapia (1990), Wand and Jones (1995) who provide a very accessible introduction to kernel smoothing methods, Simonoff (1996), Hart (1997), Efromovich (1999), and Schimek (2013).

15.1 Why Not Histograms?–Hidalgo Stamps Data

When faced with a density estimation task (i.e. the need to visualize the probability distribution underlying a set of numbers), most people construct a histogram. Many users appreciate the fact that features discovered in a histogram are impacted by choice of the binwidth. That point is illustrated in Figure 15.1 showing various histograms of the 1872 *Hidalgo Stamps* data using different binwidths. That data set was brought to the statistical literature by Izenman and Sommer (1988). It consists of $n = 485$ paper thicknesses of the Hidalgo postage stamp. The question of interest to philatelists is how many paper thicknesses appear in the data, which is unknown as the relevant records have been lost. This question is sensibly rephrased as: how many modes are in the density?

Figure 15.1 makes it clear this is not a simple issue, and in fact each panel reflects node numbers that have been suggested in the literature. In particular, at least two modes was the conclusion using bootstrap ideas in Section 16.5 of Efron and Tibshirani (1993). Three modes were found by Walther (2002), four were flagged as statistically significant by Sommerfeld et al. (2017), 3–5 by Chaudhuri and Marron (1999) and five by Minnotte (2010). Seven modes in this data were concluded by Izenman and Sommer (1988), Basford et al. (1997), and Fisher and Marron (2001). Minnotte and Scott (1993) suggested there might be up to 10 modes.

Note that several different choices for the vertical axis are commonly used. One is simple bin counts, another is the relative frequency, while a third is to make the total area of the bars equal to 1. For investigating a single data set, the choice usually does not matter much as sensible scaling gives the same impression as long as the bins all have the same width. However for comparing data sets of different sizes, the second and third options are generally preferred. For density estimation, it is only the third (used in all histograms here) that results in an actual probability density. The difference becomes visually important in the case of unequally spaced bins (for example equally spaced in terms of distribution quantiles) where it is most interpretable to choose *areas* to correspond to amounts of probability mass because areas are most naturally perceived by the human visual system.

Figure 15.1 makes it clear that too large a histogram binwidth can result in missing important discoveries in data, and a small binwidth appears to be strongly driven by random sampling variation. However, as pointed out by Mats Rudemo, the human perceptual system is rather good at visually smoothing. So in the hands of an experienced analyst who understands the issue of spurious small scale spikes, a small binwidth can lead to good results in terms of data discovery. It may be worth keeping in mind that this notion does *not* seem to be a factor in the default histogram choice of most common software packages. For example, as

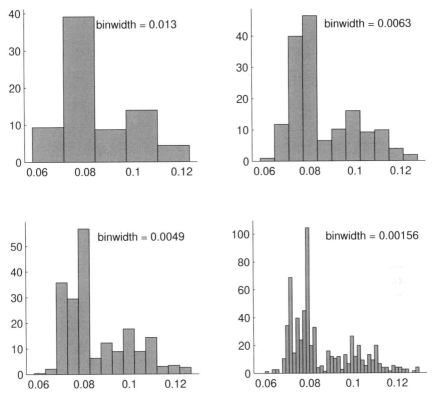

Figure 15.1 *Histograms of the Hidalgo Stamps data for various binwidths. Shows that the number of modes is strongly driven by choice of binwidth.*

noted in Section 3.3 of Scott (2015), the Sturges (1926) method tends to grossly oversmooth. Nor does Scott's recommendation give an undersmoothed binwidth because it targets the uni-modal Gaussian distribution.

While many users understand the impact of the binwidth on features discovered using histograms, as illustrated in Figure 15.1, less well understood is the *bin edge effect*, which was elegantly pointed out in Section 4.3 of Scott (2015). This is demonstrated in Figure 15.2, which shows histograms of the Hidalgo Stamps data that both have the same binwidth of 0.005 (same as in the lower left panel of Figure 15.1). The obvious difference between these histograms is entirely driven by horizontal shift, which clearly has a major impact on the number of modes (from 6 in the left panel, to 2 on the right).

One approach to handling this bin edge effect is to average the histograms over many shifts, resulting in the Average Shifted Histogram, studied in Chapter 5 of Scott (2015). This can be represented as a variant of the very popular kernel density estimator. Given a *kernel* function K with integral one (e.g. a probability

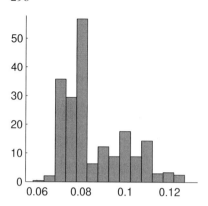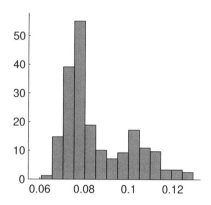

Figure 15.2 *Histograms of Hidalgo Stamps both with binwidth 0.005, but shifted to the right by half a bin in the right panel. Shows number of modes can also be very sensitive to bin location.*

density), and a *bandwidth* $h > 0$,the Kernel Density Estimator (KDE) is

$$\hat{f}(x) = n^{-1} \sum_{i=1}^{n} K_h \left(x - \widetilde{x}_i \right), \tag{15.1}$$

where $K_h(\cdot)$ denotes the area-preserving rescaling $\frac{1}{h} K \left(\frac{\cdot}{h} \right)$. Three KDEs, showing how the bandwidth h controls the smoothness, for the Hidalgo Stamps data appear in Figure 15.3. As done many times in earlier chapters, the density estimates are overlayed with the raw data as a *jitter plot* (as discussed in Section 4.1). Note that for this data set, the jitter plot gives a reasonable impression of regions of high density. The three bandwidths were manually chosen to give an impression of oversmoothing (red, $h = 0.005$,showing just two modes), undersmoothing (blue, $h = 0.0006$,with many spurious modes) and a perhaps good amount of smoothing (green, $h = 0.0016$,with 7 modes as found in several of the above references). The small kernel functions at the bottom of Figure 15.3 also give an indication of how the KDE is constructed as a sum of shifts of such functions. These kernels have been vertically rescaled so they are all visible, but essentially each KDE is a sum of such functions shifted to be centered at each data point.

Figure 15.3 shows that choice of bandwidth is critical, and more about this choice can be found in Sections 15.3 and 15.4. Here bandwidth selection was done using what Bradley Davis has termed the *Goldilocks principle*. Following a theme of the popular children's story the idea is to find bandwidths that are both clearly under and over smoothed, which then provide a context in which the one in the middle seems to be about right.

Less critical is the choice of kernel, although there are many views on that as well. Computational issues were once an important consideration, but much less so now with the ready availability of fast computation. There have been several approaches to optimal kernel choice, with the most sensible based on decoupling the bandwidth and kernel choices using the *canonical kernel* idea (see Marron

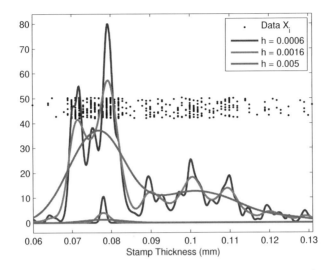

Figure 15.3 *Illustration of kernel density estimation. Raw Hidalgo Stamps data are shown using a jitter plot. Colored curves for three different bandwidths showing over (red), under (blue), and about right (green) levels of smoothing. Corresponding vertically rescaled component kernels appear at the bottom.*

and Nolan (1988)). For actual data analysis the Gaussian kernel is recommended, both because it tends to produce the fewest distracting small scale visual artifacts, and because of the variation diminishing property (it is the only kernel where the number of modes is a decreasing function of the bandwidth) as discussed in Chaudhuri and Marron (2000).

The KDE can also give clear insight into the perhaps surprising behavior of histograms highlighted in Figure 15.2. This is done in Figure 15.4, which simply adds a KDE overlay. The peaks in the KDE clearly show why there are so many modes in the histogram in the left panel (the peaks in the KDE fall nicely within histogram bins) and why there tend to be so few in the right panel (the peaks are split between bins). Since the KDE so clearly diagnoses the bin edge problem, it follows that the KDE (not the standard histogram) should be the method of first choice for visualizing univariate distributions of data. This convention, together with the jitter plot overlay as in Figure 15.3, has been followed at most points in this monograph. An exception is where each bin has a specific meaning, such as for the heat-map color distributions shown in Figures 6.9–6.12.

KDEs have already been extensively used in earlier chapters, most notably in the scatterplot matrix visualizations introduced in Figure 4.4 (for the Twin Arches data), in the marginal distribution plots described in Section 5.1, and even as early as Figure 1.4 (Spanish Mortality data). An issue deserving a little more discussion is that of subdensities as introduced in the Lung Cancer PCA example of Figure 4.7. Figure 15.5 studies an alternative to those sub-densities, which is to use full densities (recall curves with area one underneath) for each of the sub-populations

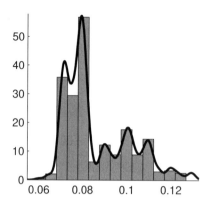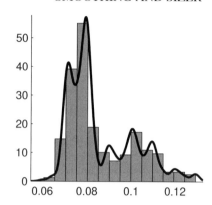

Figure 15.4 *Understanding of the histogram bin edge effect shown for the Hidalgo Stamps data in Figure 15.2, by overlaying a KDE (black curve). Shows many histogram modes when the KDE peaks fall mostly within single bins (left panel) and few histogram modes when the peaks are split between bins (right).*

of the Pan-Cancer data. The left panel is essentially a replot of the top left panel of Figure 4.10. Recall that view studied the gene expression of 6 different cancer types (highlighted using colors and symbols) compared in the direction of the projection on the DWD direction trained on HNSC (Head and Neck Squamous cell Carcinoma) vs. KIRC (KIdney Renal clear cell Carcinoma). Note that the left panel gives a clear impression of how the subpopulations combine to form the full population (whose density is the black curve). In particular, the blue KIRC group forms really all of the left-hand bump (even overlaying the black curve). While the green HNSC is most of the bump on the right, there are clearly a few magenta BLCA (Bladder Urothelial Carcinoma) cases mixed into that bump as well. It is also visually clear how the other classes make up the main bump of the black density. Plotting each colored curve as a density itself is much less useful. One problem is that, because the curves are narrower, the constraint of having area one makes them much taller than the black curve. A simple fix is to scale them to all have the same height, which is shown in the right panel of Figure 15.5. This gives a generally much less intuitive impression of how the subgroups relate to the full population. Exceptions to this principle include when some subgroups are much smaller than others (so the curves can be hard to see), and when it is desirable to think of subgroups in a relative way (independent of subgroup sizes).

A perhaps non-obvious issue in overlaying density estimates is that use of a common bandwidth h can be very important. A critical case is sub-density displays as in the left panel of Figure 15.5 which are most interpretable when the subdensities sum to the overall density (this happens naturally for a common bandwidth). Due to the high noise level inherent to bandwidth selection discussed in Section 15.3, this is generally far from the case using separate default bandwidth selectors for each. Marron and Schmitz (1992) proposed use of a sensible common bandwidth h which is the geometric mean of the individually selected bandwidths.

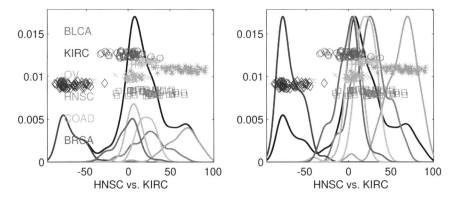

Figure 15.5 *Comparison of sub-density plots (left panel) and subset density plots (right), for the same Pan-Cancer data as in the top left panel of Figure 4.10. Indicates sub-density scale is usually more intuitive.*

A reasonable alternative (especially when subgroup sizes are very unbalanced) is to use the union of the data to choose the common bandwidth.

Some other key ideas about smoothing are reviewed in Section 15.2. Section 15.3 contains an overview of data-driven bandwidth selection. An alternate approach, based on scale-space ideas, together with a fundamental approach to statistical inference appears in Section 15.4.

15.2 Smoothing Basics–Bralower Fossils Data

As discussed above the main contexts of statistical smoothing are density estimation and nonparametric regression. An example of the latter is the Bralower Fossils data in Figure 15.6. The data set comes from Bralower et al. (1997), and was collected for the study of global climate over very long time periods. Each plus sign represents a microscopic fossil shell found in a core drilled into the floor of the ocean. The shells are dated by the surrounding material, and the ages range over about 30 million years, around a hundred million years ago. Strontium is an element that naturally occurs in two different isotopes with about 70% prevalence of one. These shells are climatic indicators because during an ice age, much of the seawater is frozen in the ice caps, so chemicals in the ocean are more concentrated. When the temperature warms, melting results in dilution. The shells absorb the strontium isotopes at differing rates under varying dilutions, which results in the observed data. This may sound like a serious stretch of the data (e.g. the numbers on the vertical axis only differ in the 4th decimal place), but note that there are clear systematic patterns in the data as opposed to being driven by pure random noise. Furthermore, there is not any natural choice of parametric model here so it is appropriate to approach the data with smoothing techniques.

The Bralower Fossil data are smoothed in Figure 15.6 using a local linear smoother, which is generally not far from a simple moving average (except that a line is fit to the data within each window). Each fit is weighted by a Gaussian window function (playing the same role as the kernel K in (15.1)). As noted in Fan and Gijbels (1996), one difference between the local linear and local average can be seen at the edges, where these smooths track the local slope quite effectively, in contrast to local averages which tend to inappropriately level off at the edges. As for the KDE, the bandwidth plays a critical role, and the Goldilocks approach to manual selection has again been used here. The red curve, using $h = 4$, seems clearly oversmoothed. In particular it does not follow the data well through the very large valley around 115 million years ago, but is also somewhat low at the peak around 105 million years ago. The blue curve, for $h = 0.4$, has many wiggles that seem to follow fine-scale noise-driven variation, and thus appears undersmoothed. This curve also reveals another property of the local linear smoother: it can leave the vertical range of the data, e.g. at around 96 million years ago. This is a consequence of sparsity in the data together with tail properties of the Gaussian kernel. In large horizontal gaps between data points, the local linear smoother tends to follow the lines determined by the two closest data points on each side, and to smoothly connect those regions in the middle of the gap. The green curve, based on $h = 1.3$ may be about the right amount of smoothing, in terms of following the overall trends in the data, while not responding to apparent measurement error. Note that this value of $h = 1.3$ is between $h = 4$ and $h = 0.4$ in a *multiplicative* (not additive) sense, i.e. it is close to the geometric mean. Generally smoothing parameters are best thought of in a multiplicative way as discussed around Figure 15.8. An important question is which of these observed features are "really there", in the sense of statistical significance? This critical inferential issue is addressed in Section 15.4.

As noted in the above-referenced monographs, many papers have been written that analyze smoothing methods using asymptotics as the sample size $n \to \infty$. Appealing lessons from them include insightful quantification of how excessive sampling variation (i.e. wiggliness of the curves) is a consequence of a small bandwidth, while bias which appears as under-estimation of peaks and valleys is exacerbated when the bandwidth is too large. When these effects are jointly studied, it is natural to combine them into a trade-off resulting in an optimality theory, which then characterizes smoothing methods in terms of rates of convergence. Early on it was noticed that for a given amount of smoothness, various methods had the same rate of convergence. This motivated the idea that these rates may be the best possible, which has been elegantly established using *lower bound* theory. Landmark papers of that type include Farrell (1972), Stone (1982), and Pinsker (1980). Particularly insightful was the geometric view of optimal rates discovered by Donoho and Liu (1991).

An interesting twist on these results was started in the statistical community by Donoho and Johnstone (1994); Donoho et al. (1995), who essentially showed that conventional smoothing methods had poor spatial adaptivity properties. They went on to show that wavelet thresholding provided an elegantly straightforward way to overcome this problem. This generated a large amount of research, including the Bayesian wavelet approach of Kohn et al. (2000).

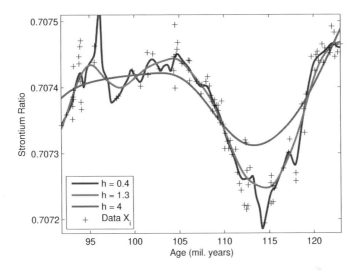

Figure 15.6 *Local linear analysis of the Bralower Fossils data. Plus symbols are ratios of strontium isotopes as a function of age. Three curves are estimates of the regression function, with bandwidths that are oversmoothed (red), undersmoothed (blue), and about right (green) for seeing important underlying structure in the data, in a situation where no parametric model is available.*

An important and sometimes controversial aspect of such mathematical analyses has been choice of error criteria. For nonparametric regression the most commonly used criteria tend to involve squared error such as the L^2 norm, which is natural for a number of reasons including the fact that the regression function itself minimizes the theoretical squared error. Another advantage of squared error measures is that their Hilbert space inner product structure allows for straightforward analysis, including direct quantification of the smoothing trade-off in terms of variance and bias. Less clear is the case of density estimation. As noted above, Devroye and Györfi (1985) present a number of compelling reasons why the L^1 norm is a more sensible criterion. However that comes at substantial cost in terms of technical tractability, which has meant that most discoveries such as performance of bandwidth selectors have first been discovered using L^2 and then with much more effort adapted to L^1. A weakness of most error criteria is that their optimal bandwidths can be quite far from what human experts would select as discussed around Figure 7.9.

While asymptotic results have been very popular and insightful, they do leave open the question of how large a sample size is needed before they take effect (some of which can be dubious such as the so-called *higher-order kernels*). A number of such questions were answered using *exact risk analysis* in the context of univariate density estimation by Marron and Wand (1992), and in the multivariate case by Wand and Jones (1993). See Marron et al. (1998) for exact risk analysis of wavelets.

With the advent of massive modern computer resources, computational time is no longer the consideration that it once was. However for large data sets it can still sometimes be an issue worth considering. In the days of slower computers, substantial effort was invested in this issue. A path-breaking approach was Silverman's Fast Fourier Transform computation of the KDE, see Silverman (1982). Another very interesting and effective approach was the *updating* idea of Seifert et al. (1994). Fan and Marron (1994) analyzed all of these, and discovered that surprisingly the speed of Silverman's method was more due to the *binning* pre-processing than to the Fast Fourier Transform computation.

15.3 Smoothing Parameter Selection

Given the importance of the bandwidth to what is seen in a KDE as demonstrated using the Hidalgo Stamps data in Figure 15.3 (parallel issues exist for all other types of smoothing method), it is natural to believe that the fundamental issue in smoothing is selection of the tuning parameter (the bandwidth h for the KDE). That idea led to a large amount of research on data-based bandwidth selection. For surveys of that area see Marron (1989), Jones et al. (1996a,b), and many of the monographs referenced above, as well as Cao et al. (1994) and Park and Turlach (1992).

As is clear from several of the contributions to the discussion edited by Härdle and Schimek (1996), there was never a consensus among researchers in the area as to how smoothing parameters should be chosen. Using terminology from Jones et al. (1996a), opinions were divided mostly between *first-generation* methods which included familiar ideas such as cross-validation and AIC and more sophisticated *second-generation* methods, such as the recommended Sheather and Jones (1991) Plug-In method, and the bootstrap approach of Jones et al. (1991).

Strong motivation for the development of second-generation methods came from deep asymptotic analysis revealing a surprisingly large amount of sampling variation, first discovered by Hall and Marron (1987a,b). For example, using the notation $\hat{h}$ for a first-generation data-driven bandwidth, and h_0 as some type of optimal bandwidth such as a minimizer of the Integrated Squared Error, $\int \left[\hat{f}(x) - f(x)\right]^2 dx$, the bandwidth variation is quantified as

$$\frac{\hat{h} - h_0}{h_0} \sim n^{-1/10},$$

in the limit as $n \to \infty$. Many variants of this were later developed. This very slow rate of convergence was very discouraging, yet did confirm an empirically known fact that methods such as cross-validation occasionally result in unacceptably poor bandwidths, which has also been observed in many other contexts. This also motivated the development of second-generation methods. However the latter sometimes required very large sample sizes for their benefits to be felt, and seemed to be less useful in other contexts such as spline smoothing. This apparently led to the lack of consensus discussed above.

A perhaps surprising property of cross-validation and other first-generation methods is that they have a tendency to actually be *negatively correlated* with various optimal bandwidths, such as the minimizer of the integrated squared error. See Chiu and Marron (1990) for a detailed analysis of this phenomenon.

For practical bandwidth choice, the recommended bandwidth selection method depends on the context. In situations where sufficient time and effort are available, the best approach is subjective visual trial and error by an experienced analyst. When there is analysis time but less expertise available, the Goldilocks approach suggested in Section 15.1 (and as applied in Figure 15.6) is recommended. In situations where insufficient human resources can be employed (e.g. when many estimates need to be constructed, such as for the marginal distribution plots introduced in Section 5.1), automatic choice is needed. Following the recommendation of Jones et al. (1996a), the plug-in method of Sheather and Jones (1991) is used here as a default here for KDEs. Similarly the Ruppert et al. (1995) plug-in method is used here for local linear regression, when a default is required.

Smoothing parameter selection issues are similarly important for the kernel approaches to classification discussed in Section 11.2, especially in the case of the radial basis function, an example of which is a Gaussian kernel. See Ahn (2010) and Oliva et al. (2016) for interesting work in that direction.

15.4 Statistical Inference and SiZer

As noted in Section 15.2, and visually clear from Figures 15.3 (Hidalgo Stamps data) and 15.6 (Bralower Fossils data), a central issue for any actual data analysis using smoothing methods is which observed phenomena (usually peaks and / or valleys) are "really there". More precisely, which represent important and reproducible underlying structure (thus real scientific discoveries), as opposed to which are spurious artifacts of the sampling variation.

A natural approach for statisticians is to attempt to generalize the idea of confidence interval for studying the uncertainty in a parameter estimate to some type of confidence bands. In the context of smoothing methods, this has three fundamental flaws:

- Variation in a bundle of curves is almost always far richer than can be understood through simply studying bands. For example think of how the rich modes of variation displayed through PCAs at many points in this monograph, such as Figures 1.1–1.6 (Spanish Mortality data), and 4.5–4.8 (Lung Cancer data), would be totally missed by simply putting bands around the set of curves in the original data.

- As noted in Härdle and Marron (1991), a major challenge to correct coverage of confidence bands in smoothing is correctly modeling the inherent bias. Doug Nychka pointed out that if one could adequately model this bias, one could just subtract it from the estimator and then be back in the same position. While a number of alternate approaches have been proposed none has been completely satisfactory. Despite this, confidence bands are still widely used. It is recommended that their inherent crudeness as an approximation be kept in mind when interpreting these.

• All existing confidence bands are based on the choice of the smoothing parameter, which is problematic as discussed in Section 15.3. This issue could be mitigated by taking the Goldilocks approach to bandwidth selection and applying confidence bands to all three choices of bandwidth, although this would result in a busy graphic that may be hard to interpret. That challenge could be handled with multiple panels, although the preference in the literature seems to be to ignore this issue.

These three challenges motivated Chaudhuri and Marron (1999, 2000) to seek a completely different approach to statistical inference in smoothing via the SiZer (SIgnificance of ZERo crossings) method. SiZer tackles the bandwidth choice issue by taking a *scale-space* approach. As elegantly detailed in Lindeberg (2013), scale-space was a model for human visual perception that was explored in the early days of the field of computer vision. The human perceptual ability to perceive both macroscopic big picture aspects as well as fine-scale details in the same visual field, was modeled as a family of Gaussian kernel smooths. In particular, the large bandwidths put the focus on large scale features while the small bandwidths focus on small scale details. Note that from this perspective bandwidth selection (as discussed in Section 15.3) is not a sensible pursuit, because it is the full collection of curves that contain the useful information, not any particular individual member.

Lindeberg (2013) was also very notable for providing a detailed exploration of the *variation diminishing property*. Stated in kernel smoothing terms, this idea is the property that the number of modes in a smooth should be a decreasing function of the bandwidth. As noted in Chaudhuri and Marron (2000), Lindeberg (2013) showed this property always holds if and only if the kernel is Gaussian, in two different ways. One approach is using total positivity ideas, and the other is based on a heat equation representation of smoothing. Parallel applications of heat equation ideas are discussed in Section 8.5 in the context of central limit theory on manifolds.

15.4.1 Case Study: British Family Incomes Data

The top panel of Figure 15.7 shows a scale-space KDE analysis of the 1985 British Family Incomes data from Marron and Schmitz (1992). As is common in income analysis, inflationary aspects are controlled by dividing by the sample mean. The sample size $n = 7145$ after 56 cases with scaled income larger than 3 have been truncated (allowing more attention to be paid to the main body of the distribution). The green points are a jitter plot of the raw data, highlighting a weakness of the jitter plot: visual impression is strongly affected by sample size, e.g. in this case population structure is obscured by too many data points. The family of blue curves are KDEs with bandwidths logarithmically equally spaced from $h = 0.01$ to $h = 1$. Logarithmic spacing is natural when working with bandwidths (note the evenly spaced visual impression among the blue curves) because bandwidths

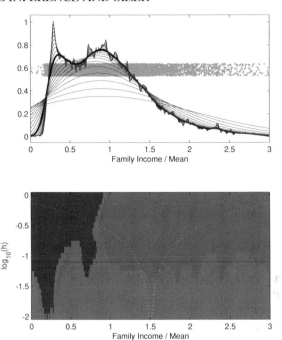

Figure 15.7 *SiZer analysis of British Family Incomes data. Reveals that both apparent large bumps represent important population structure worthy of deeper investigation, while smaller bumps may be only spurious sampling artifacts.*

work as *scale factors*. For example one more naturally considers *doubling* the bandwidth than *adding two* to the bandwidth.

As expected, the KDEs indicate that the distribution of incomes is strongly right-skewed: most people have small to moderate incomes, while relatively few have far larger incomes. However the smaller bandwidths suggest an unexpected feature: there appear to be two modes to this distribution. This was a potentially ground-breaking discovery, because the standard parametric model for incomes (there are several of these, all of which were strongly rejected by goodness of fit tests) are all entirely unimodal. The existence of this bimodal structure was confirmed in the PhD dissertation of Heinz-Peter Schmitz, by stratifying the population so that the first peak corresponded to families on fixed government incomes and the second was driven by three other income categories, each of which were reasonably fit by standard income distributions. The resulting mixture distribution then revealed that the bimodal structure is actually an inherent aspect of the population. Marron and Schmitz (1992) studied this behavior over time, revealing an interesting time trend in these peaks. As noted in Section 15.1 that paper also pointed out that when using multiple data-driven bandwidth choices the noise inherent to those choices (described in Section 15.3) can be usefully mitigated by using the common geometric mean of the selected bandwidths in each smooth.

While this example highlights the potential for scientific success using smoothing methods, it also reveals a potential risk. Had the discovered feature been a sampling artifact (not reproducible in other data sets), much analytical time and effort could be wasted. This underlines once again the importance of confirmatory analysis, i.e. of doing proper *statistical inference* when using smoothing methods to make discoveries using real data.

The scale-space way of thinking also provides a natural solution to the second drawback of confidence bands: the bias inherent to smoothing. This is done by regarding the family of data smooths as targeting not the true underlying curve (f for density estimation or m for regression), but instead targeting the *curve at the level of resolution* of the bandwidth. The natural multi-resolution scale-space representation of the underlying curve is its convolution with the kernel at each bandwidth, which is also the expected value of the estimator, so bias-free estimates are available. This comes at the price of a revised interpretation of the goal of the analysis. That interpretation has not yet been embraced by theoretical statisticians, however it is quite natural for scientists who actually analyze data and are thus quite accustomed to understanding data simultaneously at various levels of resolution.

The first flaw of confidence bands (that bands only poorly reflect variation in curves) is addressed by another shift in the focus of the analysis. The key idea is that most important scientific discoveries by smoothing result from finding unexpected peaks and/or valleys in data, so that should be the direct focus of the inference. There is a mode testing literature, started by Good and Gaskins (1980), and including Silverman (1981), Müller and Sawitzki (1991), Fisher and Marron (2001), and a number of others.

But the SiZer approach proposed by Chaudhuri and Marron (1999) directly connects the statistical inference with the scale-space visualization. The main idea is that a peak is determined by the curve going up on one side, and down on the other. Hence SiZer focuses on the derivative, and showing when it is significantly different from 0. Zero crossings between both types of significant regions indicate significant peaks and valleys. This type of statistical inference is shown in the SiZer map in the bottom panel of Figure 15.7. Each row corresponds to one scale of resolution (bandwidth), and colors across each row flag statistical inference (in terms of a hypothesis test whose null is a zero derivative). In particular, red is used for significantly decreasing, blue for significantly increasing, and the intermediate color of magenta when the hypothesis test is indeterminate. One more color used is gray, which indicates not enough (essentially < 5) points in the Gaussian window for reliable inference. The top rows of the bottom panel show blue (increasing) followed by red, so both the up and the down are statistically significant. At the finest scales (wiggliest blue curves) many peaks and valleys are present, yet the corresponding rows of the SiZer map are almost all magenta, indicating that those bumps can not be distinguished from spurious sampling variation. The black line across the SiZer map indicates the bandwidth $h = 0.08$, chosen to focus attention on the apparent bimodal structure, which also gives the thick black KDE among the family in the top panel. That row shows blue–red–blue–red indicating that both peaks are statistically significant. While the magenta indicates the small scale bumps are spurious, a common question is: "does this analysis show there are one or two bumps in the underlying distribution?" The answer is "yes", because from the scale-space perspective modality depends entirely on the level of resolution. At the coarsest scales there is one mode, at medium scales two significant modes, and at the finest scales many insignificant modes.

Of course careful attention needs to be paid to multiple comparison issues because SiZer is based on many simultaneous hypothesis tests. The first version accomplished this using some relatively crude approximations, that resulted in somewhat anti-conservative inference. Much more precise inference was developed using deeper probability theory in Hannig and Marron (2006), which is the current default and used in all examples here.

Insight as to the amount of smoothing that is done at each scale, i.e. row of the SiZer map, is provided by the dashed white curves. These indicate ±2 standard deviations of a centered Gaussian kernel at each scale.

The choice of red for down and blue for up is arbitrary. That scheme makes intuitive sense for say economic applications. However, for climatological applications where red is associated with warm and blue with cooler trends, it makes sense to reverse the colors, as done by Holmström and Erästö (2002); Erästö and Holmström (2005).

15.4.2 Case Study: Bralower Fossils Data

SiZer has also been developed for nonparametric regression, as demonstrated in Figure 15.8, which is a SiZer analysis of the Bralower Fossils data from Section 15.2. This time at the coarsest scales the blue curves are not far from linear, but are slightly steeper on the left. The top rows of the SiZer map reveal that the decrease is statistically significant on the left side. The level of resolution $h = 1.3$ is highlighted using the horizontal line in the bottom panel, and the thick black curve in the top. The SiZer colors show a significant increase on the left, and also flag the importance of the large valley that bottoms out around 115 million years ago. While the corresponding smooth green curve in Figure 15.6 seductively suggests a second valley around 98 million years ago, the gray in the SiZer map indicates that the data in this region are too sparse to draw any such conclusion.

15.4.3 Case Study: Mass Flux Data

The Mass Flux data set was introduced in Figure 12.10. Recall that the PC1 scores indicated 3 bumps in the distribution, that suggested 3 clusters. A SiZer analysis of the scores is shown in Figure 15.9. As in the above analyses the input data are shown as green in a jitter plot. But this time instead of dots, green circles are used to be more consistent with Figure 12.10. The blue-red patterns in the SiZer map show that all three modes are strong underlying features of the data, which is consistent with the three cloud types as discussed in Section 12.3. Note also that different bumps actually are significant at different bandwidths (scales). In

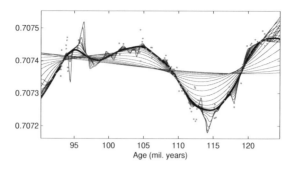

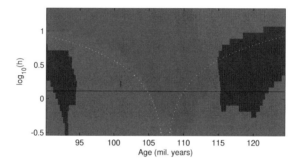

Figure 15.8 *SiZer analysis of the Bralower Fossils data from Figure 15.6. Indicates the main valley at around 115 million years ago is statistically significant, but the smaller valley around 98 might be an artifact of random variation.*

particular the center bump is only really there at bandwidths around $h = 10^{0.4} \approx$ 2.5 to $10^{0.6} \approx 4.0$, while the right bump needs $h \approx 10^{0.75} \approx 5.6$ (highlighted with the horizontal black line). Thus this example highlights how the inference done by SiZer is truly multi-scale in nature.

Of course clustering in the Mass Flux data could also be analyzed using the formal clustering methods described in Chapter 12. Both 3-means clustering and also hierarchical clustering (Euclidean distance with either average or Ward's linkage) gave very similar results. Consistent with the SiZer analysis in Figure 15.9, these are statistically significant using the SigClust method described in Section 13.2.

15.4.4 Case Study: Kidney Cancer Data

The Kidney Cancer part of the Pan-Cancer data introduced in Section 4.1.4 illustrates a variation of SiZer. Unlike the analysis there (and also in Section 13.2.1), instead of considering only a subset of $n = 50$, here the full $n = 551$ data set of KIRC cases is analyzed. The first mode of variation scores from the PCA of that data set is shown as green dots in the top panel of Figure 15.10. The pattern in these dots, as well as the overlaid kernel density estimates, suggest an apparent cluster on the right which is a potential distinct subtype of Kidney Cancer.

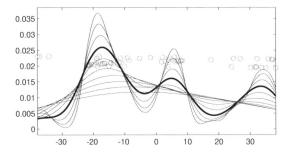

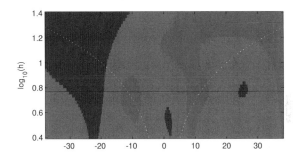

Figure 15.9 *SiZer analysis of the PC1 scores for the Mass Flux data. Shows three bumps are important underlying features that are significant at different scales.*

The importance of cancer subtypes discovered by finding clusters in gene expression was demonstrated by Perou et al. (2000), who discovered clinically relevant subtypes that since motivated the development of different treatment regimes.

The middle panel of Figure 15.10 investigates this potential subtype using a Sizer analysis. The large peak in the area of PC1 scores around -20 is clearly indicated as statistically significant by the nearby blue and red regions in the SiZer map. However, the smaller peak suggested by the cluster of PC1 scores near 100 is not indicated as statistically significant. The reason is that the density is generally sloping downwards, and the cluster is not large enough for a statistically significant upwards slope to appear.

The bottom panel of Figure 15.10 addresses this challenge by modifying the conventional SiZer analysis, from inference about slopes (first derivatives) to a parallel inference based on statistical significance of curvature (second derivatives), in an analysis first done by Tom Keefe. In particular, using the method proposed in Chaudhuri and Marron (2002), regions in the scale space map (with location on the horizontal axis and bandwidth scale, i.e. $\log_{10} h$, on the vertical axis) of significant concavity ($f'' < 0$) are colored cyan, regions of significant convexity ($f'' > 0$) are orange shaded, and inconclusive regions are colored green. The horizontal black line at the scale $h = 20$ passes through two substantial cyan regions, leading to the conclusion of two significant bumps (hence subtypes) in

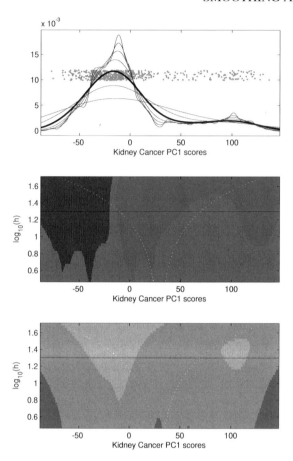

Figure 15.10 *Slope and curvature SiZer analyses of PC1 scores of the full Kidney Cancer data. Jitter plot and family of kernel density estimates in the top panel suggest bimodality. Curvature analysis (bottom panel) confirms statistical significance of bumps, suggesting subtypes that are not flagged as significant using the slope version (middle panel).*

the Kidney Cancer data, despite its ongoing downward tendency on the right side of the main peak.

This result is consistent with the SigClust discovery of two clusters in a random subsample of $n = 50$ of these data points in Section 13.2.1. An important point is that since SigClust directly targets modes it is a more powerful test of modality. In particular it found the subtypes in the Kidney Cancer data with only a small subset ($n = 50$) of the full data ($n = 551$). Subtypes in a larger set of KIRC data from TCGA have been studied in detail by Ricketts et al. (2018).

15.4.5 Additional SiZer Applications and Variants

There are a number of variants of Sizer, for example the inference can be based on the 2nd derivative, instead of the 1st (i.e. curvature not slope). The curvature version of SiZer found important population structure (modality) that was not available from the slope version in the analysis of tree-structured data objects of Shen et al. (2014). SiZer has been modified to handle the specific case of jumps (i.e. change-points) by Kim and Marron (2006). Li and Marron (2005) developed a local likelihood version of SiZer. A smoothing spline version can be found in Marron and Zhang (2005). A wavelet version was invented by Park et al. (2007). An interesting application to establish the existence of *code decay* in software engineering can be found in Eick et al. (2001). Analysis of internet traffic data motivated variations of SiZer that can handle dependent data, see Hannig et al. (2001), Park et al. (2004) and Rondonotti et al. (2007). SiZer ideas have been elegantly integrated with the manifold data object ideas of Chapter 8 by Huckemann et al. (2016).

SiZer type ideas have also been adapted to the two-dimensional case, which thus includes significance of features in images to give a methodology called *Statistical Significance in Scale-space* by Godtliebsen et al. (2002, 2004). The underlying inferential techniques are basically the same, but the visual interface is much different, mostly because of the challenges of extending the basic notions of increasing and decreasing to more than one dimension. SiZer ideas have been elegantly integrated with the manifold data object ideas of Chapter 8 by Huckemann et al. (2016).

The scale-space ideas that underpin SiZer are not unrelated to the persistent homology filtrations that are the basis of Topological Data Analysis as discussed in Section 10.1.4. This connection has been studied in Sommerfeld et al. (2017), who use it to propose an alternate approach to inference for bumps, based on the bootstrap ideas of Fasy et al. (2014), that is seen to be slightly more sensitive than SiZer.

CHAPTER 16

Robust Methods

As discussed at many points in previous chapters, *robustness* issues arise frequently in OODA. Good overview of a large body of research done in this area can be found in the monographs Hampel et al. (2011), Huber and Ronchetti (2009), and Staudte and Sheather (1990), with a more recent survey in Clarke (2018). A general view of the area is that it studies, and recommends remedies for, violation of various classical statistical modeling assumptions.

The great bulk of this work has been motivated by the fact that under the very typical assumption of a Gaussian error distribution *outliers* occur with vanishingly small probability. Yet they can be rather common in real data, and have strong potential to seriously disrupt classical statistical methods which tend to ignore them, as illustrated in Figure 16.1. The left panel shows a toy example of $n = 10$ simulated realizations from the $N_2 \left(\begin{bmatrix} 2 \\ -2 \end{bmatrix}, \begin{bmatrix} 0.49 & 0.49 \\ 0.49 & 0.03 \end{bmatrix} \right)$ distribution shown as + signs, with one point deliberately moved far away. Note that the sample mean, shown as the thick + sign with a circle around it, is a poor approximation of the center of the remaining point cloud, because the outlier has a strong influence on it (in particular with weight $\frac{1}{n} = 0.1$). This pulls it far outside the elliptical contour of the Gaussian density that contains say 95% of the probability mass, hence the sample mean is a very poor estimate of $\mu = \begin{bmatrix} 2 & -2 \end{bmatrix}^t$.

A parallel impact of an outlier on PCA is shown in the right panel of Figure 16.1. This time $n = 40$ points were drawn from the $N_2 \left(\begin{bmatrix} -2 \\ -2 \end{bmatrix}, \begin{bmatrix} 1 & 1 \\ 1 & 0.15^2 \end{bmatrix} \right)$ distribution (also + signs), one of which was similarly moved away. While the sample mean (at the intersection of the heavy dashed line segments) is somewhat affected, it at least lies within the point cloud of the bulk of the data. However the first PC direction (indicated by the longer dashed line segment) clearly does not reflect the dominant mode of variation in the overwhelming majority of the data set. That is because PC1 finds the direction of maximal projected variance, and sample variance (which is driven by squared distances, recall how that had a strong impact on the toy data set on S^2 in Figure 8.3) is notoriously impacted by outlying data points. Here this happened to the extent of actually being larger than the sample variance along the major axis of the remaining data cloud, thus pulling off the PC1 direction as seen. Scores plots (first seen in Figure 1.4 and used in many different ways above) provide a very effective diagnostic for detecting undue influence from outliers, as seen in the right-hand column of Figure 16.2. The second PC in Figure 16.1 is shown as the shorter dashed line segment to give an impression of these as axes of the PC coordinate system. In robust statistics many

approaches to dealing with such challenges have been considered from a number
of interesting perspectives.

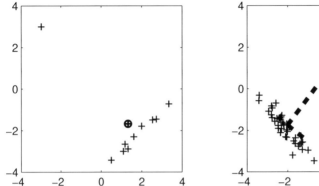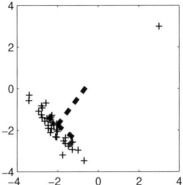

Figure 16.1 *Toy examples illustrating impact of outliers on the sample mean and PCA. The
single outlier in the left panel pulls the mean (circled + sign) actually outside of the convex
hull of the remaining Gaussian data points. On the right the single outlier pulls the PC1
direction (longer dashed line) almost orthogonal to the major axis of the rest of the again
Gaussian point cloud.*

While the study of outliers has dominated research in the area, robustness
against other violations has been studied as well, such as the violation of the typical *independence* assumption, as studied in Beran (1994), Beran et al. (2013),
and Clarke and Hall (2009). Another important violation of typical (i.e. Gaussian
error model) assumptions is treated in the area of robustness against *data heterogeneity*, as discussed in Section 11.4. The fundamental idea here is that modern
large data sets are often created by combining smaller data sets, that are frequently
not identically distributed. In such cases a Gaussian mixture error model can be far
more appropriate than the common single Gaussian. See Marron (2017a) for more
discussion, and Bühlmann and Meinshausen (2015); Meinshausen and Bühlmann
(2015) for leadership research in that area.

An overview of some controversial robustness issues is in Section 16.1. Specific
discussion of robust methods that have proven to be useful in OODA, e.g. techniques well suited for handling high-dimensional data, appears in Section 16.2.
Two specific case studies, involving dealing with surprisingly different types of
outliers in high dimensions are presented in Sections 16.2.1 and 16.2.2. A few
other robustness topics are discussed in Section 16.3.

16.1 Robustness Controversies

Perhaps because of the many potential ways to approach it, the field of robust statistics has seen its share of controversy and very robust debate over the
years. One topic is *outlier deletion* versus *robust methodology*. From the former

viewpoint, it is the responsibility of the data analyst to find outliers, and after careful consideration to delete them from the data set when that is sensible. A number of approaches to this can be found in Barnett and Lewis (1994). The latter approach has been to shift away from explicit identification of outliers toward the development of methods that are simply far less sensitive to them. The large majority of researchers in robust statistics have chosen this latter approach for a number of reasons, including:

- Outliers can be quite challenging to even define, making them difficult to delete in an objective way.

- Outliers can also be difficult to discover, especially in high-dimensional and complex data situations such as arise in OODA.

- Outlier deletion is a time-consuming process in a world where all too many analysts tend to be increasingly pressed for time, as is apparent from the fact that so little data visualization is frequently done in many modern data analyses.

- Outlier deletion may involve just too large a part of the data set, including some data objects that still contain useful information, as illustrated using the Cornea data in Section 16.2.1.

There has also been strong controversy even among those who advocate development of robust statistical methods over outlier deletion. The key to understanding the two sides in the debate is to consider the potential genesis of the outliers. One natural source is recording errors, which can easily happen while writing down numbers (e.g. missing a decimal point), or else through unnoticed failure of a measuring device. In that case, outliers are simply bad data (i.e. contain no useful information), so the statistical methods should aim to completely ignore them (i.e. give them weight 0) as done by outlier deletion. But outliers also naturally appear as just atypically large values, that are not a mistake but are an unusual yet genuine part of the data being studied, e.g. as illustrated using the Drug Discovery data in Figure 5.3 and the Pan-Can data in Figure 5.16. In such cases the outliers contain useful information, but robust methods should keep them from unduly dominating the analysis (as happened for both Gaussian toy examples in Figure 16.1). The canonical example of such a method is the sample median in $\mathbb{R}^1$, which feels a small but appropriate influence from every data point, even those that are very far from the rest of the data. The former position of implicit outlier removal was strongly advocated by the Hampel school (e.g in early editions of Hampel et al. (2011)), while the latter of accommodating yet downweighting outliers was equally strongly pushed by Huber (e.g. in early editions of Huber and Ronchetti (2009)). Of course from a broader perspective, each approach can be appropriate depending on the context. The challenge for the data analyst is to determine which to use in any particular case.

16.2 Robust Methods for OODA

As noted above the challenges of OODA have motivated the re-thinking of some robust statistical ideas. An important example is the notion of *mean outliers* versus *shape outliers* in FDA, as introduced by Dai and Genton (2018, 2019); Dai et al. (2020) (who give several interesting examples clarifying how challenging it can be to even define outliers). The latter concept is illustrated in Figure 16.2. The toy data set there starts with the $n = 50$ 10-d Tilted Parabolas in Figures 4.1 and 4.2 shown in black. The additional red curve that has been added demonstrates the concept of shape outlier. Note that the values of the red curve lie entirely in the vertical range $[0, 30]$ while the black curves occupy a larger range of $[-10, 40]$. However, instead of having that generally very smooth parabolic shape it has a much higher frequency (it is actually a coarsely discretized sinusoid) which puts it into a far different region of the space of curves. Perhaps as expected from the right panel of Figure 16.1, this red outlier has a strong impact on PCA (especially compared to Figure 4.2). A careful look at the sample mean in the top center panel of Figure 16.2 shows that the outlier has some influence there, in terms of some quite angular corners corresponding to the valleys in the red curve compared to the mean of the Tilted Parabolas data in Figure 4.1. The first mode of variation is generally similar to the original analysis in terms of mostly being an overall up and down mode, although the red outlier clearly has a strong influence on the shape of the loadings vector. The second mode of variation is completely dominated by the outlier. There is hardly any variation among the black curves in the loadings plot in the left panel of the third row. The outlier also stands out starkly in the scores plot on the right of the third row, highlighting how far the outlier is from the Gaussian bulk of the data in the 10-d feature space. As noted above, in general such scores plots are a good diagnostic for indicating the presence of the outlier. The third mode of variation (fourth row) is essentially the random tilt mode which appears as the second mode in Figure 4.2, although again there has been some corruption by the outlier. As in the bottom row of the earlier analysis, the residuals from the first three modes (bottom center panel) are relatively much smaller (recalling the lesson of vertical axis choice conveyed by Figures 4.1 and 4.2).

Among the many approaches to mitigation of outliers that have been proposed in the literature, two approaches that are intuitively appealing (and are seen in Section 16.2.1 to have good high-dimensional properties) are illustrated in Figure 16.3. The toy data sets in the two panels are the same as in Figure 16.1. The left panel shows a robust median that, as noted in Section 7.1, appears to have been re-discovered and re-named several times. Haldane (1948) named it the geometric median. It was shown to be unique in the case $d > 1$ (for $d = 1$ this is the ordinary median which of course is not unique for n even) by Milasevic and Ducharme (1987). That paper used the name spatial median, which is common in the multivariate rank statistics literature, see Möttönen and Oja (1995) and Oja (2010). In the robustness literature it is usually called Huber's L^1 M-estimate, using terminology from Huber and Ronchetti (2009). From that perspective it is the $p = 1$ special case of the L^p M-estimate defined for random vectors $\widetilde{x}_1, \cdots, \widetilde{x}_n$ centered at $\theta \in \mathbb{R}^d$ as

$$\arg \min_{\theta \in \mathbb{R}^d} \sum_{i=1}^{n} \|\widetilde{x}_i - \theta\|_2^p. \tag{16.1}$$

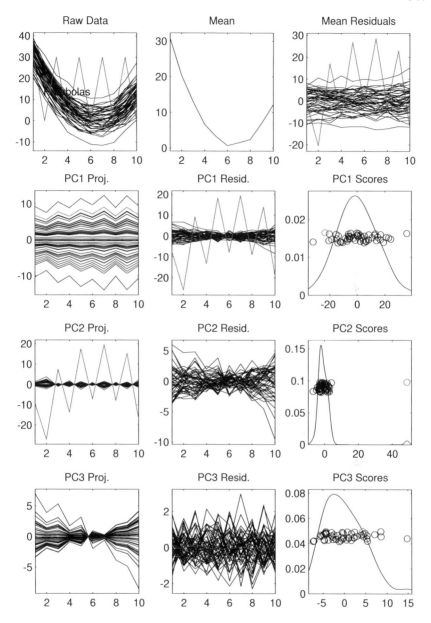

Figure 16.2 *Parabolas and Outlier toy example in the same format as the PCA of Figure 4.2. Tilted Parabola data from there are shown in black, with a red shape outlier added, illustrating how outliers can have a major impact on PCA, and how scores plots can provide useful outlier diagnostics.*

Note that for $p = 2$ this is the Fréchet mean defined at (7.5) with respect to Euclidean distance. Another important special case is $p = 1$, the Fréchet median from (7.6).

A simple iterative algorithm due to Gower (1966), provides insight into the workings of this estimator of centrality and why it is robust. In the case $p = 1$, it is based on setting the vector of d partial derivatives of the sum in (16.1) to 0. Using the notation $\boldsymbol{\theta} = \begin{bmatrix} \theta_1 & \cdots & \theta_d \end{bmatrix}^t$ and $\widetilde{\boldsymbol{x}}_i = \begin{bmatrix} \widetilde{x}_{i1} & \cdots & \widetilde{x}_{id} \end{bmatrix}^t$ this gives, for $j' = 1, \cdots, d$

$$0 = \frac{\partial}{\partial \theta_{j'}} \sum_{i=1}^{n} \left(\sum_{j=1}^{d} (\widetilde{x}_{ij} - \theta_j)^2 \right)^{1/2} = \sum_{i=1}^{n} \frac{1}{2} \left(\sum_{j=1}^{d} (\widetilde{x}_{ij} - \theta_j)^2 \right)^{-1/2} 2 (\widetilde{x}_{ij'} - \theta_{j'}) .$$

Multiplying by $\frac{1}{n}$ and putting these back into vectors gives

$$0 = \frac{1}{n} \sum_{i=1}^{n} \frac{\widetilde{\boldsymbol{x}}_i - \boldsymbol{\theta}}{\|\widetilde{\boldsymbol{x}}_i - \boldsymbol{\theta}\|_2} . \tag{16.2}$$

Insight into this quantity comes from the example in the left panel of Figure 16.3. The + signs are the points $\widetilde{\boldsymbol{x}}_1, \cdots, \widetilde{\boldsymbol{x}}_n$, and a candidate value of $\boldsymbol{\theta}$ is the bold x sign at $(1, 2)$. Note that each $\widetilde{\boldsymbol{x}}_i - \boldsymbol{\theta}$ is the vector pointing from $\boldsymbol{\theta}$ to $\widetilde{\boldsymbol{x}}_i$, and dividing by its norm (i.e. length) projects it onto the circle of radius 1 centered at $\boldsymbol{\theta}$. For the case of $\boldsymbol{\theta}$ at that x sign, these $\boldsymbol{\theta}$ centered projections are shown as small circles lying on the large dashed circle centered at $\boldsymbol{\theta}$. The equation (16.2) is solved by the choice of $\boldsymbol{\theta}$ that makes the average of these projections, shown as a bold circle, equal to $\boldsymbol{\theta}$. That is not achieved by $\boldsymbol{\theta}$ at that x sign. Gower's algorithm is iterative movement of $\boldsymbol{\theta}$ to the bold circle at each step, until convergence. The final result is shown using the bold circled x sign, and its corresponding large dashed circle with projections is also shown. This gives a much more sensible notion of center of this data set than the sample mean (shown as in Figure 16.1 using the bold circled + sign).

The key to the robustness of this notion of center is that the outlier becomes downweighted by projection onto the circle. Note that this estimator is independent of the radius of the circle (equation (16.2) can be arbitrarily rescaled), and in fact the circle shown in the figure has a radius larger than 1 to give a better visual impression of the projected small circles. As noted in Section 7.1, a version of this robust notion of center–the Fréchet median–that is especially adapted to data objects on a manifold as discussed in Chapter 8, can be found in Fletcher et al. (2009).

The right panel of Figure 16.3 shows how this basic idea can be simply extended to give a robust version of PCA. Recall that straightforward PCA applied to this data set gave a first PC direction pointing toward the first quadrant in Figure 16.1 while the bulk of the data objects (the + signs) clearly suggest a nearly orthogonal trend. That direction can be found by using the same type of projection used to understand Huber's L^1 M-estimate in the left panel, in particular onto a sphere centered there. Again the small circles on the dashed circle are projections

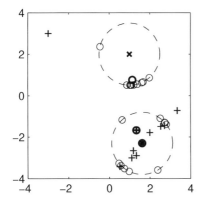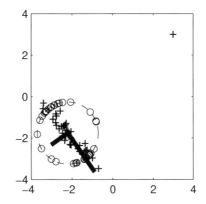

Figure 16.3 *Illustration of Huber's L^1 M-estimate of the mean (left panel) and Spherical PCA (right panel). Toy Gaussian data sets are the same as Figure 16.1.*

of all of the data points. Note that the main sausage of data is projected to ice caps on this one-dimensional sphere. The outlier now has much reduced influence, and is comparable to the projections of a few points near the center. PCA on these projections finds a first direction driven by the ice caps, thus following the main body of the data, as shown by the long solid line. Again the PC2 direction is the shorter orthogonal line indicating the PC coordinate system. This method was called *spherical PCA* by Locantore et al. (1999), and was the basis of a very effective robust analysis of cornea image data objects as discussed in Section 16.2.1. Its perhaps surprisingly poor performance in very high dimensions is explored using Genome-Wide Association data in Section 16.2.2. This same approach to robust PCA was independently discovered and called *PCA on the signs* by Möttönen and Oja (1995).

Figure 16.4 shows the results of a spherical PCA of the Parabolas and Outlier toy data set in Figure 16.2. The first mode of variation continues to be mostly curves moving up and down together, but now the loadings vector is less influenced by the outlier, hence closer to the first mode of variation of the original Tilted Parabolas data in Figure 4.2 (i.e. to that of the bulk of the data). The strong robustness properties of spherical PCA are apparent in the second mode of variation which is now driven by the random tilt, again an important property of the bulk of the data as seen in PC2 of Figure 4.2. As the outlier still represents interesting variation, it is appropriate that it now appears in the third mode of variation, as it is clearly more important than the small variation which remains. Some of the outlier remains in the PC3 Residuals (bottom center panel) which is an artifact of the nonlinear analysis done by spherical PCA. The robustness of Huber's L^1 M-estimate in the top center panel appears as less angularity at the corners than in the sample mean in Figure 16.2.

One difference between Figure 16.4 and Figures 16.2 and 4.2 is the scree plot in the upper right panel. Recall from Figure 3.5 that those showed a red curve

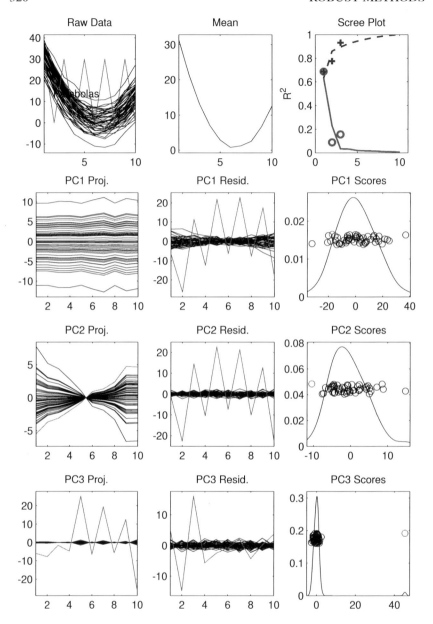

Figure 16.4 *Spherical PCA of the Parabolas and Outlier data in (nearly) the format of Figure 16.2. First and second modes of variation are now much closer to those of the bulk of the data as shown in Figure 4.2. Lessened influence of the outlier is reflected by it only appearing in the third mode.*

and circles indicating the fraction of the energy (measured using sums of squares) explained by each mode of variation, i.e. PC component. The blue dashed curve and + signs show the cumulative version of these fractions. For spherical PCA as studied in Figure 16.4 there are two such energy decompositions available and of interest. The first is the conventional PCA of the data projected to the sphere (e.g. the small circles in the right panel of Figure 16.3). This is shown as the red curve decomposition, which is monotone decreasing because it reflects the eigen analysis underlying this PCA (which determines the directions being used in the modes of variation). But perhaps of even more interest may be the energy in these spherical modes of variation in terms of the original data objects, e.g. the + signs in the right panel of Figure 16.3. These are also the sums of squares of the curves in the left columns in Figure 16.3. This decomposition is shown using red circles for the fractions, and blue + signs for the corresponding cumulatives. These are quite different, and not even monotone. That is because spherical PCA puts the outlier into the third mode of variation, even though its projected variance is larger than the variance of the random tilt mode of variation. Recall this was the reason that the robust spherical PCA is useful for this data set.

Spherical PCA was analyzed using the HDLSS asymptotic methods of Section 14.2 by Zhou and Marron (2015). That paper first showed that under classical Gaussian assumptions, spherical PCA has the same good consistency/strong inconsistency properties as conventional PCA that were quantified in (14.8). Next they showed that for a sufficiently strong sequence of outliers, PCA could be asymptotically pulled off in the fashion shown in the right panel of Figure 16.1, while spherical PCA gave the correct robust solution in that limit.

16.2.1 Case Study: Cornea Curvature Data

Spherical PCA was actually developed for the analysis of the Cornea Curvature data set described in Locantore et al. (1999). The cornea is the outer surface of the eye, and its curvature is critical to vision because most of the refraction of light entering the eye occurs there. The data objects in that study were images as shown in Figure 16.5. Color codes radial curvature (i.e. along rays emanating from the center) which is the curvature component with the most impact on visual acuity. Stronger radial curvature is represented by warmer colors. Representation of these functions on the disk was done using the *Zernike orthogonal basis*. That is a special case of orthogonal basis data object representation as discussed in Section 3.3. The Zernike basis for functions on the disc is defined in polar coordinates as a tensor product of the Fourier basis in the angular direction, and a system of Jacobi polynomials (carefully chosen to avoid a singularity at the origin) in the radial direction. See Schwiegerling et al. (1995) and Born and Wolf (2013) for further details. The 9 images shown in Figure 16.5, which is Figure 3 of Locantore et al. (1999), were selected to give an impression of the full set of $n = 43$. These are reconstructions of least squares Zernike fits, based on the data object choice of feature vectors having size $d = 66$ (which gives good noise reduction while maintaining image aspects of clinical interest). The three cases on the bottom row

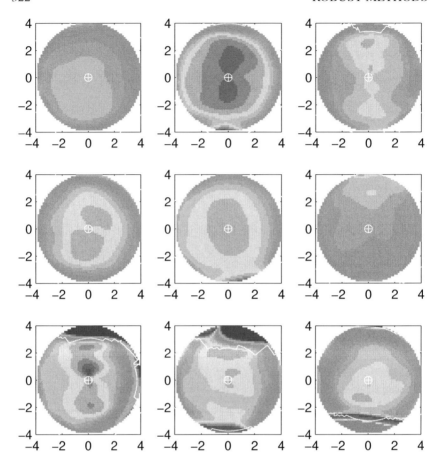

Figure 16.5 *Zernike basis representation of 9 data object images in the Cornea Curvature data from Locantore et al. (1999). Warmer colors are stronger radial curvature. Note strong edge effects are quite common. Circle plus signs denote the image (hence basis) center point, and white curves indicate boundaries of missing data.*

show extreme edge effects. These are artifacts of the data acquisition, which is done by collecting reflected light from the cornea. This can be somewhat blocked by eyelids, resulting in some missing data near the outer edge of the image. The boundary of the missing regions is indicated by the thin white curves. The wild red and blue values are the consequence of extrapolation in the Zernike fits.

As at many places in this book, PCA provides useful population-level insights through decomposition into modes of variation. The object feature space terminology of Section 3.1 is useful here, as the images are the data objects, and each is represented by a feature vector of $d = 66$ Zernike coefficients. Hence, PCA is done in $\mathbb{R}^{66}$, but the resulting modes of variation are interpreted in terms of images. This is done in Figure 16.6, which is from Figures 6–8 of Locantore et al.

(1999), using an approach similar to that of Figure 1.11, where the modes of variation are back-projected into the space of the original objects. The top row shows the first mode of variation with the sample mean in the center, the reconstructed image at two standard deviations below the mean is shown on the left, and on the right is two standard deviations above the mean. Note that the overall color is hotter on the left and cooler on the right, which is consistent with the first optometric measurement of overall curvature. The light orange figure 8 pattern in the mean is a consequence of astigmatism, which is a ridge of high curvature (the 8 shape is the radial curvature view of such ridges). The astigmatism is stronger when there is more overall curvature, indicating a correlation that was not previously known by the clinician co-authors of Locantore et al. (1999).

The second and third of these modes of variation are visualized on the second and third rows of Figure 16.6, using the same format (mean between minus and plus two standard deviations). Both the second and third mode have strong artifacts that look similar to those in the bottom row of Figure 16.5. In particular, the case on the lower right appears to be impacting the PC2 direction in $\mathbb{R}^{66}$, in a manner similar to that illustrated in the right panel of Figure 16.1. This was confirmed by looking at the PC2 scores plot, which looked quite similar to the PC2 scores of the Parabolas and Outlier data in Figure 16.2, verifying that this direction was indeed driven by that single outlying case. A natural approach was to delete that case, which indeed eliminated the impact of that outlier, but then PC2 was dominated by another outlier (not surprising from the bottom row of Figure 16.6. Even after sequentially deleting 4 such outliers, there were still strong outlier effects on the PCA, which is a concern because that is almost 10% of the data. That motivated a robust approach, i.e. an analog of PCA that downweighted the effect of the outliers, while still leaving them in the data set. This is also sensible since the central parts of all images contain useful information.

As discussed in Section 17.1 there are a number of approaches to PCA. Most of these can motivate approaches to robust PCA. One is via the characterization of PCA as eigen analysis of the sample covariance matrix (3.5), with the variance and covariance estimates replaced by robust versions. A major challenge to this approach is that the resulting estimated covariance is typically no longer positive definite, resulting in hard to interpret negative eigenvalues. The approach of Li and Chen (1985) followed the idea of PCA as directions of maximal projected variation, doing an iterative search over directions employing more robust measures of spread on the projections. Local optima are a challenge to that approach, and there do not seem to be successful implementations for dimensions as large as $d = 66$. Another intuitively appealing approach is the minimum volume ellipsoid approach of Rousseeuw and Leroy (1987), but that has the drawback of requiring $d < n$, which is sensible for affine invariance reasons, but not workable for this Cornea Curvature data set with $d = 66$ and $n = 43$. This dearth of widely-known robust PCA methods for high-dimensional situations motivated Locantore et al. (1999) to invent spherical PCA.

While spherical PCA gave a much improved analysis, the outliers still had a substantial influence. This was due to some of the Zernike coefficients being

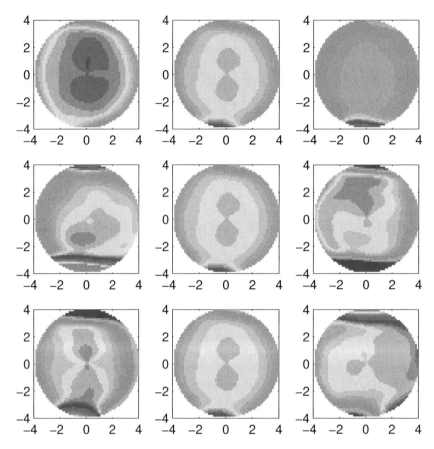

Figure 16.6 *Conventional PCA of the Cornea Curvature data from Locantore et al. (1999).*
Rows are first 3 modes of variation, center column shows the sample mean (same for each
row). Left (right, respectively) column is the mean minus (plus) 2 standard deviations.
First mode is clinically interesting. Second and third appear to be dominated by outliers,
motivating a robust approach.

orders of magnitude different from others (a common issue with orthogonal basis
representations). Locantore et al. (1999) tackled that problem by extending the
spherical PCA idea illustrated in Figure 16.3, to *elliptical PCA*. The main idea is
to replace projection onto the sphere by projection onto an ellipse whose axes are
parallel to the coordinate axes, thus effectively handling the wildly different scales
of the Zernike coefficients. The elliptical PCA analysis of the Cornea Curvature
data is shown in Figure 16.7 (from Figures 19–21 of Locantore et al. (1999)), using the same format as Figure 16.6. Note that the first mode of variation is quite
similar to that of the first PC above, with the same interpretation, but without the
small influence of outliers. The second mode now clearly appears as the known (to
the clinicians) mode of variation of steeper on top versus steeper on the bottom.

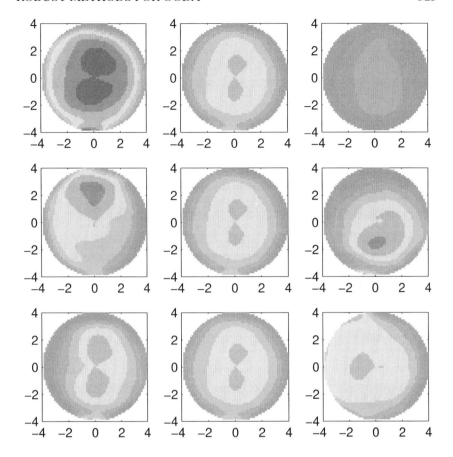

Figure 16.7 *Elliptical PCA of the same data (and in the same format, again from Locantore et al. (1999)) as Figure 16.6. Shows outlier effects have been removed. All three modes of variation are now clinically interpretable.*

The third mode is also clinically interpretable in terms of direction of astigmatism. While astigmatism can appear in any direction, most of it is approximately vertical, called "with the rule" in optometry. Most of the rest is horizontal, and is called "against the rule". The third mode is with the rule versus against the rule variation, although only one lobe of the figure 8 in the lower right plot actually appears above the yellow to orange color threshold.

As noted in Section 6.11.2 of Maronna et al. (2019), spherical PCA has since become a standard approach to robust PCA.

16.2.2 Case Study: Genome-Wide Association Data

Genome-Wide Association Studies (GWAS) were discussed in Section 14.2. As introduced in Klein et al. (2005), the data objects are vectors representing many

L2PCA

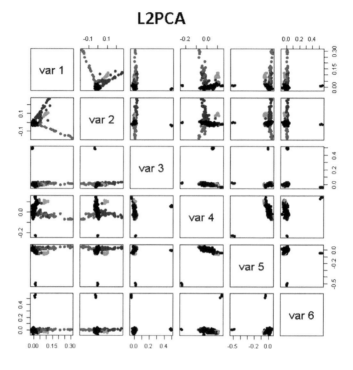

Figure 16.8 *Conventional L^2 PCA of a subset of the Cystic Fibrosis GWAS data set, from Zhou and Marron (2016). Shows major ethnic groups in the first two modes of variation. Others are driven by pairs of outliers.*

(up to 5 million with technology available at this writing) local variants in the DNA of humans at genomic locations called Single Nucleotide Proteins (SNPs). The entries of such vectors are typically binary, yet as seen in Figure 16.8, which is part of Figure 1 in Zhou and Marron (2016), PCA still shows a large amount of interesting population structure. This is an extreme example of how PCA is useful even for non-Gaussian data. In fact these data objects all lie at vertices of a very high-dimensional unit cube.

Colors are used in Figure 16.8 to represent some ethnic groups, which are clearly important aspects of the variation in this population. The data come from the cystic fibrosis analysis done by Wright et al. (2011). That paper did careful preprocessing resulting in binary SNP vectors of length $d = 21,000$ (chosen to give a representative set of SNPs). The data set is quite large, so for convenient display only a subset of $n = 347$ people chosen as described in Zhou and Marron (2016) is shown here. While the ethnic groups stand out strongly, note that some of the PC directions (PCs 3–6) are clearly dominated by just a few outliers. The above robust PCA ideas suggest analyzing this data set using spherical PCA, which is done in Figure 16.9.

Based on the ideas in Figure 16.3, it is natural to expect that the influence of the outliers would be greatly reduced by using spherical PCA. Yet that clearly has not happened in Figure 16.9 (the rest of Figure 1 in Zhou and Marron (2016)), which looks very similar to Figure 16.8, particularly in terms of the impact of the outliers. This perhaps surprising behavior can be understood using the HDLSS geometric representation ideas of Section 14.2. Those concepts suggest that all the data should be equidistant around the population expected value, i.e. essentially lying on the surface of a sphere of radius $d^{1/2}$. Furthermore pairwise angles (from the population center) are all approximately $90°$. A strong exception to this overall pairwise orthogonality is first-degree relatives (e.g. siblings or parents and children) who share about half of their SNPs. This results in a pairwise angle of approximately $\cos^{-1}\left(\frac{1}{2}\right) = 60°$, i.e. much smaller than the typical $90°$, which is enough to make just a pair of points drive a PC direction. It is not easy to comprehend that the clearly visible ethnic arms visible in Figure 16.8 are the consequences of pairwise angles only slightly smaller than $90°$, caused by groups of people having a larger number of SNPs in common. As these data objects already lie near the surface of the sphere centered at the population expected value, it is actually not surprising that the spherical PCA device of projecting to a sphere has no meaningful impact, i.e. Figure 16.8 is so similar to Figure 16.9.

An effective solution to the first degree relative outlier problem in GWAS data illustrated in Figures 16.8 and 16.9 is the Visual L^1 PCA proposed by Zhou and Marron (2016). That analysis is shown in Figure 16.10. The first two PC modes of variation clearly capture the ethnic groups visible in the first two components of the earlier analyses. But the remaining modes appear to be much more useful, as they shift the focus from outlying pairs to separate groups of people that appear to represent various ethnic subgroups.

Visual L^1 PCA is based on the robust L^1 PCA of Brooks et al. (2013). As noted in Section 8.6, that method uses a clever backwards implementation. While L^1 PCA gives a robust and useful set of directions for understanding population variation, the corresponding L^1 scores tend to be quite hard to interpret as noted in Zhou and Marron (2016), due to the lack of rotation invariance of the L^1 norm. The Visual L^1 approach is based on the same robust direction vectors of maximal L^1 variation, but computes scores using the more interpretable L^2 projections. This gives the needed insensitivity to outliers demonstrated in Figure 16.10, which is part of Figure 6 in Zhou and Marron (2016).

16.3 Other Robustness Areas

As discussed in Section 7.1, there are many notions of multivariate median. These all have the property that in the case $d = 1$ they are just the conventional sample median. Good access to this area can be found in Small (1990) and Chakraborty and Chaudhuri (1999). Especially appealing are the *data depth* methods, see e.g. Liu et al. (1999), Vardi and Zhang (2000), and López-Pintado and Romo (2006). A median type estimate of scale is the Median Absolute Deviation (from the median), defined in (13.2).

Spherical PCA

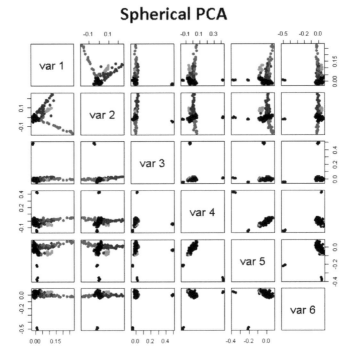

Figure 16.9 *Spherical PCA of the same data (also from Zhou and Marron (2016)) as in Figure 16.8. Shows essentially no impact of projecting data onto a sphere and then computing PCA. The strong influence of the outlying pairs, and the perhaps surprising lack of robustness, is understood using HDLSS ideas from Section 14.2.*

A novel approach to finding outliers in RNAseq data, of the type illustrated in Sections 4.1.3 and 13.2.1 can be found in Choi et al. (2018). See Ahn et al. (2019) for interesting outlier detection methods in HDLSS contexts.

Some discussion of robustness in landmark shape analysis has been given by Dryden and Mardia (2016, Section 13.6). In particular distinctions are made between three different types of robustness: resistance to landmark outliers (some of the d variables are unusual within a particular object); resistance to object outliers (some of the n objects are unusual in the sample); and robustness to model misspecification. Dryden and Walker (1999) discuss methods for landmark outlier resistance and adapt the S-estimator (Rousseeuw and Yohai (1984); Rousseeuw and Leroy (1987)) and the Least Median of Squares estimator (Rousseeuw (1984)) for landmark shape matching. It is important to balance having a high breakdown with high statistical efficiency, where *breakdown* is the minimum percentage of points that can be moved arbitrarily to achieve maximum discrepancy. It is also crucial that any approach to robust shape analysis be equivariant, for example the same match should be obtained regardless of the arbitrary rotations of the objects.

V-L1PCA

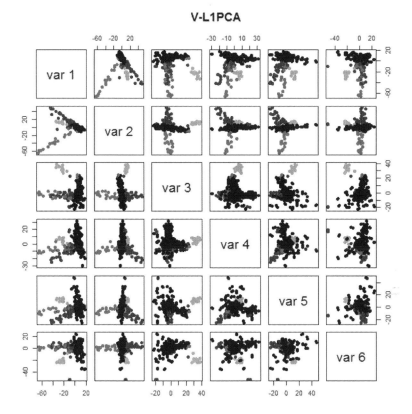

Figure 16.10 *Visual L^1 analysis (also from Zhou and Marron (2016)) of the same GWAS data in Figures 16.8 and 16.9. This view is much more robust against the pairs of outlying first-degree relatives, instead shifting the focus to additional groups of people (different ethnic subgroups).*

Dryden and Walker (1999) used such an S-estimator with 25% breakdown, which has a good trade-off between breakdown and efficiency. Least Median of Squares has almost 50% breakdown, and an example of its use for matching electrophoresis gels is given by Dryden and Mardia (2016, Section 13.6). These resistant shape analysis techniques can be helpful for identifying outlier landmarks, which are highlighted by very large residuals after resistant matching.

CHAPTER 17

PCA Details and Variants

PCA (recall from Section 1.1, this is an acronym for Principal Component Analysis) has been very useful at many points in preceding chapters. A good source for many aspects of PCA (including the history, such as the name itself going back to Hotelling (1933)) is Jolliffe (2002). Common terminology for PCA is "dimension reduction", which appears to be rooted in the era of pre-computational statistics, when indeed dimension reduction was a critical task. In the current era of abundant computational capabilities, more relevant and specific ways of thinking about PCA are:

- *Data visualization*. For example where scores distribution plots clearly show the *relationships* between data objects. This was done in preceding chapters at many points, starting with the Spanish Mortality data in Figure 1.6 and the Twin Arches toy data in Figure 4.4. Further insights came from decomposition into easily interpretable one-dimensional modes of variation (each revealing both loadings and scores), starting again with Spanish Mortality in Figures 1.4–1.5, and with the 2-d Toy data in Figures 3.2–3.4.

- *Denoising*. Useful in situations where a low-rank approximation of the data can retain most of the signal and thus PCA may reduce noise in the data.

- *Efficient data representation*. Which comes up in HDLSS situations where $d > n$ as discussed in Section 14.2. The $(n-1) \times n$ matrix that contains all nonzero PC scores is essentially a rotation of the mean centered data into the lower dimensional space $\mathbb{R}^n$, giving an $(n-1)$ dimensional (exact) representation of the data, as detailed in Section 17.1.4.

Several quite different but useful ways of thinking about PCA are given in Section 17.1. Section 17.2 discusses methods that parallel PCA in the sense of shifting the focus from variation in a single data block to analyzing *joint* variation between two different but *linked* blocks of data.

Often really good ideas are discovered several times independently, usually under a different name. PCA is a clear example of this, with alternate names appearing in other fields as listed in Table 17.1.

DOI: 10.1201/9781351189675-17

Name	Field
Factor Analysis	Social Sciences
Karhunen Loeve Expansion	Probability Theory / Electrical Engineering
Empirical Orthogonal Functions (EOF)	Geo and Climate Sciences
Proper Orthogonal Decomposition (POD)	Applied Mathematics

Table 17.1 *Alternate names for concepts essentially similar to PCA in other fields.*

Not all of these are direct synonyms for PCA. For example, as discussed in Section 17.1.3, in psychometrics factor analysis is not exactly PCA, as it includes simultaneous likelihood estimation of the residual variance. However, the name factor analysis is also used in a number of other fields where it actually is a synonym for PCA. Also, the Karhunen-Loeve expansion is the same idea, but typically applied to probability distributions instead of to data sets. Of course these can be related by thinking of a data set in terms of its empirical discrete probability distribution, using the idea of *empirical probability measure*, that puts probability mass n^{-1} on each data object.

17.1 Viewpoints of PCA

As discussed above there are many ways of considering PCA, most of which provide differing useful insights. As noted in Section 6.3 *data centering* issues tend to be downplayed, but actually have a surprisingly important and sometimes non-intuitive role (see for example the combined view analysis of the Twin Arches data in Figures 6.9 and 6.10). This is true for both discussions about rows or columns of the data matrix being viewed as the data objects (recall the discussion in Section 3.1), and also in data visualizations, as studied below in Section 17.1.1. Good insight into many aspects of PCA, including the decomposition of the column object-centered data matrix into insightful modes of variation, comes from studying the Singular Value Decomposition (SVD) in Section 17.1.2. A traditional and important approach to PCA is through a low-rank factor model whose parameters are estimated by Gaussian likelihood estimation (which can be deceptive) as reviewed in Section 17.1.3. Relevant computational and graphical issues are discussed in Section 17.1.4.

Recalling notation from Section 3.3, and the use of tildes to indicate random quantities from Table 7.1, write a $d \times n$ data matrix as

$$\widetilde{X} = \begin{bmatrix} \widetilde{x}_{1,1} & \cdots & \widetilde{x}_{1,n} \\ \vdots & \ddots & \vdots \\ \widetilde{x}_{d,1} & \cdots & \widetilde{x}_{d,n} \end{bmatrix} = \begin{bmatrix} \widetilde{x}_1 & \cdots & \widetilde{x}_n \end{bmatrix}, \qquad (17.1)$$

where the data object column vectors (sometimes called "cases" or even "samples") are

$$
\widetilde{\boldsymbol{x}}_j = \begin{bmatrix} \widetilde{x}_{1,j} \\ \vdots \\ \widetilde{x}_{d,j} \end{bmatrix} \in \mathbb{R}^d,
$$

and where each number $\widetilde{x}_{i,j}$ is the trait value (sometimes called "variable" or "feature") for $j = 1, \cdots, n$, and $i = 1, \cdots, d$. Linear algebra provides a useful mathematical backbone for understanding many aspects of OODA, and an important issue is the space in which this is done. Many mathematical treatments focus on the vector spaces $\mathbb{R}^d$ and/or $\mathbb{R}^n$, but in this chapter we take the unconventional approach of developing the ideas in the vector space $\mathbb{R}^{d \times n}$. Recall that space from (3.4) where the symbol $\times$ in the exponent is used to indicate the vector space of matrices, in contrast with $\mathbb{R}^{dn}$, which denotes the corresponding vector space of long vectorized column vectors of length dn. Subspaces and projections in $\mathbb{R}^{d \times n}$ provide a simple framework for understanding the relationships between many OODA operations. This approach is a simplification of the ideas that Mammen et al. (2001) developed to understand how kernel smoothing (discussed in Sections 15.1 and 15.2) can be viewed as a projection.

Recall from (4.3) that $\boldsymbol{I}_d$ denotes the $d \times d$ identity matrix, from (10.1) that $\boldsymbol{1}_{d,n}$ is the $d \times n$ matrix of ones, and from (4.2) that $\boldsymbol{0}_{d,n}$ indicates the $d \times n$ matrix of 0s.

The operation of *projection* is fundamental here. Given any metric space with distance δ, a subset $\mathcal{S}$, and an element $\boldsymbol{x}$, the projection of $\boldsymbol{x}$ onto $\mathcal{S}$ is the closest point in $\mathcal{S}$ to $\boldsymbol{x}$, i.e.

$$
P_{\mathcal{S}}(\boldsymbol{x}) = \arg \min_{\boldsymbol{s} \in \mathcal{S}} \delta(\boldsymbol{s}, \boldsymbol{x}).
$$

Centering operations are studied in Section 17.1.1, using projections in $\mathbb{R}^{d \times n}$, with respect to the Frobenius norm $\|\cdot\|_F$, defined in (7.9). Additional useful linear algebra comes from the corresponding Frobenius inner product

$$
\langle \boldsymbol{M}, \boldsymbol{N} \rangle_F = \sum_{i=1}^{d} \sum_{j=1}^{n} M_{i,j} N_{i,j}, \tag{17.2}
$$

on $\mathbb{R}^{d \times n}$.

Two relevant subspaces of $\mathbb{R}^{d \times n}$ are based on the concept of *flat vectors* whose entries are all the same. For example in $\mathbb{R}^d$, the set of flat vectors is the subspace

$$
\left\{ \begin{pmatrix} u \\ \vdots \\ u \end{pmatrix} : u \in \mathbb{R} \right\},
$$

which is sometimes called "the 45-degree line", although that only makes sense in the case $d = 2$. The flat vectors are also sometimes called "constant vectors", but the "flat" terminology seems easier to keep in mind in the context of the matrix space $\mathbb{R}^{d \times n}$.

The concept of flat vectors allows interpretation of conventional univariate means as projections. First note that a unit basis vector (i.e. with norm 1) of the subspace of flat vectors in $\mathbb{R}^n$ is $\frac{1}{\sqrt{n}}\mathbf{1}_{n,1}$. Hence given $\boldsymbol{x} = \begin{pmatrix} x_1 \\ \vdots \\ x_n \end{pmatrix} \in \mathbb{R}^n$,

the $\mathbb{R}^n$ projection coefficient of $\boldsymbol{x}$ onto the flat subspace is the inner product $\frac{1}{\sqrt{n}}\mathbf{1}_{1,n}\boldsymbol{x} = \sqrt{n}\bar{x}$. Multiplying that coefficient by the basis vector results in the $\mathbb{R}^n$ projection of $\boldsymbol{x}$ onto the subspace of flat vectors as

$$\frac{1}{\sqrt{n}}\mathbf{1}_{n,1}\left(\frac{1}{\sqrt{n}}\mathbf{1}_{1,n}\boldsymbol{x}\right) = \frac{1}{n}\mathbf{1}_{n,n}\boldsymbol{x} = \mathbf{1}_{n,1}\bar{x}, \tag{17.3}$$

which is the flat vector whose common entry is $\bar{x}$.

To understand data centering as projection in the matrix space $\mathbb{R}^{d \times n}$, useful notation is $\mathcal{S}_{FT}$ for the subspace (of $\mathbb{R}^{d \times n}$) in which all of the row trait vectors are flat, where FT stands for *Flat Traits*, i.e.

$$\mathcal{S}_{FT} = \left\{ \boldsymbol{u}\mathbf{1}_{1,n} : \boldsymbol{u} \in \mathbb{R}^d \right\}.$$

Similarly define the subspace (of $\mathbb{R}^{d \times n}$) consisting of matrices composed of flat column object vectors (*Flat Objects*) to be

$$\mathcal{S}_{FO} = \left\{ \mathbf{1}_{d,1}\boldsymbol{v}^t : \boldsymbol{v} \in \mathbb{R}^n \right\}.$$

Applying the transpose of (17.3) to each row of the data matrix $\widetilde{\boldsymbol{X}}$ gives the projection

$$P_{\mathcal{S}_{FT}}\left(\widetilde{\boldsymbol{X}}\right) = \widetilde{\boldsymbol{X}}\left(\frac{1}{\sqrt{n}}\mathbf{1}_{n,1}\right)\left(\frac{1}{\sqrt{n}}\mathbf{1}_{1,n}\boldsymbol{x}\right) = \widetilde{\boldsymbol{X}}\left(\frac{1}{n}\mathbf{1}_{n,n}\right) = \bar{\boldsymbol{x}}_{CO}\mathbf{1}_{1,n},$$

which essentially extends the $d \times 1$ column object mean $\bar{\boldsymbol{x}}_{CO}$ from (6.1) into a $d \times n$ element of $\mathcal{S}_{FT}$. The column object centered version of the data matrix can also be written as a projection onto the orthogonal (with respect to the Frobenius inner product (17.2)) complementary subspace $\mathcal{S}_{FT}^{\perp} = \left\{ \boldsymbol{M} \in \mathbb{R}^{d \times n} : \boldsymbol{M} \perp \mathcal{S}_{FT} \right\}$ as,

$$P_{\mathcal{S}_{FT}^{\perp}}\left(\widetilde{\boldsymbol{X}}\right) = \widetilde{\boldsymbol{X}} - \bar{\boldsymbol{x}}_{CO}\mathbf{1}_{1,n} = \widetilde{\boldsymbol{X}}\left(\boldsymbol{I}_n - \frac{1}{n}\mathbf{1}_{n,n}\right).$$

Similarly, row trait mean centering as defined in (6.2) can be studied in terms of projections onto the flat object subspace $\mathcal{S}_{FO}$

$$P_{\mathcal{S}_{FO}}\left(\widetilde{\boldsymbol{X}}\right) = \mathbf{1}_{d,1}\bar{\boldsymbol{x}}_{RT}^t = \left(\frac{1}{d}\mathbf{1}_{d,d}\right)\widetilde{\boldsymbol{X}},$$

and its orthogonal complement $\mathcal{S}_{FO}^{\perp}$

$$P_{\mathcal{S}_{FO}^{\perp}}\left(\widetilde{\boldsymbol{X}}\right) = \widetilde{\boldsymbol{X}} - \mathbf{1}_{d,1}\bar{\boldsymbol{x}}_{RT}^t = \left(\boldsymbol{I}_d - \frac{1}{d}\mathbf{1}_{d,d}\right)\widetilde{\boldsymbol{X}}.$$

17.1.1 Data Centering

Some mean centering issues were discussed in Sections 6.3 and 7.3.3. As noted in Prothero et al. (2021), while centering seems like a simple issue, it can be surprisingly slippery. Here a deeper view based on the above ideas of projections onto $\mathbb{R}^{d \times n}$ is based on the Sine Wave toy data set, with $d = 20$ rows and $n = 10$ columns, shown in Figure 17.1. The data were generated (for $i = 1, \cdots, 20$ and $j = 1, \cdots, 10$) as

$$x_{i,j} = T_1 + T_2 + T_3 - 3 + N(0, 10^{-6}) \qquad (17.4)$$

where

$$T_1 = \sin\left(5 \cdot \pi \cdot (i - 1)/19\right),$$
$$T_2 = 0.3 \cdot (j - 5.8)^2,$$
$$T_3 = 0.005 \cdot (i - 10.5) \cdot (j - 5.5).$$

Let T_1, T_2, and T_3, denote the matrix versions of the first three terms on the right-hand side of (17.4). Note that each of these is a *mode of variation* as defined in Section 3.1. These modes illustrate how column object and row trait mean centering are understood through projections in $\mathbb{R}^{d \times n}$. In particular, note that $T_1 = u1_{1,n}$ ($u \in \mathbb{R}^d$ generates the sine wave) has flat rows, so $T_1 \in \mathcal{S}_{FT}$ (i.e. is a flat trait mode) and that the variation across rows follows a sin wave. Similarly $T_2 = 1_{d,1}v^t$ ($v \in \mathbb{R}^n$ gives the parabola) has flat columns, so $T_2 \in \mathcal{S}_{FO}$ and its (flat object) variation across columns follows a parabola. Both of these modes are clearly visible in the views of this Sine Wave data set shown in Figure 17.1. The third mode T_3 is a product of two (both mean 0) linear vectors and has a much smaller coefficient to make it a much smaller contribution to the overall variation. The linear factors have both been carefully centered, which entails $T_3 \in \mathcal{S}_{FT}^{\perp} \cap \mathcal{S}_{FO}^{\perp}$ which will leave T_3 remaining after both centering operations. Finally a very low level of independent Gaussian variation is also added.

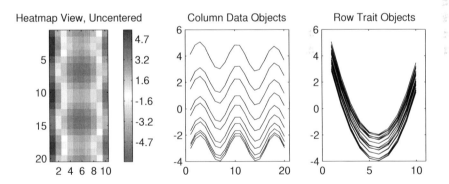

Figure 17.1 *Combined view of the Sine Wave raw data matrix. Heat-map in the left panel. Curve views of same data set, with columns as data objects in the center panel and row trait vectors on the right. Visual connections between the three different views take some effort to make.*

Figure 17.1 explores the Sine Wave data set of (17.4) using a different formatting of the combined data matrix view introduced in Section 6.3. A heat-map view (as discussed in Sections 6.1) is shown in the left panel. As expected from the form of (17.4), the heat-map contains both vertical sine wave and horizontal parabolic patterns.

Another view of the Sine Wave data matrix (in the spirit of the top right panels of Figures 6.9–6.12) is of the column data objects (essentially sine waves), shown as $n = 10$ curves in the center panel. That view suggests that vertical shift of the (common) sine pattern is the dominant mode of variation. A third view highlights the $d = 20$ row trait vectors in $\mathbb{R}^n$ shown as curves in the right panel. That graphic is in the spirit of the lower left panels of Figures 6.9–6.12. It gives the quite different impression that the dominant mode of variation is instead about vertical shift of the parabolic curve. A careful look at the matrix view in the left panel reveals what is happening: the variation in the height of the sine waves follows the parabola, and the variation in the heights of the parabola follows the sine wave pattern. An important point is *neither* is a single mode of variation, as defined in Section 3.1.4, as they are not rank-one matrices. In fact both are the sum of *two* modes shown as $T_1 + T_2$ in (17.5). The first mode $T_1 = u1_{1,n}$ can be thought of as either extending the sine wave u in a flat horizontal way (giving the common sine wave in the center panel) or else providing sinusoidal variation to the flat line $1_{1,n}$. The second mode $T_2 = 1_{d,1}v^t$ similarly plays a dual role. It both provides parabolic variation (modeled by v^t) to the vertical flat line $1_{d,1}$ to drive the variation in the center panel, and also gives a vertical flat extension of the parabola v^t apparent in the right panel.

The effect of column object mean centering, i.e. $P_{S_{FT}^\perp}\left(\widetilde{X}\right)$, is studied in Figure 17.2. Because $T_1 \in S_{FT}$ that subtraction of the column data object mean essentially removes the vertical sine wave from the heat map (left panel of Figure 17.2) making the horizontal parabolic structure of the mode T_2 even more apparent. The conventional column data object view in the middle panel shows that the common structure of the curves in the middle of Figure 17.1, i.e. the sine wave structure, has been removed, leaving mostly just the flat vertical shift mode of variation T_2 (with heights determined by the parabola). Also note that the linear T_3 mode of variation begins to be visible in these curves as well. This same type of removal of mean structure was very useful in the Spanish Mortality data analysis of Figure 1.3 where the column object mean nicely captured the mortality structure that is common over time, while the mean residuals contained the variation across time. It also was demonstrated using the Tilted Parabolas data in the top row of Figure 4.1. The alternate row trait view (right panel of Figure 17.2) instead shows that column object mean centering removes most of the T_1 driven variation visible in the right panel of Figure 17.1, again since that variation was driven by the sinusoidal structure in the data, ending up with the set of parabolas that are nearly all the same in T_2.

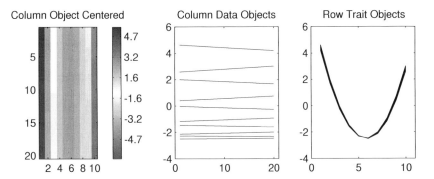

Figure 17.2 *Effect of column object centering for the Sine Wave data of Figure 17.1, using the same panel format. Shows removal of vertical sine wave component, leaving nearly constant columns whose heights follow the parabola.*

Figure 17.3 shows the effect of complementary row trait mean centering, i.e. $P_{\mathcal{S}_{FO}^{\perp}}\left(\widetilde{X}\right)$, on the Sine Wave data set. The heatmap view in the left panel shows that this time the horizontal parabola effect T_2 (defined in (17.4)) has been removed, leaving mostly horizontal stripes following the T_1 vertical sine wave pattern. In the conventional column data object curve view in the center panel, this projection has the effect of removing the dominant and important T_2 vertical shift mode of variation (since $T_2 \in \mathcal{S}_{FO}$). The vertical axes in the center and right panels of Figures 17.1, 17.2, and 17.3 all use a common scale to show the T_2 parabolic variation has the larger magnitude relative to the T_1 sine wave component. A major point illustrated here is that the type of centering used can have a strong visual impact, although frequently too little attention is paid to this point. Finally the right panel shows that this operation removes the T_2 parabolic structure from the corresponding view of the original data on the right of Figure 17.1. Note that the T_3 tilted linear component is now a little more apparent (actually in both plots).

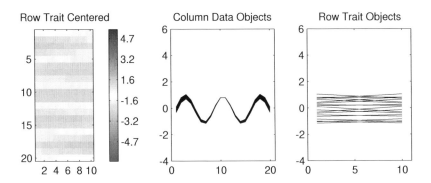

Figure 17.3 *Row trait centering of Sine Wave data from Figure 17.1, with the same panel format. Shows removal of T_2 horizontal parabolic component.*

Another data centering option is *double centering* shown in Figure 17.4. Double centering can be viewed in a number of ways. The simplest is through simultaneous removal of both types of mean via a composition of projections $P_{\mathcal{S}_{FT}^{\perp}}\left(P_{\mathcal{S}_{FO}^{\perp}}\left(\widetilde{X}\right)\right)$. Other ways of thinking about (or even recipes for computation of) double centering include

$$P_{\mathcal{S}_{FT}^{\perp}}\left(P_{\mathcal{S}_{FO}^{\perp}}\left(\widetilde{X}\right)\right) = \left(\widetilde{X} - P_{\mathcal{S}_{FO}}\left(\widetilde{X}\right)\right) - P_{\mathcal{S}_{FT}}\left(\widetilde{X} - P_{\mathcal{S}_{FO}}\left(\widetilde{X}\right)\right) =$$

$$= \widetilde{X} - P_{\mathcal{S}_{FO}}\left(\widetilde{X}\right) - P_{\mathcal{S}_{FT}}\left(\widetilde{X}\right) + P_{\mathcal{S}_{FT}\cap\mathcal{S}_{FO}}\left(\widetilde{X}\right).$$

Writing this in matrix form gives the double centered version of the data

$$\overline{\overline{X}} = \widetilde{X} - \bar{x}_{CO}\cdot\mathbf{1}_{1,n} - \mathbf{1}_{d,1}\cdot\bar{x}_{RT}^{t} + \bar{x}_{AA}\cdot\mathbf{1}_{d,n} = \qquad (17.5)$$

$$= \left(I_d - \frac{\mathbf{1}_{d,d}}{d}\right)\widetilde{X}\left(I_n - \frac{\mathbf{1}_{n,n}}{n}\right).$$

where the *grand mean* (i.e. mean over all of the entries of the data matrix) is the scalar

$$\bar{x}_{AA} = d^{-1}\sum_{i=1}^{d}\bar{x}_{i,A} = n^{-1}\sum_{j=1}^{n}\bar{x}_{A,j} = (nd)^{-1}\sum_{i=1}^{d}\sum_{j=1}^{n}\widetilde{x}_{i,j} = (nd)^{-1}\mathbf{1}_{1,d}X\mathbf{1}_{n,1}.$$

One way of understanding why the grand mean $\bar{x}_{AA}$ should be added back in is that an overall vertical shift in the entire data matrix (e.g. the fourth term in (17.4)) will be subtracted twice from the previous terms, so it should be added back in to give the correct overall impact.

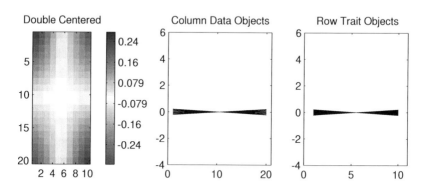

Figure 17.4 *Double centered version of the Sine Wave data in Figure 17.1. Shows most structure has been removed leaving just the product of sloping lines.*

Note that in Figure 17.4 both the horizontal and vertical effects that dominated Figure 17.1 have been removed. That leaves mostly just the product of linear components generated by $T_3 \in \mathcal{S}_{FT}^{\perp}\cap\mathcal{S}_{FO}^{\perp}$ showing up as a distinctive pattern that results from a set of lines whose slopes change linearly from positive to negative.

Note that in the heat-map in the left panel, the color bar has been changed to reflect a finer color scale (than in Figure 17.1), since use of the original scale results in very light colors that are harder to interpret. However, the vertical axes in the center and right panels have been kept the same, so one can still visually gauge the relatively small magnitude of the tilted lines that are a consequence of the T_3 product linear mode of variation. Further discussion of these different notions of mean centering of a data matrix, and their impact on PCA type decompositions of variation can be found in Zhang et al. (2007a).

Next, this Sine Wave data set defined in (17.4) is used to understand how these centering operations interact with classical functional PCAs, such as illustrated using the Tilted Parabolas example in Figure 4.1. In particular, Figure 17.5 shows an analysis of the Sine Wave data, using the format of Figure 4.1. The top left panel shows the column object raw data curves using a rainbow color scheme (starting with magenta for the left-hand column of the heat map through red for the right-hand column, i.e. over indices $j = 1, \cdots, 10$). These are colored versions of the curves in the top center panel of Figure 17.1, where the color order is best understood using the scores plot in the bottom right panel of Figure 17.5. The top center panel shows the data object mean curve, $\overline{x}_{CO}$, which can be thought of in an $\mathbb{R}^{d \times n}$ way as an overlay of the n identical columns (as curves) of $\overline{x}_{CO} 1_{1,n} = P_{\mathcal{S}_{FT}}\left(\widetilde{X}\right)$. The top right panel shows the residuals from subtracting that mean, i.e. $P_{\mathcal{S}_{FT}^{\perp}}\left(\widetilde{X}\right)$. The latter are colored versions of those in the center panel of Figure 17.2. This makes it clear that the classical FDA mean centering operation should be thought of as removing a mode of variation from the data (in this case the T_1 mode). The PC1 mode of variation in the bottom left panel of Figure 17.5 is essentially $T_2 = 1_{d,1} v^t$, i.e. a product of the flat loadings vector $1_{d,1}$ with parabolic scores v^t. The latter are highlighted in the scores plot in the bottom right, where both the rainbow colors and the parabolic pattern are clear. An important point is that the appearance of the flat trait mode T_2 as the PC1 mode of variation should not be thought of as typical, and instead is just a consequence of the design of this particular example. However, first PC modes frequently have at least some influence from the flat trait mode $P_{\mathcal{S}_{FO}}\left(\widetilde{X}\right)$, see for example the Spanish Mortality data in Figure 1.4. The residuals in the bottom center plot of Figure 17.5 are linear with slopes again following the rainbow colors, which capture the small variation in the slopes of the line segments in the top right panel, thus displaying the T_3 component of the variation that is a colored version of the curves in the center panel of Figure 17.4.

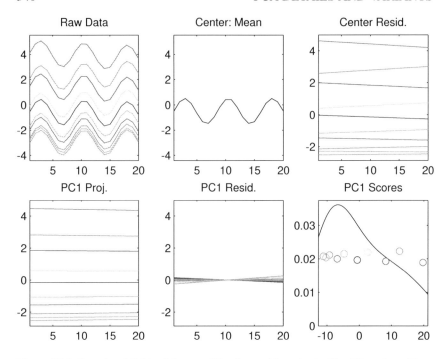

Figure 17.5 *Functional PCA of the* $n = 10$ *column objects in the Sine Wave data. Shows the sine wave component is common structure appearing in the column object mean (top center panel). PC1 mode of variation is essentially a constant function (flat mode), as seen in the bottom left loadings plot. Residuals from this mode show up as lines with ordered slopes (bottom center). Rainbow color scheme follows the ordering of the columns, clearly understood in the scores plot (lower right).*

The dual version of FDA, based on row trait vectors as data objects, is shown in Figure 17.6. This time a very different heat scale color scheme is used, running from black through red to yellow following the indices $i = 1, \cdots, 20$, to indicate that now color indicates rows of the data matrix $\widetilde{X}$. The top left panel shows the colored parabolic version of the curves in the left panel of Figure 17.1. This time it is the T_2 parabolic mode of variation that shows up as common structure captured by the row trait mean (top center). The sine wave component T_1 is now much harder to see in the first PC mode in the bottom left panel, which again are essentially flat lines. This is because the sine wave appears in the ordering of the curves, as reflected in the scores shown in the lower right panel. The residuals in the bottom center panel are again lines with linearly varying slope driven by T_3, where the indices i and j have swapped roles relative to the bottom right panel of Figure 17.5.

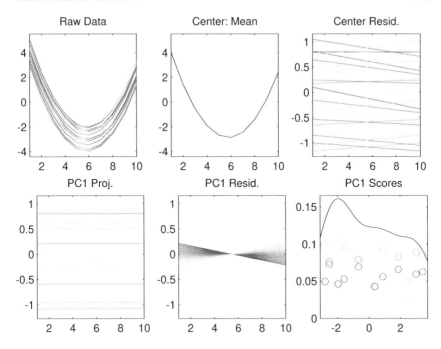

Figure 17.6 *Functional PCA of the $d = 20$ row trait vectors in the Sine Wave data. Heat color scheme is used to highlight row ordering. From this viewpoint the parabola appears as mean structure, and the sine wave component generates the variation of the flat curves in the first mode of variation. Again the residuals are lines with ordered slopes.*

It can be challenging to connect the structure seen in the column object PCA in Figure 17.5 in an intuitive fashion with the row trait PCA shown in Figure 17.6. Attention to where the terms of (17.4) appear in Figures 17.5 and 17.6, makes clear the differences and also connections between these PCAs based on column data objects and on row trait vectors. The main point is that loadings in one correspond to scores in the other. In this sense the two types of PCA are duals to each other. That point is generally somewhat obscured by centering issues, but can be clarified by focusing on modes of variation. This time the PC1 mode of variation in the bottom left panel of Figure 17.6 is essentially $T_1 = u1_{1,n}$, which is a product of the constant direction vector $1_{1,n} \in \mathbb{R}^n$ with sine wave scores u. The scores plot in the lower right reveals the sine wave pattern following the circles from cool colors at the bottom through hotter colors at the top. Also this time the bi-linear PC1 residuals in the bottom center plot have a form similar to those in the bottom center plot of Figure 17.4, but with a more informative coloring.

A special property of this Sine Wave toy data set is that the mean mode of each is the first PC mode of the other. Hence Figures 17.5 and 17.6 are able to illustrate the important general principle that for each *common* mode of variation, the scores for one are the loadings for the other. However, in general for column object PCA, after mean object centering, the constant direction vector $1_{d,1} \in \mathbb{R}^d$ will not point in any single PC direction, so the modes of variation of the mean residuals will not be the same as for row trait object PCA. Thus exact duality will usually not hold, except for modes where both loadings and scores are orthogonal to their respective flat directions, i.e. sum to 0. When this type of duality is desirable (i.e. when it is important that the scores on one analysis be the loadings of the other), it can be guaranteed by starting the PCA at the double centered version of the data $\overline{\overline{X}}$ from (17.5). This amounts to separately accounting for both of the mean-based modes of variation, before finding the PC modes.

Another viewpoint of centering ideas is *data filtering*. Filtering a data set by row trait mean centering $P_{\mathcal{S}_{FO}^{\perp}}\left(\widetilde{X}\right)$ removes the flat mode of variation in (conventional) column object PCA (sometimes quite desirable, e.g. for removing library effects in gene expression). Similarly filtering by column object mean centering removes the flat mode of variation in row trait (sometimes called dual) PCA. Other common types of filtering include scaling (e.g. dividing either rows or columns by their standard deviation) as discussed in Section 5.2, and transformation as discussed in Section 5.3.

17.1.2 Singular Value Decomposition

A typical conception of PCA is sequential in nature through the components (modes of variation), where at each step one searches for a direction of maximal variation in the subspace orthogonal to the previously found directions, and then uses projections of the data objects on these directions to define the modes. For example this viewpoint was taken in discussing the analysis of the Spanish Mortality data in Section 1.1 and the Tilted Parabolas (Twin Arches) example in Section 4.1.1 (4.1.2, respectively). But an interesting property of PCA is that finding the full set of modes can be usefully considered to be a single matrix operation.

As noted for example in Section 3.5 of Jolliffe (2002), *Singular Value Decomposition* (SVD) provides a direct and simple approach to computing PCA, and to understanding its properties. The full matrix version of the SVD of a $d \times n$ data matrix $\widetilde{X}$ is

$$\widetilde{X} = UDV^t, \qquad (17.6)$$

where $U \in O(d)$ (the set of orthogonal matrices as defined in Section 7.3.5) is a $d \times d$ orthonormal basis matrix of $\mathbb{R}^d$, where D is a $d \times n$ diagonal matrix (i.e. all matrix entries are 0 except on the main diagonal) of *singular values* that are nonnegative and assumed to be sorted in decreasing order, and where $V \in O(n)$ is an $n \times n$ orthonormal basis matrix of $\mathbb{R}^n$.

An interesting interpretation of the SVD comes from thinking of $\widetilde{X}$ as the matrix of coefficients of a linear transformation from $\mathbb{R}^n$ to $\mathbb{R}^d$, when multiplying from the left, and also of linear transformation from $\mathbb{R}^d$ to $\mathbb{R}^n$, under right multiplication applied to the transpose of a vector. In both cases, note that the transformations corresponding to orthonormal matrices are *isometries*. These are elements of $O(d)$ (or $O(n)$), which are a little more general than rigid rotations (elements of $SO(d)$ or $SO(n)$ as defined in Section 7.3.3), since various axis flippings are also included. Note that they preserve Euclidean distances. Furthermore linear transformation by a diagonal matrix simply rescales (possibly zeroing out) each vector entry. Thus the SVD in (17.6) represents both (i.e. left and right) linear transformations implied by $\widetilde{X}$ as essentially a rotation, followed by a coordinate by coordinate rescaling (or zeroing), which is followed by another rotation.

While the full matrix version of the SVD defined above lends itself to linear transformation interpretation, it is usually numerically inefficient, in the sense that it can be written in terms of smaller matrices. When $d \neq n$ the diagonal matrix D has at most only $d \wedge n$ (where $\wedge$ denotes the minimum) nonzero elements, so there are either rows or columns of D that are all zero and thus are not actually used in the product reconstruction of $\widetilde{X}$. In particular, replacing by appropriate sub-matrices, the SVD in (17.6) can have the same form where U is $n \times (d \wedge n)$, D is $(d \wedge n) \times (d \wedge n)$, and V is $(d \wedge n) \times n$.

Further reduction is available in the lower rank case when $\widetilde{X}$ has rank $r < (d \wedge n)$, because then some of the singular values are 0, so again there are parts of the matrices that have no effect on the product. In this case, (17.6) still has the same form but now U is $n \times r$, D is $r \times r$, and V is $r \times n$. This is the most efficient *loss-less* version of SVD which gives an exact representation of $\widetilde{X}$. The columns of

$$U = [u_1 \cdots u_r], \quad V = [v_1 \cdots v_r] \tag{17.7}$$

are called the *left* and *right singular vectors,* respectively.

SVD can be thought of as providing the solution to a number of optimization problems. Two of these are in terms of optimal projections onto direction vectors in $\mathbb{R}^d$ and in $\mathbb{R}^n$. Recalling the notation (3.2) for high-dimensional spheres, the set of all direction vectors in $\mathbb{R}^d$ is the sphere $S^{d-1} = \{u \in \mathbb{R}^d : \|u\| = 1\}$. Note that S^{d-1} is one way to represent the Grassmannian manifold, $\mathbf{Gr}(1, d)$, i.e. the set of all 1-d subspaces of $\mathbb{R}^d$. The $1 \times n$ row vector whose entries are projection coefficients (scores) of each data object (column of $\widetilde{X}$) onto a given direction vector $u \in S^{d-1}$ is given as $u^t \widetilde{X}$. A useful measure of *signal power in the direction* u is the sum of squared scores, $u^t \widetilde{X} \left(u^t \widetilde{X}\right)^t = u^t \widetilde{X} \widetilde{X}^t u$. The left singular vectors $u_1, \cdots, u_d$ give (sequential) *maximal signal power* of possible projections in $\mathbb{R}^d$ in the sense that

$$u_l = \arg \max_{u \in S^{d-1}, u \perp u_1, \cdots, u_{l-1}} u^t \widetilde{X} \widetilde{X}^t u \tag{17.8}$$

for $l = 1, \cdots, d$. Furthermore the maximum signal powers are the squares of the *singular values* $s_1, \cdots, s_d$, the diagonal elements of D (using the convention $s_l = 0$ for $l > (d \wedge n)$). In the case of the appropriately centered and scaled data matrix in (17.13) below, these squared singular values drive the relative heights of the red circles in the scree plots introduced in Figure 3.5. Thus SVD simultaneously provides the full PCA decomposition without the need for sequential computation.

The right singular vectors (columns of V) solve parallel optimization problems in $\mathbb{R}^n$, in particular, given a direction vector $v \in S^{n-1} \subset \mathbb{R}^n$, the $d \times 1$ column vector of projection coefficients (loadings) of the rows of $\widetilde{X}$ onto v is $\widetilde{X}v$. Signal power in the direction v is the sum of squares $\left(\widetilde{X}v\right)^t \widetilde{X}v = v^t \widetilde{X}^t \widetilde{X}v$, which are sequentially maximized by the right singular vectors $v_1, \cdots, v_n$,

$$v_l = \arg \max_{v \in S^{n-1}, v \perp v_1, \cdots, v_{l-1}} v^t \widetilde{X}^t \widetilde{X}v$$

for $l = 1, \cdots, n$, where the maximum values are the *same* singular values $s_1, \ldots, s_n$ using appropriate 0s as needed to fill in discrepancies due to $d \neq n$.

Another important optimization problem solved by SVD (key to deep understanding of the modes of variation generated by PCA) can be seen by writing it as a sum of rank 1 matrices:

$$\widetilde{X} = \sum_{l=1}^{r} s_l u_l v_l^t, \tag{17.9}$$

where s_l is the l-th singular value and u_l and v_l are defined in (17.7). This representation gives easy insight into the good matrix approximation properties of SVD. In particular, for $k \leq r$ (the rank of $\widetilde{X}$), from the ordering of the singular values $s_1 \geq \cdots \geq s_r > 0$, it follows that $\widetilde{X}_k = \sum_{l=1}^{k} s_l u_l v_l^t$ is the best rank k approximation of $\widetilde{X}$ in the sense that

$$\widetilde{X}_k = \arg \min_{M \in \mathcal{R}_k} \left\| \widetilde{X} - M \right\|_F,$$

where $\mathcal{R}_k$ is the set (not a subspace) of matrices of rank $\leq k$. Note that $\widetilde{X}_k$ has a representation of the form (17.6), $\widetilde{X}_k = U_k D_k V_k^t$ where U_k, D_k, V_k^t are the first k columns of U, the upper $k \times k$ sub-diagonal of D, and the first k rows of V^t, respectively. This also shows how each rank k SVD can be viewed as a projection onto $\mathcal{R}_k$ in $\mathbb{R}^{d \times n}$, the space of $d \times n$ matrices, as was done in Section 17.1.1. In addition, each rank one matrix $s_l u_l v_l^t$ is the projection of the data matrix $\widetilde{X}$ onto the one-dimensional subspace generated by $u_l v_l^t$, and the rank k approximation is the projection onto the k dimensional subspace generated by $u_1 v_1^t, \cdots, u_k v_k^t$. Finally note that the representation (17.9) connects with the discussion of modes of variation in Section 3.1.4. In particular, each rank 1 matrix $s_l u_l v_l^t$ (for $l = 1, \cdots, k$) is a mode of variation, containing both the loadings (u_l) and scores ($s_l v_l^t$).

The representation (17.9) also provides a very useful connection between the SVD loadings and scores. In particular, the scores are projections of the data onto the (subspace generated by the) respective loadings vectors (in $\mathbb{R}^d$) in the sense that for $l = 1, \cdots, r$

$$u_l^t \widetilde{X} = u_l^t \sum_{l'=1}^{r} s_{l'} u_{l'} v_{l'}^t = s_l v_l^t. \tag{17.10}$$

Similarly, loadings have a parallel simple representation in terms of scores as

$$\widetilde{X} v_l = \sum_{l'=1}^{r} s_{l'} u_{l'} v_{l'}^t v_l = s_l u_l. \tag{17.11}$$

The matrix version of (17.10) is the projection of each data object (column of $\widetilde{X}$)

onto each of the singular vectors, through the matrix inner product calculation

$$U^t \widetilde{X} = U^t U D V^t = D V^t. \tag{17.12}$$

These inner products give the coefficients of the projections that are called SVD scores, which are the basis of many useful visualizations (e.g. scatterplot matrices as discussed in Section 6.4) in this book.

Perhaps the most direct mathematical understanding of PCA comes from viewing it as the SVD of the (sample size scaled) column object-centered version of the data,

$$\check{X} = n^{-1/2} \left(\widetilde{X} - \overline{x}_{CO} \mathbf{1}_{1,n} \right) = n^{-1/2} \widetilde{X} \left(I_n - \frac{1}{n} \mathbf{1}_{n,n} \right), \tag{17.13}$$

written as

$$\check{X} = \check{U} \check{D} \check{V}^t. \tag{17.14}$$

For $l = 1, \cdots, r$, the l-th set of PCA *scores*, i.e. the l-th *principal components*, are in the l-th row of the $r \times n$ matrix $\check{D} \check{V}^t$. Recall these scores provide insights into how the data objects relate to each other through scatterplots as seen in Section 6.4 and other places. As noted starting in Section 1.1, insight into the drivers of these relationships comes from the *loadings*. For $l = 1, \cdots, r$, the l-th column of $\check{U}$ is the l-th loadings vector which is the direction (recall a vector of norm 1) in $\mathbb{R}^d$ of l-th largest variation in the data. For $i = 1, \cdots, d$, the i-th entry of the loadings vector reflects direction and magnitude of the influence of the i-th variable on the l-th direction. The PC scores and loadings are intimately related to each other via projections. In particular, since

$$\check{U}^t \check{X} = \check{U}^t \check{U} \check{D} \check{V}^t = \check{D} \check{V}^t, \tag{17.15}$$

the l-th row of the scores matrix $\check{D} \check{V}^t$ is the $n \times 1$ vector of inner products (essentially projection coefficients) of the l-th loadings vector with the centered data objects (columns of $\check{X}$). Similarly

$$\check{X} \left(\check{D} \check{V}^t \right)^t \check{D}^{-2} = \check{U} \check{D} \check{V}^t \check{V} \check{D} \check{D}^{-2} = \check{U} \tag{17.16}$$

shows that the loadings can be represented as a normalization (rescaling by the inverse variances) of the inner products of the centered data matrix and the scores. In summary, (17.15) and (17.16) show that both loadings and scores can be derived from the other by an appropriate product with the scaled, centered data matrix.

The column centering operation in (17.13) provides enhanced interpretability in several ways, including connection with the sequential view of PCA. First, using $\check{X}$ instead of $\widetilde{X}$ in (17.8) gives the interpretation of PC directions first discussed in Section 1.1: $\check{u}_1$ is the direction vector (based at the sample mean) that *maximizes sample variance* of the projections. Note that it is the centering operation and division by $n^{1/2}$ that changes the criterion from sum of squares to sample variance. Since the columns of $\check{U}$ are orthonormal, each $\check{u}_l$ (for $l = 2, \cdots, r$) is the direction vector in the subspace orthogonal to $\check{u}_1, \cdots, \check{u}_{l-1}$ that maximizes the sample variance of the data projections. A second enhancement resulting from column object centering was illustrated in the bottom panels of Figure 6.8, comparing uncentered SVD with PCA in a toy example. In particular, the column object mean centering used in PCA (together with the orthogonality of the rows of $\breve{V}^t$, i.e. the columns of $\breve{V}$) results in more easily interpretable *uncorrelated* scores scatterplots. As noted in Section 6.3, PCA scores scatterplots all have correlation 0 because otherwise the direction vector could be rotated to result in a larger sample variance of the scores. Another view of this phenomenon comes from calculating the $r \times r$ matrix of inner products of the scores vectors, again using the orthonormality of the columns of $\breve{V}$ and the fact that $\breve{D}$ is diagonal,

$$\breve{D}\breve{V}^t \left(\breve{D}\breve{V}^t \right)^t = \breve{D}\breve{V}^t\breve{V}\breve{D}^t = \breve{D}^2. \tag{17.17}$$

Orthogonality of the scores vectors (recall rows of $\breve{D}\breve{V}^t$), and hence 0 sample correlation of the entries of those vectors as seen in scores scatterplots, follows from the fact that $\breve{D}^2$ is also a diagonal matrix, whose off-diagonal entries are thus all 0.

Without column object mean centering, there is generally some (i.e. non-zero) sample correlation of SVD scores plots, again as demonstrated in the lower left panel of Figure 6.8. However, in many situations, this correlation is not distracting, because frequently one of the singular vectors (usually the first) happens to point approximately in the direction of the sample column object mean. In particular, many data sets tend to have an important flat mode of variation, for example T_2 as illustrated in (17.4). That flat mode is often roughly in the first SVD direction, so the SVD modes of variation are frequently similar to those of a standard PCA.

Again a major lesson of Figure 6.8 is that column mean object centering results in uncorrelated PC scores. The symmetry of the SVD in terms of rows and columns, readily apparent from (17.6) and (17.14), suggests a parallel relationship: row trait mean centering implies that such pairwise scatterplots of the PCA loadings will be similarly uncorrelated. While column object mean centering is routine for PCA, row trait mean centering (or double mean centering as defined in (17.5)) is not. As shown in Figure 6.8 appropriate centering will ensure that scores are uncorrelated. This does not seem to have been widely noticed for two reasons. First, with the noticeable exception of *biplot* data views, as defined by Gabriel (1971), and more completely developed in Gower and Hand (1996); Gower (1966), there has not been strong interest in visualizing PC loadings. While biplots can be insightful in low dimensional contexts, a version adapted to high dimensions has not yet been developed. The second reason is that in typical analyses, the column mean object centering removes the grand mean $\bar{x}_{AA}$, which often is a large factor in the lack of correlation in the scores. An exception is shown in Figure 17.7, based on the Spanish Mortality data studied in Section 1.1.

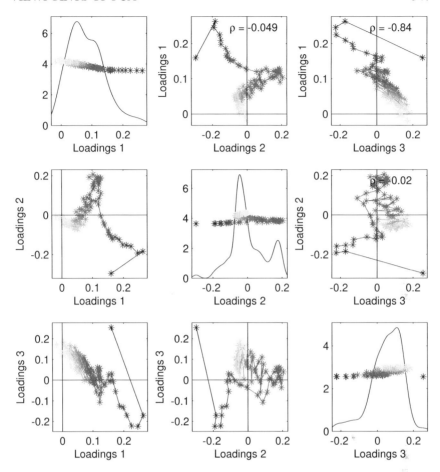

Figure 17.7 *PCA loadings for the Spanish Mortality data studied in Section 1.1. Heat coded coloring indicates age from 0 (black) to 98 (yellow). Shows correlations in loadings caused by the lack of row trait mean centering.*

Note that the correlations between the first and second as well as the second and third loadings vectors are small, but the first and third are strongly negatively correlated. Furthermore the first loadings do not have mean 0. Thus, in situations where it is important to simultaneously consider both scores and loadings, such as the multi-block analyses studied in Section 17.2, it can be important to use an SVD of the double centered version of the data $\overline{\overline{X}}$ defined in (17.5).

Also note that there are strong connections with the column object-oriented analysis in Section 1.1. In particular the pattern in the first loadings can be seen in the top center panel by tracing (through the ages on the horizontal axis) of the upper magenta curve in the left panel of Figure 1.4. There is an initial relatively small value (black) with an initial jump followed by slow decrease with later wiggles due to the age rounding (red and yellow). In that same panel of Figure 17.7

the pattern in the second loadings (left panel of Figure 1.5) also follows the visible magenta curve, initially negative swinging to positive (for the 20 to 45 year olds) and then back to negative. This is natural because as noted in Section 17.1.1 the usual PCA (with columns as data objects) scores and loadings defined above essentially (modulo centering) become the loadings and scores, respectively when the rows are taken as data objects, which is now easy to understand from taking the transpose of (17.14).

It is also insightful to formulate these quantities in terms of the sample covariance matrix $\widehat{\Sigma}$ defined at (3.5). Given a data matrix $\widetilde{X}$, the entries of $\widehat{\Sigma}$ are the sample variances

$$\widehat{var}_i = \frac{1}{n}\sum_{j=1}^{n}(\widetilde{x}_{i,j} - \overline{x}_{i,A})^2 \qquad (17.18)$$

and the sample covariances

$$\widehat{cov}_{i,i'} = \frac{1}{n}\sum_{j=1}^{n}(\widetilde{x}_{i,j} - \overline{x}_{i,A})(\widetilde{x}_{i',j} - \overline{x}_{i',A}) \qquad (17.19)$$

for $i, i' = 1, \cdots, d$. The sample covariance matrix is related to PCA through representing it as the outer product

$$\widehat{\Sigma} = \breve{X}\breve{X}^t. \qquad (17.20)$$

Using the singular value representation (17.14) gives

$$\widehat{\Sigma} = \breve{U}\breve{D}\breve{V}^t\breve{V}\breve{D}^t\breve{U}^t = \breve{U}\breve{\Lambda}\breve{U}^t \qquad (17.21)$$

where $\breve{\Lambda} = \breve{D}\breve{D}^t$ is a diagonal matrix whose entries are the squares of the entries of $\breve{D}$, i.e. the squares of the singular values $\breve{s}_l$. The representation (17.21) is called an *eigenvalue decomposition* or *eigen analysis* at many points above. This reveals that in addition to the (17.14) representation of PCA as a Singular Value Decomposition of the column object-centered data matrix, it can also be thought of as an eigenvalue analysis of the sample covariance matrix. The entries of $\breve{\Lambda}$ are the eigenvalues, which divided by their sum (i.e. the total energy in the centered data matrix $\breve{X}$) are the heights of the scree plots shown e.g. in Figure 3.5. Such eigenvalues were also fundamental to the *random matrix theory* studied in Section 14.1.

17.1.3 Gaussian Likelihood View

A common theme throughout this book has been a *nonparametric* or data centric development of analytical methods for OODA. This is a contrast to many introductions to multivariate analysis where the focus is instead on underlying parametric statistical models and resulting likelihood approaches. Low-rank *latent factor* models provide a traditional approach of this latter type to PCA. In particular, assume the $d \times n$ data matrix can be written in the form

$$\widetilde{X} = \mu\mathbf{1}_{1,n} + LS + \widetilde{E},$$

where for some low-rank r, L is a $d \times r$ orthonormal loadings matrix, S is an $r \times n$ scores matrix whose rows are orthogonal and sum to 0 (for identifiability) and $\widetilde{E}$ is a $d \times n$ matrix of errors whose entries are often assumed to be independent $N\left(0, \sigma^2\right)$. Rows of S are often thought of as unobserved latent factors, and their estimation is the primary goal of *factor analysis*. Standard calculations show that for a given r, using notation from (6.1) and (17.14), the maximum likelihood estimates of $\boldsymbol{\mu}$, L and S are $\overline{\boldsymbol{x}}_{CO}$, $\breve{U}$ and $\breve{D}\breve{V}^t$, respectively. This is a sense in which standard PCA can be thought of from the likelihood viewpoint.

While this approach is mathematically elegant, as noted in Section 4.1 it has the serious downside that it has fed a common misconception that PCA is only useful when the data approximately follow a multivariate Gaussian distribution. As seen at many points in this book, PCA provides insightful data visualizations in many non-Gaussian situations. Important real data examples of this appear in Figures 4.7 (Lung Cancer data) and 4.9 (Pan-Cancer).

As noted at the beginning of this chapter, properly defined factor analysis is closely related to PCA (as opposed to being exactly the same). The difference is that factor analysis includes estimation of the residual variance σ^2 as part of the likelihood calculation. This results in the same loadings (column vectors of $\breve{U}$), i.e. directions for visualizations, as PCA. However the scores are somewhat impacted by the simultaneous estimation of σ^2.

17.1.4 PCA Computational Issues

Good discussion of various approaches to computing PCA can be found in Section A.1 of Jolliffe (2002). Using most software packages, which contain well-optimized SVD functions, the SVD approach to computation described in Section 17.1.2 is both insightful and generally computationally useful.

As noted at the beginning of this chapter, PCA can give an efficient (exact) low dimensional representation of a $d \times n$ data matrix. In particular, it follows from (17.13) and (17.14) that

$$\widetilde{X} = \overline{\boldsymbol{x}}_{CO}\mathbf{1}_{1,n} + \breve{U}_r\left(n^{1/2}\breve{D}_r\breve{V}_r^t\right),\tag{17.22}$$

where r is the rank of $\widetilde{X} - \overline{\boldsymbol{x}}_{CO}\mathbf{1}_{1,n}$ (which is usually the rank of $\widetilde{X}$ when $n > d$ and one smaller otherwise), where the *loadings* matrix $\breve{U}_r$ contains the first r columns of $\breve{U}$ (these are the PCA directions of projection in $\mathbb{R}^d$) and where $\left(n^{1/2}\breve{D}_r\breve{V}_r^t\right)$ is the $r \times n$ matrix of PCA scores (projection coefficients), plotted at many points in this book starting with Figure 1.4. Thus the scores are a translated and orthogonal transformed (rotation and/or sign flip) version of the data that are a lower-dimensional isometric (hence loss-less) representation in $\mathbb{R}^r$.

The data representation (17.22) is closely related to the formula for simulating a $d \times n$ random matrix $\widetilde{X} \sim N_d\left(\boldsymbol{\mu}, \boldsymbol{\Sigma}\right)$ as

$$\widetilde{X} = \boldsymbol{\mu}\mathbf{1}_{1,n} + \boldsymbol{\Sigma}^{1/2}Z = \boldsymbol{\mu}\mathbf{1}_{1,n} + U\boldsymbol{\Lambda}^{1/2}U^tZ = \boldsymbol{\mu}\mathbf{1}_{1,n} + U\boldsymbol{\Lambda}^{1/2}Z^* \tag{17.23}$$

where both Z and Z^* are $d \times n$ matrices of independent standard normals since

U^t is an orthogonal matrix. To simulate Gaussians that follow the data set in any matrix $\widetilde{X}$ as well as possible, use (17.23) with μ (and $U\Lambda^{1/2}$) replaced by the sample version $\overline{x}_{CO}$ (and $n^{1/2}\check{U}_r\check{D}_r$, respectively).

17.2 Two Block Decompositions

As noted at many points above, a set of data objects that is usefully viewed as columns of a matrix is a common format (recall the OODA convention of cases as columns of the data matrix and traits, i.e. variables as rows). However, an increasing number of OODA contexts have been appearing where data objects naturally consist of several such column vectors of traits that are grouped in meaningful ways. A canonical example is the TCGA data set, introduced in Sections 4.1.3 and 4.1.4. In addition to the gene expression studied in those sections, around a dozen other types of measurements were made for each patient, including protein expression and mutation status. Most measurement types result in long vectors of numbers, so the resulting data objects are naturally considered to be sets of vectors. For clear study of not just the measurements but also the relationships between the groups, it is convenient to group the vectors into several matrices, called data *blocks* here. In particular, for $b = 1, \cdots, B$ let the $d_b \times n$ data block be

$$\widetilde{X}^{(b)} = \left[\widetilde{x}_1^{(b)} \cdots \widetilde{x}_n^{(b)}\right]. \tag{17.24}$$

An important aspect of multi-block data is that for each $j = 1, \cdots, n$ the corresponding measurement vectors $\widetilde{x}_j^{(1)}, \cdots, \widetilde{x}_j^{(B)}$ are all made on the same experimental unit, e.g. tissue sample in the case of the TCGA data. There are several synonyms for "multi-block", such as *multi-view* which is common in the machine learning literature. An accessible overview of multi-block methods can be found in De Bie et al. (2005).

Time-honored approaches to multi-block data, also called data integration, are *Partial Least Squares* (PLS) studied in Section 17.2.1 and *Canonical Correlation Analysis* (CCA) discussed in Section 17.2.2. As discussed there, these powerful methods can be approached from several viewpoints, and have a number of variants. Some of these variants are usefully understood and contrasted through their corresponding modes of variation.

A simple starting point for this is the case $B = 2$ and the joint covariance matrix

$$\begin{bmatrix} \widehat{\Sigma}^{(1)} & \widehat{\Sigma}^{(1,2)} \\ \widehat{\Sigma}^{(1,2)} & \widehat{\Sigma}^{(2)} \end{bmatrix}, \tag{17.25}$$

where $\widehat{\Sigma}^{(1)}$ and $\widehat{\Sigma}^{(2)}$ are the respective within block sample covariance matrices (defined at (3.5) and usefully represented as an outer product in (17.20)) of the two blocks, and where $\widehat{\Sigma}^{(1,2)}$ is the *cross-covariance matrix* whose entries are sample covariances of the form (17.19) with $i = 1, \cdots, d_1$ (and $i' = 1, \cdots, d_2$) representing the index of a trait (variable) in $\widetilde{X}^{(1)}$ (and $\widetilde{X}^{(2)}$, respectively). In the spirit of (17.20), using block indexed versions of the column object-centered data matrices (17.13), the $d_1 \times d_2$ cross-covariance matrix has an outer product representation as $\widehat{\boldsymbol{\Sigma}}^{(1,2)} = \breve{\boldsymbol{X}}^{(1)} \left(\breve{\boldsymbol{X}}^{(2)} \right)^t$, where $\breve{\boldsymbol{X}}^{(b)}$ is the column object-centered version of the data in block b for $b = 1, 2$.

Additional extensions of PLS and CCA ideas include kernel (in the classification sense of Section 11.2) versions in Bach and Jordan (2002), the sparsity approaches of Gao et al. (2015) and the deep learning implementations proposed by Andrew et al. (2013). For more than two blocks there are dozens of extensions as seen in Tenenhaus and Tenenhaus (2011). For example, CCA alone has 20 generalizations to multiple blocks: Kettenring (1971); Nielsen (2002); Asendorf (2015).

17.2.1 Partial Least Squares

The terminology Partial Least Squares (PLS) has been used for a large number of methods for multi-block analysis. The most important of these are a set of methods for deriving joint modes of variaftion from the cross-covariance matrix $\widehat{\boldsymbol{\Sigma}}^{(1,2)}$. Good overview appears in the survey paper Wegelin (2000). The name PLS was coined by Wold (1975, 1985).

A straightforward version, labeled PLS-SVD by Wegelin (2000), develops joint modes of variation using ideas parallel to PCA (as described in Section 17.1.2). It starts with the SVD of the cross-covariance matrix

$$\widehat{\boldsymbol{\Sigma}}^{(1,2)} = \widetilde{\boldsymbol{U}} \widetilde{\boldsymbol{D}} \widetilde{\boldsymbol{V}}^t. \tag{17.26}$$

For $l = 1, \cdots, d_1 \wedge d_2$ the columns $\widetilde{\boldsymbol{u}}_l$ of $\widetilde{\boldsymbol{U}}$ and $\widetilde{\boldsymbol{v}}_l$ of $\widetilde{\boldsymbol{V}}$ are direction vectors in $\mathbb{R}^{d_1}$ and $\mathbb{R}^{d_2}$, (respectively) that sequentially maximize the covariance of projections of $\widetilde{\boldsymbol{X}}^{(1)}$ onto $\widetilde{\boldsymbol{u}}_l$ and of $\widetilde{\boldsymbol{X}}^{(2)}$ onto $\widetilde{\boldsymbol{v}}_l$, subject to orthogonality of columns of both $\widetilde{\boldsymbol{U}}$ and $\widetilde{\boldsymbol{V}}$. Each pair of direction vectors generates *joint modes of variation* in $\mathbb{R}^{d_1 \times n}$ and $\mathbb{R}^{d_2 \times n}$ by computing corresponding scores vectors using the data matrix multiplication principle of (17.15). The resulting *cross-covariance scores* are projection coefficients of the data objects onto the direction loading vectors as in (17.10). Hence these are computed as $(d_1 \wedge d_2) \times n$ *PLS scores* matrices

$$\widetilde{\boldsymbol{S}}^{(1)} = \widetilde{\boldsymbol{U}}^t \left(\widetilde{\boldsymbol{X}}^{(1)} - \overline{\boldsymbol{x}}_{CO}^{(1)} \mathbf{1}_{1,n} \right), \ \widetilde{\boldsymbol{S}}^{(2)} = \widetilde{\boldsymbol{U}}^t \left(\widetilde{\boldsymbol{X}}^{(2)} - \overline{\boldsymbol{x}}_{CO}^{(2)} \mathbf{1}_{1,n} \right).$$

Scatterplot matrix plots, as in e.g. Figure 4.4 (Twin Arches data), of scores (projections on the loadings vectors) within each block gives insights (parallel to PCA scatterplot matrix plots) into object relationships that are maximally related to the other block. Also of interest is to study the amount of dependence between the blocks through, for $l = 1, \cdots, d_1 \wedge d_2$, viewing the scatterplots of the scores $\widetilde{\boldsymbol{S}}_l^{(1)}$ vs. $\widetilde{\boldsymbol{S}}_l^{(2)}$. As for PCA, the entries of the vectors $\widetilde{\boldsymbol{u}}_1$ and $\widetilde{\boldsymbol{v}}_1$ are both called *loadings* in this context. These loadings can be plotted as in Figure 4.11 to give insights as to the drivers of the *joint variation* between data blocks. When the loadings are interpretable as smooth curves, such as for the Spanish Mortality data studied in

Section 1.1, useful insights come from multiplying the loadings vectors by scores to produce mode of variation plots, such as those in Figures 1.4 and 1.5, and in many other places.

While the PLS-SVD approach is quite straightforward to compute and gives practically useful modes of variation, as demonstrated in Singh et al. (2010) and other places noted in Wegelin (2000), it does have some limitations. In particular these modes of variation do not share the *bi-orthogonality* property of PCA modes, which was that *both* the columns of the matrix $\breve{U}$ and also the columns of $\breve{V}$ (i.e. the rows of $\breve{V}^t$) are orthonormal bases (of $\mathbb{R}^d$ and $\mathbb{R}^n$, respectively). In particular, the rows of the block 1 scores matrix $\widetilde{S}^{(1)}$ are generally *not* orthogonal to each other, and similarly for the rows of $\widetilde{S}^{(2)}$. Because each vector of scores has sample mean 0 of its entries, these non-orthogonalities entail that the sample correlations between vector entries are not 0, i.e. there is some correlation between these vectors, which can be quite apparent and even disconcerting in scores scatterplot matrix views. However, there is some orthogonality *across* the blocks, that follows from a calculation similar to (17.17). In particular the parallel $(d_1 \wedge d_2) \times (d_1 \wedge d_2)$ inner product matrix is

$$\widetilde{S}^{(1)} \left(\widetilde{S}^{(2)} \right)^t = \widetilde{U}^t \left(\widetilde{X}^{(1)} - \overline{x}_{CO}^{(1)} 1_{1,n} \right) \left(\widetilde{V}^t \left(\widetilde{X}^{(2)} - \overline{x}_{CO}^{(2)} 1_{1,n} \right) \right)^t =$$

$$= \widetilde{U}^t \left(\widetilde{X}^{(1)} - \overline{x}_{CO}^{(1)} 1_{1,n} \right) \left(\widetilde{X}^{(2)} - \overline{x}_{CO}^{(2)} 1_{1,n} \right)^t \widetilde{V} =$$

$$= \widetilde{U}^t \widehat{\Sigma}^{(1,2)} \widetilde{V} = \widetilde{U}^t \left(\widetilde{U} \widetilde{D} \widetilde{V}^t \right) \widetilde{V} = \widetilde{D}. \qquad (17.27)$$

Hence the diagonality of the singular value matrix $\widetilde{D}$ shows that the scores in each block are orthogonal to the non-corresponding scores in the other block. This does not seem to be widely acknowledged, perhaps because there does not appear to be motivation to construct such cross block scores scatterplots.

The above (within blocks) lack of orthogonality of the PLS-SVD scores can be viewed as motivating a number of variants. The best known of these comes from Wold (1975, 1985) who proposed *PLS regression* as an improvement of *PCA regression*. The latter is useful in multiple linear regression contexts with many and/or collinear predictors, which tend to cause numerical instability in classical least squares algorithms. The main idea of PCA regression is to replace those predictors with a reduced set of PCA scores, which often provide a good low-rank approximation (thus containing most of the relevant information of the original predictors while hopefully eliminating noise), sometimes called *latent variables*. That also avoids numerical instability in the regression formulas because of the orthogonality of the PCA scores vectors. An obvious drawback is that PCA is driven completely by the variation in the predictors, which has no explicit connection to the responses. The intuitively attractive improvement of PLS regression is to replace the low-rank approximation based on SVD of (17.14) inherent to PCA with a more directly relevant approximation based on SVD of (17.25). That gives modes of variation that focus on the connection between predictors and responses, instead of merely maximizing variation among the predictors. However, as discussed above, the scores from the SVD of (17.25) are not entirely suitable as predictors because of their non-orthogonality. This can be thought of as the motivation for Wold's PLS which does have uncorrelated scores. Those give both enhanced numerical performance of the regression algorithm, and also more interpretable scores in latent variable terms.

There are various approaches to Wold's version of PLS. An algorithm-centered approach (focused on sequentially finding appropriate directions) can be found in Wegelin (2000). The idea is to find a first mode of variation using (17.26) as above. Additional modes are found by iteratively solving (17.26) with the added constraint of orthogonality of the scores vectors. Many authors prefer a model-based latent factor way of thinking along the lines of that approach to PCA described in Section 17.1.3, with the same type of sequential implementation. Other approaches are more optimization-based, see for example Rosipal and Krämer (2005) and Xu et al. (2013).

The orthogonality of the scores in Wold's PLS comes at a price of loss of orthogonality of the loadings vectors. The orthogonality of scores across blocks shown in (17.27) is similarly lost in Wold's PLS. There do not seem to be two block decompositions into modes of variation that share the bi-orthogonality property of PCA.

A question that does not appear to be addressed in the literature is how far are the scores resulting from PLS-SVD from being orthogonal, i.e. how far are the PLS-SVD modes of variation from being bi-orthogonal? This is investigated in Table 17.2 which summarizes results from a small simulation study based on 1000 realizations of pairs of matrices with i.i.d. Gaussian entries of dimensions $d \times n$. Each pair was double centered as described in Section 17.1.1 and scores from an SVD of the cross-covariance matrix $\widehat{\Sigma}^{(1,2)}$ were computed. The angles between the first and second scores vectors for both matrices were computed, and the minimums and averages have been summarized in Table 17.2.

d	n	Min. Angle	Average Angle
10000	10	88.70	89.67
1000	100	86.59	89.19
100	1000	84.65	88.83
10	10000	87.85	89.55

Table 17.2 *Summary of simulation study, exploring angles between PLS-SVD scores vectors, in the case of i.i.d. Gaussian data. Shows scores vectors tend to be orthogonal across a wide range of settings.*

Note that broadly over this wide range of contexts, these angles are all perhaps surprisingly close to orthogonal, i.e. $90°$. In the HDLSS case ($d \gg n$) this is not surprising, given the fundamental ideas from Section 14.2 that random directions tend to be orthogonal in those cases. Less obvious is why the angles are also very close to $90°$ in the other cases. An open theoretical problem seems to be an asymptotic analysis that quantifies these observations. Of course, the results of Table 17.2 only apply to pure noise contexts, and some types of low-rank signal can easily lead to much smaller angles between scores vectors.

An important limitation of PLS is that because it is driven by covariance it does not always focus completely on joint structure but instead also feels variation within each block, as demonstrated in Figure 17.8. For example when one variable has a much larger variance than the others, it will tend to unduly influence PLS. A solution to this problem is given in Section 17.2.2.

17.2.2 Canonical Correlations

As noted in Section 17.2.1, PLS-SVD provides modes of joint variation that maximize covariance between scores. As discussed in the context of correlation PCA in Section 5.2, a potential drawback of covariance is that it feels the units of the data. This can create challenges in several contexts, an important example being data blocks whose traits are measured in non-commensurate units. In those situations it is very sensible to replace the covariance criterion with the unit free Pearson's correlation coefficient. That goal results in CCA, which was named by Hotelling (1936), who as noted at the beginning of this chapter also coined the term PCA.

CCA starts with directions $\widetilde{u} \in \mathbb{R}^{d_1}$ and $\widetilde{v} \in \mathbb{R}^{d_2}$. Corresponding vectors of projections of the ($\frac{1}{n}$ rescaled) column object-centered data (17.13) are $\left(\check{X}^{(1)}\right)^t \widetilde{u}, \left(\check{X}^{(2)}\right)^t \widetilde{v} \in \mathbb{R}^n$. Their sample correlation is

$$\widehat{\rho}\left(\left(\check{X}^{(1)}\right)^t \widetilde{u}, \left(\check{X}^{(2)}\right)^t \widetilde{v}\right) = \frac{\widehat{cov}\left(\left(\check{X}^{(1)}\right)^t \widetilde{u}, \left(\check{X}^{(2)}\right)^t \widetilde{v}\right)}{\left(\widehat{var}\left(\left(\check{X}^{(1)}\right)^t \widetilde{u}\right) \widehat{var}\left(\left(\check{X}^{(2)}\right)^t \widetilde{v}\right)\right)^{1/2}} =$$

$$= \frac{\left(\left(\check{X}^{(1)}\right)^t \widetilde{u}\right)^t \left(\check{X}^{(2)}\right)^t \widetilde{v}}{\left(\left(\left(\check{X}^{(1)}\right)^t \widetilde{u}\right)^t \left(\check{X}^{(1)}\right)^t \widetilde{u} \left(\left(\check{X}^{(2)}\right)^t \widetilde{v}\right)^t \left(\check{X}^{(2)}\right)^t \widetilde{v}\right)^{1/2}} =$$

$$= \frac{\widetilde{u}^t \widehat{\Sigma}^{(1,2)} \widetilde{v}}{\left(\widetilde{u}^t \widehat{\Sigma}^{(1)} \widetilde{u} \widetilde{v}^t \widehat{\Sigma}^{(2)} \widetilde{v}\right)^{1/2}}.$$

Assuming both sample covariance matrices $\widehat{\Sigma}^{(1)}$ and $\widehat{\Sigma}^{(2)}$ are full rank and hence invertible, this sample correlation can be rewritten in a form allowing calculation by a direct SVD using the change of variables

$$\check{u} = \left(\widehat{\boldsymbol{\Sigma}}^{(1)} \right)^{1/2} \widetilde{\widetilde{u}}, \ \check{v} = \left(\widehat{\boldsymbol{\Sigma}}^{(2)} \right)^{1/2} \widetilde{\widetilde{v}}. \tag{17.28}$$

Note that to arrive at directions that maximize correlation it is not enough to just make each variable commensurate (i.e. to standardize each variable by dividing by its standard deviation as done to compute the correlation matrix discussed in Section 5.2). Instead standardization by full root inverse covariance matrices (sometimes called "sphering", as done in Figure 11.2) is needed, which essentially results in projected correlations over all directions in $\mathbb{R}^{d_1}$ and $\mathbb{R}^{d_2}$.

The change of variables (17.28) results in the (reparameterized) sample correlation becoming

$$\widehat{\rho}\left(\check{u}, \check{v} \right) = \frac{ \check{u}^t \left(\widehat{\boldsymbol{\Sigma}}^{(1)} \right)^{-1/2} \widehat{\boldsymbol{\Sigma}}^{(1,2)} \left(\widehat{\boldsymbol{\Sigma}}^{(2)} \right)^{-1/2} \check{v} }{ \left(\check{u}^t \check{u} \check{v}^t \check{v} \right)^{1/2} }.$$

But the SVD optimization seeks unit (i.e. direction) vectors which have norm 1. Hence to maximize $\widehat{\rho}\left(\check{u}, \check{v} \right)$, it is enough to consider $\check{u}$ and $\check{v}$ with $\| \check{u} \| = \| \check{v} \| = 1$, and thus to compute the SVD

$$\left(\widehat{\boldsymbol{\Sigma}}^{(1)} \right)^{-1/2} \widehat{\boldsymbol{\Sigma}}^{(1,2)} \left(\widehat{\boldsymbol{\Sigma}}^{(2)} \right)^{-1/2} = \check{U} \check{D} \check{V}^t, \tag{17.29}$$

where as above $\check{U}$ and $\check{V}$ are orthonormal matrices and $\check{D}$ is diagonal. Furthermore, their columns are sets of orthogonal maximizers of $\widehat{\rho}\left(\check{u}, \check{v} \right)$. Inverting the variable change (17.28) returns those sets of coefficients to the original data scale. In particular, the matrices of *CCA loadings* are

$$\widetilde{\widetilde{U}} = \left(\widehat{\boldsymbol{\Sigma}}^{(1)} \right)^{-1/2} \check{U}, \ \widetilde{\widetilde{V}} = \left(\widehat{\boldsymbol{\Sigma}}^{(2)} \right)^{-1/2} \check{V}. \tag{17.30}$$

Note that while the columns of $\check{U}$ and $\check{V}$ are orthonormal, this generally does not hold for the CCA loadings $\widetilde{\widetilde{U}}$ and $\widetilde{\widetilde{V}}$. However, the latter does provide very useful modes of variation. In particular, treating the columns of $\widetilde{\widetilde{U}}$ and $\widetilde{\widetilde{V}}$ as direction vectors and finding the corresponding projection coefficient matrices (again in the spirit of multiplying by the data blocks as in (17.15)) give the *CCA scores*

$$\widetilde{\widetilde{S}}^{(1)} = \widetilde{\widetilde{U}}^t \check{X}^{(1)}, \ \boldsymbol{S}^{(2)} = \widetilde{\widetilde{V}}^t \check{X}^{(2)}.$$

An interesting property of these scores is that

$$\begin{aligned} \widetilde{\widetilde{S}}^{(1)} \left(\widetilde{\widetilde{S}}^{(1)} \right)^t &= \widetilde{\widetilde{U}}^t \check{X}^{(1)} \left(\check{X}^{(1)} \right)^t \widetilde{\widetilde{U}} \\ &= \check{U}^t \left(\widehat{\boldsymbol{\Sigma}}^{(1)} \right)^{-1/2} \widehat{\boldsymbol{\Sigma}}^{(1)} \left(\widehat{\boldsymbol{\Sigma}}^{(1)} \right)^{-1/2} \check{U} = I_n, \end{aligned} \tag{17.31}$$

i.e. the rows of the CCA scores matrix turn out to be orthonormal vectors in $\mathbb{R}^n$. A very parallel calculation shows that the rows of $\widetilde{\widetilde{S}}^{(2)}$ are also orthonormal. Thus the transformation (17.28) results in an interesting contrast between CCA and the PLS-SVD introduced at the beginning of Section 17.2.1. In particular, PLS-SVD has orthonormal loadings (not scores), while CCA has orthonormal scores and not loadings. This again suggests that the bi-orthogonality property of PCA modes of variation is indeed a special property. An interesting variation of CCA that deliberately produces bi-orthogonal (in a particular sense) modes of variation has been proposed by Shu et al. (2020).

Various rescalings of the CCA scores appear in different software packages. A common variation multiplies both sides (17.31) by n, which gives a scores matrix that is interpretable as having an identity covariance matrix. When variables are not commensurate, as discussed in Section 5.2, it can be useful to replace the co-variance matrices that are input to CCA with corresponding correlation matrices.

The case of covariance matrices not of full rank is sensibly handled by an application of the methodology called *generalized SVD*. That also conveniently allows for lower rank versions of CCA. However, it should be kept in mind that in high-dimensional cases, such methods generally lead to drastic overfitting. This is caused by the existence of pairs of directions that spuriously give perfect correlation of the projections, which happens with probability one in high-dimensional cases. Note that such pairs will appear in the analysis ahead of pairs of directions representing important underlying joint variation. In such cases PLS is a simple approach that can give results that are much less sensitive to spurious noise artifacts than CCA. There are also many other approaches to this problem that use sparsity ideas and other types of regularization. In some elegant unpublished work, Iain Carmichael uses the generalized SVD approach to formulate a general framework that encompasses, PLS, CCA, and many more. That paper also includes access to a number of extensions of PLS to the multi-block (multi-view) case of $B > 2$. An earlier survey of multi-block methods can be found in Kettenring (1971).

A special case of CCA, useful in classification as discussed in Chapter 11 is Canonical Variate Analysis. This generalizes LDA (see Section 11.1) to the multi-class case, by computing CCA with one data block consisting of row indicators (zeros and ones) for each class.

A toy example highlighting the difference between CCA and PLS-SVD is given in Figure 17.8. This is a two-block example, with the first data block containing data object vectors of the form $\begin{pmatrix} x_1 & x_2 \end{pmatrix}^t \in \mathbb{R}^2$. The second block consists of scalars $y \in \mathbb{R}$. The top panel displays the relationship between the data blocks using two rotated views, where each of the $n = 100$ data objects appears as a point of the form $\begin{pmatrix} x_1 & x_2 & y \end{pmatrix}^t \in \mathbb{R}^3$ indicated with a plus sign. There is a very strong relationship between data blocks in the sense that the points lie entirely in the cyan plane. As shown in the lower right panel, the data are distributed within the cyan plane to have a strong first PC direction shown in red in all panels. That red PC1 direction has been chosen to also lie within the first data block $\mathbb{R}^2$ space, shown as the yellow plane. Both PLS and CCA seek optimal pairs of directions, one in each space. Because the second data block lies in $\mathbb{R}$ there is only one possible direction, so the optimizing direction is just the vertical axis. More informative are the PLS and CCA directions in the second block (which thus lie in the yellow plane), shown in red and green, respectively. The CCA direction points in the direction of steepest increase of the cyan plane, which is natural because that will maximize the projected correlation (the target of CCA). On the other hand, the PLS direction is a compromise between the CCA and PC 1 directions, because it optimizes covariance which responds also to the very strong variance in the PC 1 direction.

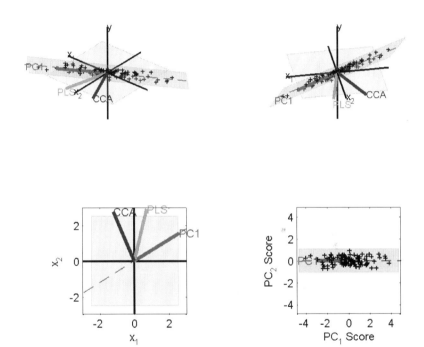

Figure 17.8 *Toy example highlighting the difference between PLS (green) and CCA (blue) directions. The concatenated ($\mathbb{R}^3$) data points all lie in the cyan plane. The $\check{X}^{(1)}$ data block lies in the yellow plane, as do the PLS-SVD and CCA directions. This shows how CCA maximizes correlation with the $\check{X}^{(2)}$ data that lie in the vertical axis, while the PLS direction is strongly influenced by the red PC1 direction.*

A range of compromise directions between PCA, PLS and CCA have been proposed in the unpublished PhD dissertation of Lee (2007), based on the *continuum regression* ideas of Stone and Brooks (1990).

CCA also has a strong relationship to classical linear regression. Using a small circle above symbols to distinguish that notation, the linear regression model is usually written as

$$\mathring{y} = \check{X}\mathring{\beta} + \mathring{\epsilon},$$

where $\mathring{y}$ is an $n \times 1$ vector of responses, $\check{X}$ is an $n \times d$ design matrix whose columns are thought of as multivariate predictors, $\mathring{\beta}$ is a $d \times 1$ vector of regression coefficients, and $\mathring{\epsilon}$ is an $n \times 1$ vector of residuals. Direct connection with the above CCA context comes from assuming the data have been centered in the sense that the entries of $\mathring{y}$ have sample mean 0, as do the columns of $\check{X}$. That allows using the column object centered notation (17.13) to write $\check{X}^{(1)} = \frac{1}{\sqrt{n}}\check{X}^t$ and $\check{X}^{(2)} = \frac{1}{\sqrt{n}}\mathring{y}^t$. This gives inputs to CCA as

$$\widehat{\Sigma}^{(1)} = \check{X}^{(1)}\left(\check{X}^{(1)}\right)^t = \frac{1}{n}\check{X}^t\check{X},$$

$$\widehat{\Sigma}^{(1,2)} = \check{X}^{(1)}\left(\check{X}^{(2)}\right)^t = \frac{1}{n}\check{X}^t\mathring{y},$$

$$\widehat{\Sigma}^{(2)} = \check{X}^{(2)}\left(\check{X}^{(2)}\right)^t = \frac{1}{n}\mathring{y}^t\mathring{y} = \hat{\sigma}_Y^2,$$

where the scalar $\hat{\sigma}_Y^2$ is an estimate of the standard deviation of the entries of $\mathring{y}$. The corresponding SVD to compute CCA as in (17.29) is thus applied to the $d \times 1$ matrix

$$\left(\frac{1}{n}\check{X}^t\check{X}\right)^{-1/2}\frac{1}{n}\check{X}^t\mathring{y}\hat{\sigma}_Y^{-1}.$$

That SVD results in a $d \times 1$ vector of norm 1

$$\check{U} = \frac{\left(\frac{1}{n}\check{X}^t\check{X}\right)^{-1/2}\frac{1}{n}\check{X}^t\mathring{y}\hat{\sigma}_Y^{-1}}{\left\|\left(\frac{1}{n}\check{X}^t\check{X}\right)^{-1/2}\frac{1}{n}\check{X}^t\mathring{y}\hat{\sigma}_Y^{-1}\right\|},$$

the single singular value

$$\check{D} = \left\|\left(\frac{1}{n}\check{X}^t\check{X}\right)^{-1/2}\frac{1}{n}\check{X}^t\mathring{y}\hat{\sigma}_Y^{-1}\right\|$$

and just the scalar $\check{V} = 1$. Inverting the change of variables as in (17.30) and dividing by the length of the resulting vector gives a CCA loadings vector which points in the same direction in $\mathbb{R}^d$ as the familiar least squares regression estimate $\hat{\beta} = \left(\check{X}^t\check{X}\right)^{-1}\check{X}^t\mathring{y}$ of the coefficient vector $\mathring{\beta}$. An important consequence is that the direction in the object space $\mathbb{R}^d$, where a set of data objects $\widetilde{X}$ best correspond to a set of n values in the vector $\widetilde{y}$, is proportional to the least squares estimate $\hat{\beta}$ (assuming $\widetilde{X}$ has been column object mean centered and that the mean of the $\widetilde{y}$ values is 0).

17.2.3 Joint and Individual Variation Explained

The methods described in Sections 17.2.1 and 17.2.2 all focus on the investigation of *joint variation*, i.e. finding modes of variation that reflect how data blocks vary together with each other. But in many applications there is also keen interest in *individual variation*, i.e. variation that is block specific and in some sense independent (either stochastically, or else in terms of appropriate orthogonality of modes of variation) of the other blocks. That idea was formalized as a method called Joint and Individual Variation Explained (JIVE) by Lock et al. (2013). An approach based on the simpler and more direct linear algebra method of *Principal Angle Analysis* is the Angle-based JIVE of Feng et al. (2018). Yu et al. (2017b) and Carmichael et al. (2019) used that method to demonstrate the value of careful investigation of both joint and individual variation in neuroscience and cancer genomics/imaging applications. An approach to the case of *partially shared blocks* (allowing joint variation between various subsets of data blocks) was proposed by Gaynanova and Li (2019).

CHAPTER 18

OODA Context and Related Areas

This chapter discusses the origins of the OODA terminology in Sections 18.1 and 18.2. Other related types of general statistical frameworks are described in Sections 18.3 and 18.4.

18.1 History and Terminology

The terminology *Object Oriented Data Analysis* (OODA) has a clear connection to the notion of *Object Oriented Programming* (OOP) from Computer Science. A good definition of OOP is: *Programming that supports encapsulation, inheritance, polymorphism, and abstraction*. More detail on this connection appears in Section 18.2.

The use of these concepts in a statistical context was pioneered by John M. Chambers and colleagues at the former Bell Laboratories, through the development of the statistical software package S and subsequently S-Plus. See Venables and Ripley (1994, 2013) for a good overview. An important historical point is that S was a major precursor of the currently very popular statistical software package R (R Core Team, 2020). A common misconception is that the name R was chosen as the letter before S. In fact it was in part from the first initial of the co-founders Robert Gentleman and Ross Ihaka.

OODA itself has its roots in the concept of FDA, which was pioneered by James O. Ramsay and colleagues, see the monographs by Ramsay and Silverman (2002, 2005) and Ferraty and Vieu (2006) for good overview of this area. While this use of "functional" is now quite standard in statistics, it is problematic for researchers with a strong mathematical training, because in that area a *functional* is essentially a function which maps functions into numbers (or more generally maps a vector space into its underlying field of scalars). Personal discussion with James O. Ramsay led to the realization that the notion of *data objects*, i.e. atoms of the statistical analysis as discussed in Chapter 2, provides the basis of this way of thinking, which led to the coining of the term OODA in Wang and Marron (2007).

The perceived value of scientific naming is an interesting cultural issue. Computer scientists seem to enjoy coining many names, trying them out for a while and then frequently abandoning most of them, except for the few that are viewed as having "gained traction". In contrast statisticians have a noticeable tendency to be very careful, in fact are usually quite conservative, about applying new names. Some have observed that at statistical meetings there tends to be too strong a focus on a rather few fashionable areas. At the time of this writing *sparsity* and *FDA* are the over-represented areas, in the past the perhaps overly dominant areas included

DOI: 10.1201/9781351189675-18

kernel smoothing and *robustness*. A natural question at this point is whether this apparent narrowness of fashionable research is a consequence of the reluctance to seek new names.

The terminology OODA itself has raised objections on occasion. For example, Lu et al. (2014b) contains an example demonstrating the value of the OODA viewpoint. The example came from the desire to automate the basic biological science practice of growing cells in *wells* on a plate. A challenging part of that automation was making the decision of when to move a subset of the cells to a new well based on digital images, because they have grown to fill the capacity of the current well. The issue of what should be the data objects, between features summarizing aspects of the whole well (e.g. cell counts) and features of individual cells (e.g. shape and size aspects), turned out to be pivotal to the investigation and even led to some interesting theoretical work discussed in that paper. An early submission of that paper was rejected by a well-known journal on the grounds that the terminology of "data objects" did not bring added value over the more traditional "experimental units". This point made sense for that particular project, but is limited in the context of the larger data analytic picture. In particular, generally choice of data objects includes not only experimental units, but also data representation issues, for example the choice of original versus log scale illustrated in Figure 1.1 of Section 1.1, the choice to focus on amplitude and/or phase variation in Section 2.1, the choice of shape or tree representation discussed in Sections 1.2 and 2.2, and which aspect of sounds an analysis should be centered upon in Section 2.3.

The discussion of the overview paper by Marron and Alonso (2014) covers quite a few other interesting aspects of OODA.

In some situations, there have been variations on the name OODA. For example, in 2010–2011 the Statistics and Applied Mathematical Sciences Institute hosted a program on OODA under the name Analysis of Object Data. That version of the name is also prominently featured in the monograph Patrangenaru and Ellingson (2015), which provides an important overview of statistical analysis for data lying in manifolds and stratified spaces. However, Piercesare Secchi has pointed out that inclusion of Oriented in OODA is quite appropriate in the sense that careful consideration of the data objects indeed tends to *orient* the analysis.

18.2 OODA Analogy with Object-Oriented Programming

In Object-Oriented Programming (OOP) there are many important concepts, including classes, objects, inheritance, methods, modularity, abstraction, encapsulation, extensibility, and polymorphism (Pitt-Francis and Whiteley, 2012). An object is an instance of a class in OOP, and so we can consider the different types of data spaces in OODA as being analogous to OOP classes, and the data sets that lie in the spaces as being analogous to OOP objects. The classes in OOP often have a tree-like relationship, with a base class at the root and different levels of derived classes below the root. Each class in OOP has functionality (including variables and methods) and this functionality is inherited by the derived classes. From

Pitt-Francis and Whiteley (2012, p9) inheritance is probably the most important aspect of OOP for the facilitation of interoperability.

To highlight the connection between OODA and OOP, we show how OODA concepts can naturally be put into an OOP type framework in Figure 18.1.

Object Oriented Programming vs. Object Oriented Data Analysis Examples

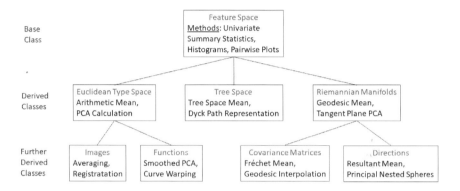

Figure 18.1 *Relationships between OOP and OODA.*

The *base class* in our example consists of all types of data spaces in OODA. *Methods* in the base class could be basic statistical operations such as univariate summary statistics, histograms, pairwise scatterplots, which are building blocks that are appropriate to calculate for any type of data set. Then the *derived classes* in the example are a subdivision of the types of data space, for example Euclidean-type data spaces, Riemannian manifolds, tree data space, etc. which require different methods of calculation for similar concepts. The methods at this level could be calculation of a mean or PCA calculation, for example. In OOP the derived classes *inherit the functionality* of the base class, in particular the base class methods are all applicable exactly in an unaltered form to the derived classes (they are inherited). This is true in our example for OODA, where the calculation of a histogram for example can be appropriate for any type of data.

There is a *further* level of derived class, e.g. different classes derived from Euclidean data, e.g. vectors, matrices, images, functions which all have the same method of mean and PCA calculation (which is inherited from the Euclidean class) but some other methods/calculations might need to be specified differently at this level.

The strict OOP analogy is that a sample of data objects in the OODA sense would be an *object* in the OOP sense. The choice of data object and its space in OODA drives the type of analysis and methods in a similar way that an OOP program is constructed.

The analogy can be extended. In OOP *polymorphism* is where the same code can be used for a variety of objects, and so for example computing PCA based on a Euclidean covariance matrix is useful not just for computing the Euclidean data-type PCA, but also for computing PCA in shape analysis using tangent space coordinates. The same code is used in both methods, but the inner workings are kept hidden from the user (OOP *encapsulation*). That code can be extended, e.g. PCA in Euclidean data can be extended to PCA for shape data using a tangent space (OOP *extensibility*). The essential features and methods of a class are put in one place but the inner details are easily ignored by the user (OOP *abstraction*).

We should stress that OODA is not OOP, and so it is *not* our intention to build an actual single computer program to carry out OODA. Rather the building blocks of OODA are analogous to those of OOP, particularly with the choice of data objects and classes driving the appropriate methodology and analysis.

Users of R are familiar with aspects of OOP, where the objects are instances of classes, e.g. objects from the linear model fitting command `lm` or `aov` are instances of S3 classes which can then have methods applied to them like `summary` or `coef`. In our example the object is a data set which is a member of a class (Euclidean, Shape, Tree etc.) and then a generic method that can be applied is `mean` or `pca` for example, although the implementation details may be different "under the hood" for data sets in different classes.

18.3 Compositional Data Analysis

The field of statistical *compositional data analysis* goes back at least to Aitchison (1982) and has connections to OODA. The original motivation was the study of variation in geological composition, in terms of vectors of proportions.

Data objects in that context were typically vectors $x \in \mathbb{R}^d$, each entry of which is the proportion of a given material in the geological sample. Note that each such object $x = (x_1, \cdots, x_d)$ is a point on the unit simplex in $\mathbb{R}^d$, i.e. $x_j \geq 0$ for $j = 1, \cdots, d$ and $\sum_{j=1}^{d} x_j = 1$.

Good insight can come from considering such data objects to be *probability vectors*, as is common in Markov Chains, see e.g. Hastings (1970). In econometrics terminology, such data objects are sometimes called *fractional responses*, see Papke and Wooldridge (1993, 2008) and Murteira and Ramalho (2016).

Data objects restricted to the unit simplex create some serious statistical challenges. For example, standard Euclidean analysis methods such as PCA (see Sections 1.1 and 3.1 and Chapter 17), or even use of the Gaussian distribution for statistical inference become clumsy at best, because such methodologies tend to leave the unit simplex. Other examples of naive analyses tending to leave the feature space include covariance matrices as data objects in Section 7.3.5 and phase data objects as discussed around Figure 9.13.

An often advocated choice of data object in this context is the *log-ratio* method, developed by Aitchison and Shen (1980) and Aitchison (1982, 1986).

This approach has worked well in many analyses, and is especially appropriate when the primary focus is on *ratios* of different amounts. However, in other situations, there can be a cost of some distortion, particularly when some entries are 0 or near 0. This has motivated other data objects choices for compositional data analysis, such as the square root transform which moves the data from the unit simplex to the unit sphere, in e.g. Scealy and Welsh (2011). Other power transformations have been proposed and studied by Tsagris et al. (2011) and Scealy et al. (2015). Butler and Glasbey (2008) address this issue using a latent Gaussian modeling approach, while Stewart and Field (2011) took a mixture modeling approach. Scealy and Welsh (2014) provide a fascinating historical discussion of major controversy that has occurred over such data object choices.

See Section 11.4 and Xiong et al. (2015) for a quite different example of data objects on the unit simplex, in the context of virus hunting using DNA methods. That paper also considered unit sphere versus simplex data object representations and found the best performance in that case came from working directly on the unit simplex.

18.4 Symbolic Data Analysis

Another statistical area related to OODA is *Symbolic Data Analysis*, see the books Bock and Diday (2012) and Billard and Diday (2006). The goal of that area is to find intuitive summaries of various aspects of relational databases. These summaries are called *symbols*, which are distributional summaries, such as ranges (intervals), frequencies (for categorical variables), histograms or quantiles. There are at least two levels of relationship between Symbolic Data Analysis and OODA. First symbols (e.g. probability densities) are often the data objects of interest. Second, given any set of data objects, the large and well-developed set of Symbolic Data Analysis ideas can provide a number of types of useful summarizations of object-oriented data via symbols of the data set. In particular, many specialized methods and software suites from that literature are very useful for analyzing symbols as data objects. For good recent access see Diday (2016) and Verde et al. (2016).

There are several different types of symbolic data, and in particular there is *native* symbolic data and *aggregated* symbolic data. Native symbolic data may be recorded in original form, e.g. as numbers or even as an interval $[0.2 - 0.4]$ perhaps as a range of values capturing uncertainty in an expert's opinion, Ellerby et al. (2020). Aggregated data may be a summary of data that were recorded at a finer stage, e.g. a histogram of values. Billard and Diday (2006) and Diday (1986, 2016) outline some of the benefits of treating data as symbolic data rather than classical data.

An example of some symbolic data is given below from the RSDA package in R, Rodriguez et al. (2020). The approach to analyzing symbolic data has analogies with OODA. In particular it is important to consider what are the most appropriate data objects. For example should interval data be considered as uniform probability distributions, or a type of indicator functional data (1 if in the interval and 0 otherwise, as defined at (11.2)), or perhaps a Gaussian random variable centered at the middle of the interval and standard deviation a quarter of the interval width? What distances should be used for comparing intervals, for example perhaps the Euclidean distance between end points, the L^2 norm or Hausdorff distance?

```
# A Symbolic Data Table :   7  x  7
        F1        F2                       F3     F4         F5              F6          F7
Case1   2.8    [1,2]  M1:10% M2:70% M3:20%    6  {e,g,i,k}     [0,90]     [9,24]
Case2   1.4    [3,9]  M1:60% M2:30% M3:10%    8  {a,b,c,d}   [-90,98]    [-9,9]
Case3   3.2   [-1,4]  M1:20% M2:20% M3:60%   -7  {2,b,1,c}    [65,90]   [65,70]
Case4  -2.1    [0,2]  M1:90% M2:0%  M3:10%    0  {a,3,4,c}    [45,89]   [25,67]
Case5    -3  [-4,-2]  M1:60% M2:0%  M3:40% -9.5  {e,g,i,k}    [20,40]    [9,40]
Case6   0.1  [10,21]  M1:0%  M2:70% M3:30%   -1  {e,1,i}       [5,8]     [5,8]
Case7     9   [4,21]  M1:20% M2:20% M3:60%  0.5  {e,a,2}  [3.14,6.76]    [4,6]
```

The example data set above consists of seven cases (in each of the rows) and there are four different types of data: F1, F4 continuous data; F2, F6, F7 interval data; F3 histogram data; F5 set-valued data. Variables F1, F4 are simply standard classical continuous data. F2, F6, F7 contain interval data for which we need to have appropriate methods for carrying out analysis such as computing means, PCA, and cluster analysis. Similarly the data types F3 (Histogram) and F5 (set-valued) also require special methods for statistical analysis. Combining the analysis of different data types would be a challenge, although generalizations of the multi-block methods of Section 17.2 are an appropriate way forward.

Let us consider some further interval data from the data set Lynne2 from the RSDA package, which is a data set of 15 individuals with interval measurements on pulse rate, systolic blood pressure, and diastolic blood pressure. We treat the intervals as uniform distributions and the interval-valued means are calculated from the arithmetic averages of the lower and upper end-points of each interval. This method of calculation results from the use of interval arithmetic, see Moore et al. (2009). The sample interval means are as follows: for pulse rate $[63.47, 81.73]$, systolic pressure $[125.8, 154.7]$, diastolic pressure $[85.8, 105.53]$. This interval mean estimator is also equivalent to the Fréchet mean (discussed in Section 7.1) with respect to Euclidean distance between the 2-vector of lower and upper end-points. In Figure 18.2 we see symbolic histograms and symbolic scatter plots of the data, using the same visually linked graphics matrix format as in Figure 4.4 (for the Twin Arches data) and many others. The symbolic histograms are obtained by superimposing uniform distributions on each interval, and then binning. The symbolic scatter plots include a rectangle for each case, where the corners of the rectangles are at the coordinates of the combinations of lower and upper end-points. These plots are obtained using the RSDA package (Version 2.0). The bottom and left panels all reveal a clear outlier in diastolic pressure. The middle right panel indicates a positive mostly linear association between systolic and diastolic pressures, while the middle left panel suggests a quadratic association between pulse rate and systolic pressure.

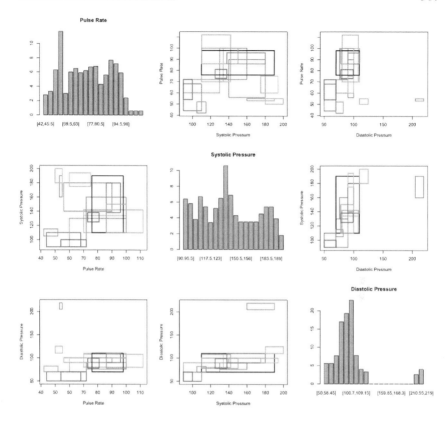

Figure 18.2 *Symbolic histograms and scatterplots of the three-dimensional* Lynne2 *blood measurement interval data. Finds interesting relationships between variables and a clear outlier.*

Further analysis, much in the spirit of other analyses given in this book can be carried out using the RSDA package.

An important historical note is that the terminology Symbolic Data Analysis came well before OODA, going back at least to Diday (1986).

18.5 Other Research Areas

There are several other areas, not discussed in detail here, where OODA ideas and terminology are potentially very useful, mostly because of the many complicated research questions that are typically addressed there.

One is *Object Oriented Spatial Statistics*, reviewed by Menafoglio and Secchi (2017). In this area a number of the tasks and approaches considered in this book are extended to the important case of spatial data. These include data sets where location plays a key role, and must be properly included in competent analyses.

Another such area, which has had a major impact on both neuroscience and also the study of many aspects of human behavior is *Functional Magnetic Resonance Imaging*, see e.g. Huettel et al. (2004). This method involves brain imaging over time, using blood flow as a surrogate for brain activity, measured at a set of voxels (recall the three-dimensional version of pixels). Many choices of data objects have been made in this area. In some studies, the focus is on a particular voxel (thus one brain region), so the time series at that point is the data object choice. In other studies the behavior over time is summarized by a single number, so the data objects can be three-dimensional sets of voxels. Still other studies treat the full 3-d movies over time as data objects. An example, showing joint analysis of how imaged brain function jointly interacts with behavioral scores is discussed in Yu et al. (2017b).

One more research area with close links to OODA is *Deep Learning*, which aims to provide computational methods that work in ways parallel to the human brain. Main methods in this area are based on *neural networks*, which go back at least to McCulloch and Pitts (1943). That area was quite popular in the 1990s, but seems to have been over-advertised at the time, with many attempted applications apparently failing to live up to their promise. However, more recently there has been a very strong resurgence, perhaps fueled by much larger typical data sets, together with much more powerful computing capabilities. These ideas have created research revolutions in areas such as computer vision. See Hagan et al. (2014) for important ideas in this area. Bengio et al. (2013) suggest that much of the success of deep learning methods comes from the ability of neural networks to provide a type of *automatic data representation*. For example, in classification tasks, the last step is typically a classical method (of the type discussed in Chapter 11), while the preceding neural layers are usefully viewed as providing inputs, via a search over a very large potential feature space. This can be viewed as an interesting way of automating the step of data object representation, as discussed in Section 3.1.

An example of such deep learning derived features in an image analysis context can be found in Carmichael et al. (2019) in the context of cancer data. The goal of that paper was to jointly analyze histology images (still the gold standard for cancer diagnosis) with genomic data (fundamental to recent breakthroughs in cancer research). This was accomplished with data integration using the AJIVE method discussed in Section 17.2.3.

A further area with OODA connections is *Natural Language Processing*, an example of which was studied in Section 10.2.4. This area aims to develop algorithms for the computational extraction of meaning from text. One part of that field is called *latent semantic analysis*, see e.g. Martin and Berry (2007), where the key idea is Singular Value Decomposition (essentially PCA without mean centering, as noted in Section 17.1.2) of some variation of an *occurrence matrix*, which summarizes appearance of words in large collections of documents. As noted in Berry and Browne (2005), there are many data object choices to be made, in terms of both how to summarize word/phrase occurrences and also how to weight various aspects of the decomposition. Baroni et al. (2014) showed that in many cases neural network based word embedding algorithms gave better performance than traditional matrix factorization based approaches, for a variety of standard measures.

However, Levy et al. (2015) demonstrated that these performance gains are likely due to data object choices that can be easily carried over to make the traditional matrix factorization approaches achieve state of the art performance.

We anticipate that in the future many additional applications and research areas will arise where OODA ideas will be useful.

Bibliography

Ahn, J. (2010). A stable hyperparameter selection for the Gaussian RBF kernel for discrimination. *Stat. Anal. Data Min.*, 3(3):142–148. pages 229, 303

Ahn, J., Lee, M. H., and Lee, J. A. (2019). Distance-based outlier detection for high dimension, low sample size data. *J. Appl. Stat.*, 46(1):13–29. pages 328

Ahn, J. and Marron, J. S. (2010). The maximal data piling direction for discrimination. *Biometrika*, 97(1):254–259. pages 221

Ahn, J., Marron, J. S., Muller, K. M., and Chi, Y.-Y. (2007). The high-dimension, low-sample-size geometric representation holds under mild conditions. *Biometrika*, 94(3):760–766. pages 287

Aitchison, J. (1982). The statistical analysis of compositional data. *J. Roy. Statist. Soc. Ser. B*, 44(2):139–177. With discussion. pages 364

Aitchison, J. (1986). *The statistical analysis of compositional data*. Monographs on Statistics and Applied Probability. Chapman & Hall, London. pages 364

Aitchison, J. and Shen, S. M. (1980). Logistic-normal distributions: Some properties and uses. *Biometrika*, 67(2):261–272. pages 364

Alfaro, C. A., Aydın, B., Valencia, C. E., Bullitt, E., and Ladha, A. (2014). Dimension reduction in principal component analysis for trees. *Comput. Statist. Data Anal.*, 74:157–179. pages 199

Amari, S. (1985). *Differential-geometrical methods in statistics*, volume 28 of *Lecture Notes in Statistics*. Springer-Verlag, New York. pages 185

Amenta, N., Datar, M., Dirksen, A., de Bruijne, M., Feragen, A., Ge, X., Pedersen, J. H., Howard, M., Owen, M., Petersen, J., Shi, J., and Xu, Q. (2015). Quantification and visualization of variation in anatomical trees. In Leonard, K. and Tari, S., editors, *Research in Shape Modeling*, pages 57–79. Springer International. pages 197

An, H. (2017). *Gaussian centered l-moments*. PhD thesis, The University of North Carolina at Chapel Hill. pages 85

An, H., Marron, J. S., Schwartz, T. A., Renner, J. B., Liu, F., Lynch, J., Lane, N., Jordan, J. M., and Nelson, A. E. (2016). Novel statistical methodology reveals that hip shape is associated with incident radiographic hip osteoarthritis among african american women. *Osteoarthr. Cartil.*, 24(4):640–646. pages 68

Anderson, C. R. (1997). *Object recognition using statistical shape analysis*. PhD thesis, University of Leeds. pages 139

Anderson, T. W. (2003). *An introduction to multivariate statistical analysis.* Wiley Series in Probability and Statistics. Wiley-Interscience [John Wiley & Sons], Hoboken, NJ, third edition. pages 85

Andrew, G., Arora, R., Bilmes, J., and Livescu, K. (2013). Deep canonical correlation analysis. In *International conference on machine learning*, pages 1247–1255. PMLR. pages 351

Aoshima, M., Shen, D., Shen, H., Yata, K., Zhou, Y.-H., and Marron, J. S. (2018). A survey of high dimension low sample size asymptotics. *Aust. N. Z. J. Stat.*, 60(1):4–19. pages 281, 290

Arlot, S. and Celisse, A. (2010). A survey of cross-validation procedures for model selection. *Stat. Surv.*, 4:40–79. pages 241

Arsigny, V., Fillard, P., Pennec, X., and Ayache, N. (2006). Log-euclidean metrics for fast and simple calculus on diffusion tensors. *Magn. Reson. Med.*, 56(2):411–421. pages 144, 145, 172

Arsigny, V., Fillard, P., Pennec, X., and Ayache, N. (2007). Geometric means in a novel vector space structure on symmetric positive-definite matrices. *SIAM J. Matrix Anal. Appl.*, 29(1):328–347. pages 144, 145, 172, 173

Asendorf, N. A. (2015). *Informative data fusion: Beyond canonical correlation analysis.* PhD thesis, University of Michigan. pages 351

Aston, J. A. D., Pigoli, D., and Tavakoli, S. (2017). Tests for separability in nonparametric covariance operators of random surfaces. *Ann. Stat.*, 45(4):1431 – 1461. pages 174

Aïzerman, M. A., Braverman, E. M., and Rozonoèr, L. I. (1964). A probabilistic problem on automata learning by pattern recognition and the method of potential functions. *Avtomat. i Telemeh.*, 25:1307–1323. pages 226

Aydin, B., Pataki, G., Wang, H., Bullitt, E., and Marron, J. S. (2009). A principal component analysis for trees. *Ann. Appl. Stat.*, 3(4):1597–1615. pages 198

Aydın, B., Pataki, G., Wang, H., Ladha, A., and Bullitt, E. (2011). Visualizing the structure of large trees. *Electron. J. Stat.*, 5:405–420. pages 199

Aylward, S. R. and Bullitt, E. (2002). Initialization, noise, singularities, and scale in height ridge traversal for tubular object centerline extraction. *IEEE Trans. Med. Imaging*, 21(2):61–75. pages 23, 24, 199

Bach, F. R. and Jordan, M. I. (2002). Kernel independent component analysis. *J. Mach. Learn. Res.*, 3(Jul):1–48. pages 351

Bai, Z. and Saranadasa, H. (1996). Effect of high dimension: By an example of a two sample problem. *Statist. Sinica*, 6(2):311–329. pages 262

Bai, Z. and Silverstein, J. W. (2010). *Spectral analysis of large dimensional random matrices.* Springer Series in Statistics. Springer, New York, second edition. pages 277

Bandulasiri, A. and Patrangenaru, V. (2005). Algorithms for nonparametric inference on shape manifolds. In *Proc. Joint Statistical Meetings. Minneapolis, MN*, pages 1617–22. American Statistical Association. pages 134

Banks, D. and Constantine, G. M. (1998). Metric models for random graphs. *J. Classification*, 15(2):199–223. pages 197

Barden, D. and Le, H. (2018). The logarithm map, its limits, and Fréchet means in orthant spaces. *Proc. Lond. Math. Soc. (3)*, 117(4):751–789. pages 202

Barden, D., Le, H., and Owen, M. (2018). Limiting behaviour of Fréchet means in the space of phylogenetic trees. *Ann. Inst. Statist. Math.*, 70(1):99–129. pages 202

Barnett, V. and Lewis, T. (1994). *Outliers in statistical data*. Wiley Series in Probability and Mathematical Statistics: Applied Probability and Statistics. John Wiley & Sons, Ltd., Chichester, third edition. pages 315

Baroni, M., Dinu, G., and Kruszewski, G. (2014). Don't count, predict! a systematic comparison of context-counting vs. context-predicting semantic vectors. In *Proceedings of the 52nd Annual Meeting of the Association for Computational Linguistics (Volume 1: Long Papers)*, pages 238–247, Baltimore. Association for Computational Linguistics. pages 368

Basford, K., McLachlan, G., and York, M. (1997). Modeling the distribution of stamp paper thickness via finite normal mixtures: The 1872 hidalgo stamp issue of mexico revisited. *J. Appl. Stat.*, 24(2):169–180. pages

Basser, P. J., Mattiello, J., and LeBihan, D. (1994). Mr diffusion tensor spectroscopy and imaging. *Biophys. J.*, 66(1):259–267. pages 172

Basser, P. J. and Pierpaoli, C. (1996). Microstructural and physiological features of tissues elucidated by quantitative-diffusion-tensor MRI. *J. Magn. Reson. Series B*, 111:209–219. pages 173

Becker, R. A. and Cleveland, W. S. (1987). Brushing scatterplots. *Technometrics*, 29(2):127–142. With discussion and a reply by the authors. pages 57

Becker, R. A., Cleveland, W. S., and Shyu, M.-J. (1996). The visual design and control of trellis display. *J. Comput. Graph. Stat.*, 5(2):123–155. pages 51

Beg, M. F., Miller, M. I., Trouvé, A., and Younes, L. (2005). Computing large deformation metric mappings via geodesic flows of diffeomorphisms. *INT. J. COMPUT. VISION.*, 61(2):139–157. pages 152

Bendich, P., Marron, J. S., Miller, E., Pieloch, A., and Skwerer, S. (2016). Persistent homology analysis of brain artery trees. *Ann. Appl. Stat.*, 10(1):198–218. pages 69, 203, 204

Bengio, Y., Courville, A., and Vincent, P. (2013). Representation learning: A review and new perspectives. *IEEE Trans. Pattern. Anal. Mach. Intell.*, 35(8):1798–1828. pages 241, 368

Benito, M., García-Portugués, E., Marron, J., and Peña, D. (2017). Distance-weighted discrimination of face images for gender classification. *Stat*, 6(1):231–240. pages 28

Benito, M., Parker, J., Du, Q., Wu, J., Xiang, D., Perou, C. M., and Marron, J. S. (2004). Adjustment of systematic microarray data biases. *Bioinformatics*, 20(1):105–114. pages 235, 237, 238

Benjamini, Y. and Hochberg, Y. (1995). Controlling the false discovery rate: A practical and powerful approach to multiple testing. *J. Roy. Statist. Soc. Ser. B*, 57(1):289–300. pages 68

Beran, J. (1994). *Statistics for long-memory processes*, volume 61 of *Monographs on Statistics and Applied Probability*. Chapman and Hall, New York. pages 314

Beran, J., Feng, Y., Ghosh, S., and Kulik, R. (2013). *Long-memory processes: Probabilistic properties and statistical methods*. Springer, Heidelberg. pages 314

Bernardi, M., Sangalli, L. M., Secchi, P., and Vantini, S. (2014a). Analysis of juggling data: An application of k-mean alignment. *Electron. J. Stat.*, 8(2):1817–1824. pages 256

Bernardi, M., Sangalli, L. M., Secchi, P., and Vantini, S. (2014b). Analysis of proteomics data: Block k-mean alignment [mr3273585]. *Electron. J. Stat.*, 8(2):1714–1723. pages 21, 256

Berry, M. W. and Browne, M. (2005). *Understanding search engines: Mathematical modeling and text retrieval*. SIAM. pages 368

Besag, J. (1986). On the statistical analysis of dirty pictures. *J. Roy. Statist. Soc. Ser. B*, 48(3):259–302. pages 28

Bhattacharya, A. (2008). Statistical analysis on manifolds: A nonparametric approach for inference on shape spaces. *Sankhyā*, 70(2, Ser. A):223–266. pages 134

Bhattacharya, A. and Bhattacharya, R. (2012). *Nonparametric inference on manifolds, with applications to shape spaces*, volume 2 of *Institute of Mathematical Statistics (IMS) Monographs*. Cambridge University Press, Cambridge. pages 167

Bhattacharya, R. and Patrangenaru, V. (2003). Large sample theory of intrinsic and extrinsic sample means on manifolds. I. *Ann. Statist.*, 31(1):1–29. pages 168

Bhattacharya, R. and Patrangenaru, V. (2005). Large sample theory of intrinsic and extrinsic sample means on manifolds. II. *Ann. Statist.*, 33(3):1225–1259. pages 168

Bhattacharya, R. N., Buibas, M., Dryden, I. L., Ellingson, L. A., Groisser, D., Hendriks, H., Huckemann, S., Le, H., Liu, X., Marron, J. S., Osborne, D. E., Patrangenaru, V., Schwartzman, A., Thompson, H. W., and Wood, A. T. A. (2013). Extrinsic data analysis on sample spaces with a manifold stratification. In Beznea, L., Brizanescu, V., Iosifescu, M., Marinoschi, G., Purice, R., and Timotin, D., editors, *Advances in Mathematics, Invited Contributions at the Seventh Congress of Romanian Mathematicians, Brasov, 2011, The Publishing House of the Romanian Academy*, pages 227–240. pages 131

Billard, L. and Diday, E. (2006). *Symbolic data analysis: Conceptual statistics and data mining*. Wiley Series in Comput. Stat.. John Wiley & Sons, Ltd., Chichester. pages 365

Biswas, M., Mukhopadhyay, M., and Ghosh, A. K. (2014). A distribution-free two-sample run test applicable to high-dimensional data. *Biometrika*, 101(4):913–926. pages 290

Biswas, M., Mukhopadhyay, M., and Ghosh, A. K. (2015). On some exact distribution-free one-sample tests for high dimension low sample size data. *Statist. Sinica*, 25(4):1421–1435. pages 290

Biswas, M., Sarkar, S., and Ghosh, A. K. (2016). On some exact distribution-free tests of independence between two random vectors of arbitrary dimensions. *J. Statist. Plann. Inference*, 175:78–86. pages 290

Bloomfield, P. (2000). *Fourier analysis of time series: An introduction.* Wiley Series in Probability and Statistics: Applied Probability and Statistics. Wiley-Interscience, New York, second edition. pages 40, 121

Boas, F. (1905). The horizontal plane of the skull and the general problem of the comparison of variable forms. *Science*, 21(544):862–863. pages 139

Bock, H.-H. and Diday, E., editors (2012). *Analysis of symbolic data: Exploratory methods for extracting statistical information from complex data*, Berlin. Springer-Verlag. pages 365

Bookstein, F. L. (1986). Size and shape spaces for landmark data in two dimensions. *Stat. Sci.*, 1(2):181–222. pages 12

Bookstein, F. L. (2014). *Measuring and reasoning: Numerical inference in the sciences.* Cambridge University Press, Cambridge. pages 167

Borg, I. and Groenen, P. J. F. (2005). *Modern multidimensional scaling: Theory and applications.* Springer Series in Statistics. Springer, New York, second edition. pages 134

Borland, D. and Taylor, R. M. I. (2007). Rainbow color map (still) considered harmful. *IEEE Comput. Graph. Appl.*, 27(2):14–17. pages 97

Born, M. and Wolf, E. (2013). *Principles of optics: Electromagnetic theory of propagation, interference and diffraction of light.* Elsevier. pages 321

Borysov, P., Hannig, J., and Marron, J. S. (2014). Asymptotics of hierarchical clustering for growing dimension. *J. Multivariate Anal.*, 124:465–479. pages 286

Borysov, P., Hannig, J., Marron, J. S., Muratov, E., Fourches, D., and Tropsha, A. (2016). Activity prediction and identification of mis-annotated chemical compounds using extreme descriptors. *J. Chemom.*, 30:99–108. pages 74, 81, 83

Bouveyron, C., Celeux, G., Murphy, T. B., and Raftery, A. E. (2019). *Model-based clustering and classification for data science.* Cambridge Series in Statistical and Probabilistic Mathematics. Cambridge University Press, Cambridge. With applications in R. pages 243, 274

Bowley, A. L. (1920). *Elements of statistics*, volume 2. PS King and Son, London. pages 85

Box, G. E. and Cox, D. R. (1964). An analysis of transformations. *J. R. Stat. Soc. Series B Stat. Methodol.*, pages 211–252. pages 94

Bradley, R. C. (2005). Basic properties of strong mixing conditions. A survey and some open questions. *Probab. Surv.*, 2:107–144. Update of, and a supplement to, the 1986 original. pages 287

Bralower, T., Fullagar, P., Paull, C., Dwyer, G., and Leckie, R. (1997). Mid-cretaceous strontium-isotope stratigraphy of deep-sea sections. *Geol. Soc. Am. Bull.*, 109(11):1421–1442. pages 299

Breiman, L. (2001). Random forests. *Mach. Learn.*, 45(1):5–32. pages 241

Breiman, L., Friedman, J. H., Olshen, R. A., and Stone, C. J. (1984). *Classification and regression trees*. Wadsworth Statistics/Probability Series. Wadsworth Advanced Books and Software, Belmont, CA. pages 241

Brillinger, D. R. (1981). *Time series: Data analysis and theory*. Holden-Day, Inc., Oakland, Calif., second edition. pages 40, 121

Broadhurst, R. E., Stough, J., Pizer, S. M., and Chaney, E. L. (2005). Histogram statistics of local model-relative image regions. In *Deep Structure, Singularities, and Computer Vision*, pages 72–83. Springer. pages 16

Brooks, J. P., Dulá, J. H., and Boone, E. L. (2013). A pure L_1-norm principal component analysis. *Comput. Statist. Data Anal.*, 61:83–98. pages 171, 327

Bubenik, P. (2015). Statistical topological data analysis using persistence landscapes. *J. Mach. Learn. Res.*, 16:77–102. pages 204

Bubenik, P. and Kim, P. T. (2007). A statistical approach to persistent homology. *Homology Homotopy Appl.*, 9(2):337–362. pages 204

Bühlmann, P. and Meinshausen, N. (2015). Magging: Maximin aggregation for inhomogeneous large-scale data. *Proceedings of the IEEE*, 104(1):126–135. pages 239, 314

Buja, A., Swayne, D. F., Littman, M. L., Dean, N., Hofmann, H., and Chen, L. (2008). Data visualization with multidimensional scaling. *J. Comput. Graph. Statist.*, 17(2):444–472. pages 134

Bullitt, E. and Aylward, S. R. (2002). Volume rendering of segmented image objects. *IEEE Trans. Med. Imaging*, 21(8):998–1002. pages 23

Bullitt, E., Muller, K. E., Jung, I., Lin, W., and Aylward, S. (2005). Analyzing attributes of vessel populations. *Med. Image Anal.*, 9(1):39–49. pages 203

Burges, C. J. (1998). A tutorial on support vector machines for pattern recognition. *Data Min. Knowl. Discov.*, 2(2):121–167. pages 232

Butler, A. and Glasbey, C. (2008). A latent Gaussian model for compositional data with zeros. *J. Roy. Statist. Soc. Ser. C*, 57(5):505–520. pages 365

Cabanski, C. R., Qi, Y., Yin, X., Bair, E., Hayward, M. C., Fan, C., Li, J., Wilkerson, M. D., Marron, J. S., Perou, C. M., et al. (2010). Swiss made: Standardized within class sum of squares to evaluate methodologies and data set elements. *PloS one*, 5(3):e9905. pages 247

Cabassi, A., Pigoli, D., Secchi, P., and Carter, P. A. (2017). Permutation tests for the equality of covariance operators of functional data with applications to evolutionary biology. *Electron. J. Stat.*, 11(2):3815–3840. pages 174

Calissano, A., Feragen, A., and Vantini, S. (2020). Populations of unlabeled networks: Graph space geometry and geodesic principal components. Technical report, Dipartimento di Matematica Politecnico di Milano. MOX-Report No. 14/2020. pages 214

Cao, R., Cuevas, A., and Manteiga, W. G. (1994). A comparative study of several smoothing methods in density estimation. *Comput. Stat. Data Anal.*, 17(2):153–176. pages 302

Cardoso, J.-F. and Souloumiac, A. (1993). Blind beamforming for non-Gaussian signals. *IEE proceedings F (radar and signal processing)*, 140(6):362–370. pages 120

Carlsson, G. (2009). Topology and data. *Bull. Amer. Math. Soc. (N.S.)*, 46(2):255–308. pages 204

Carmichael, I., Calhoun, B. C., Hoadley, K. A., Troester, M. A., Geradts, J., Couture, H. D., Olsson, L., Perou, C. M., Niethammer, M., Hannig, J., et al. (2019). Joint and individual analysis of breast cancer histologic images and genomic covariates. *arXiv preprint arXiv:1912.00434*. pages 359, 368

Carmichael, I. and Marron, J. S. (2017). Geometric insights into support vector machine behavior using the KKT conditions. *arXiv preprint arXiv:1704.00767*. pages 234, 237

Carmichael, I. and Marron, J. S. (2018). Data science vs. statistics: Two cultures? *Japanese Journal of Statistics and Data Science*, 1:117–138. pages 1, 3, 257

Casella, G. and Hwang, J. T. a. (1982). Limit expressions for the risk of James-Stein estimators. *Canad. J. Statist.*, 10(4):305–309. pages 281

Cates, J., Fletcher, P. T., Styner, M., Shenton, M., and Whitaker, R. (2007). Shape modeling and analysis with entropy-based particle systems. In *Biennial International Inf. Process. Med. Imaging.*, pages 333–345. Springer. pages 13

Cattell, R. B. (1966). The scree test for the number of factors. *Multivariate Behav. Res.*, 1(2):245–276. pages 37

Chakraborty, B. and Chaudhuri, P. (1999). A note on the robustness of multivariate medians. *Statist. Probab. Lett.*, 45(3):269–276. pages 327

Chaney, E., Pizer, S., Joshi, S., Broadhurst, R., Fletcher, T., Gash, G., Han, Q., Jeong, J., Lu, C., Merck, D., et al. (2004). Automatic male pelvis segmentation from ct images via statistically trained multi-object deformable m-rep models. *Int. J. Radiat. Oncol. Biol. Phys.*, 60(1):S153–S154. pages 10, 16

Chang, T. (1988). Estimating the relative rotation of two tectonic plates from boundary crossings. *J. Amer. Statist. Assoc.*, 83(404):1178–1183. pages 151

Chapelle, O., Schölkopf, B., and Zien, A. (2006). *semi supervised learning*. Adaptive computation and machine learning. MIT Press, Cambridge, MA, USA. pages 256

Chaudhuri, P. and Marron, J. (2002). Curvature vs. slope inference for features in nonparametric curve estimates. Technical report, Cornell University Operations Research and Industrial Engineering. pages 309

Chaudhuri, P. and Marron, J. S. (1999). SiZer for exploration of structures in curves. *J. Amer. Statist. Assoc.*, 94(447):807–823. pages 70, 294, 304, 306

Chaudhuri, P. and Marron, J. S. (2000). Scale space view of curve estimation. *Ann. Statist.*, 28(2):408–428. pages 297, 304

Chavel, I. (2006). *Riemannian geometry: A modern introduction*, volume 98 of *Cambridge Studies in Advanced Mathematics*. Cambridge University Press, Cambridge, second edition. pages 150

Chen, S. X. and Qin, Y.-L. (2010). A two-sample test for high-dimensional data with applications to gene-set testing. *Ann. Statist.*, 38(2):808–835. pages 262

Chen, Y., Marron, J. S., and Zhang, J. (2019). Modeling seasonality and serial dependence of electricity price curves with warping functional autoregressive dynamics. *Ann. Appl. Stat.*, 13(3):1590–1616. pages 189

Cheng, W., Dryden, I. L., Hitchcock, D. B., and Le, H. (2014). Analysis of proteomics data: Bayesian alignment of functions. *Electron. J. Stat.*, 8(2):1734–1741. pages 21

Cheng, W., Dryden, I. L., and Huang, X. (2016). Bayesian registration of functions and curves. *Bayesian Anal.*, 11(2):447–475. pages 193

Cherkasov, A., Muratov, E. N., Fourches, D., Varnek, A., Baskin, I. I., Cronin, M., Dearden, J., Gramatica, P., Martin, Y. C., Todeschini, R., et al. (2014). QSAR modeling: Where have you been? where are you going to? *J. Med. Chem.*, 57(12):4977–5010. pages 74

Chiu, S.-T. and Marron, J. S. (1990). The negative correlations between data-determined bandwidths and the optimal bandwidth. *Statist. Probab. Lett.*, 10(2):173–180. pages 303

Choi, H. Y., Hayes, D. N., and Marron, J. S. (2018). Identification of rna-seq shape abnormality. *Proc. Annu. Meet. Am. Assoc. Cancer. Res. 2018*, 78:4273. pages 328

Choi, H. Y. and Marron, J. (2018). Estimation of the number of spikes using a generalized spike population model and application to rna-seq data. *arXiv preprint arXiv:1804.10675*. pages 279, 281

Clarke, B. R. (2018). *Robustness theory and application*. John Wiley & Sons. pages 70, 128, 313

Clarke, S. and Hall, P. (2009). Robustness of multiple testing procedures against dependence. *Ann. Statist.*, 37(1):332–358. pages 314

Cleveland, W. S. (1993). *Visualizing data*. Hobart Press, Summit, NJ. pages 47

Cleveland, W. S. et al. (1985). *The elements of graphing data*. Wadsworth Advanced Books and Software Monterey, CA. pages 47

Coleman, J., Aston, J., and Pigole, D. (2015). Reconstructing the sounds of words from the past. *The Scottish Consortium for ICPhS*. pages 25

Cook, R. D. and Lee, H. (1999). Dimension reduction in binary response regression. *J. Amer. Statist. Assoc.*, 94(448):1187–1200. pages 123

Cootes, T. F., Hill, A., Taylor, C. J., and Haslam, J. (1994). Use of active shape models for locating structures in medical images. *Image and vision computing*, 12(6):355–365. pages 13

Cox, T. F. and Cox, M. A. (2000). *Multidimensional scaling, second edition.* Chapman and Hall/CRC, Boca Raton. pages 134

Cristianini, N. and Shawe-Taylor, J. (2000). *An introduction to support vector machines and other kernel-based learning methods.* Cambridge university press. pages 125, 231, 233

Dai, W. and Genton, M. G. (2018). Multivariate functional data visualization and outlier detection. *J. Comput. Graph. Statist.*, 27(4):923–934. pages 316

Dai, W. and Genton, M. G. (2019). Directional outlyingness for multivariate functional data. *Comput. Statist. Data Anal.*, 131:50–65. pages 316

Dai, W., Mrkvička, T., Sun, Y., and Genton, M. G. (2020). Functional outlier detection and taxonomy by sequential transformations. *Comput. Stat. Data. Anal.*, page 106960. pages 316

Damon, J. and Marron, J. S. (2014). Backwards principal component analysis and principal nested relations. *J. Math. Imaging Vision*, 50(1-2):107–114. pages 169, 170

Darwin, C. (1859). *On the origin of species.* John Murray, London. pages 199

Daubechies, I. (1992). *Ten lectures on wavelets*, volume 61 of *CBMS-NSF Regional Conference Series in Applied Mathematics.* Society for Industrial and Applied Mathematics (SIAM), Philadelphia, PA. pages 40

Davis, B., Foskey, M., Rosenman, J., Goyal, L., Chang, S., and Joshi, S. (2005). Automatic segmentation of intra-treatment CT images for adaptive radiation therapy of the prostate. *Med. Image Comput. Comput. Assist. Interv. MICCAI 2005*, pages 442–450. pages 16

De Bie, T., Cristianini, N., and Rosipal, R. (2005). Eigenproblems in pattern recognition. In *Handbook of Geometric Computing*, pages 129–167. Springer. pages 350

de Boor, C. (2001). *A practical guide to splines*, volume 27 of *Appl. Math. Sci.*. Springer-Verlag, New York, revised edition. pages 40

Delaigle, A. and Hall, P. (2016). Approximating fragmented functional data by segments of Markov chains. *Biometrika*, 103(4):779–799. pages 106

Devroye, L. and Györfi, L. (1985). *Nonparametric density estimation: The L_1 view.* Wiley Series in Probability and Mathematical Statistics: Tracts on Probability and Statistics. John Wiley & Sons, Inc., New York. pages 41, 293, 301

Devroye, L., Györfi, L., and Lugosi, G. (1996). *A probabilistic theory of pattern recognition*, volume 31 of *Applications of Mathematics (New York)*. Springer-Verlag, New York. pages 224

Diaconis, P., Goel, S., and Holmes, S. (2008). Horseshoes in multidimensional scaling and local kernel methods. *Ann. Appl. Stat.*, 2(3):777–807. pages 133, 135, 136, 179

Diday, E. (1986). New kinds of graphical representation in clustering. In *Compstat*, pages 169–175. Springer. pages 365, 367

Diday, E. (2016). Thinking by classes in data science: The symbolic data analysis paradigm. *Wiley Interdiscip. Rev. Comput. Stat.*, 8(5):172–205. pages 365

Dobriban, E. (2015). Efficient computation of limit spectra of sample covariance matrices. *Random Matrices Theory Appl.*, 4(4):1550019, 36. pages 281

Domingos, P. and Pazzani, M. (1997). On the optimality of the simple Bayesian classifier under zero-one loss. *Mach. Learn.*, 29(2-3):103–130. pages 217

Donoho, D. L. and Johnstone, I. M. (1994). Ideal spatial adaptation by wavelet shrinkage. *Biometrika*, 81(3):425–455. pages 40, 300

Donoho, D. L., Johnstone, I. M., Kerkyacharian, G., and Picard, D. (1995). Wavelet shrinkage: Asymptopia? *J. Roy. Statist. Soc. Ser. B*, 57(2):301–369. With discussion and a reply by the authors. pages 40, 300

Donoho, D. L. and Liu, R. C. (1991). Geometrizing rates of convergence. II, III. *Ann. Statist.*, 19(2):633–667, 668–701. pages 300

Dryden, I. L. (2021). `shapes` *package*. R Foundation for Statistical Computing, Vienna, Austria. Contributed package, Version 1.2.6. pages 139, 142, 225

Dryden, I. L., Hill, B., Wang, H., and Laughton, C. A. (2017). Covariance analysis for temporal data, with applications to DNA modelling. *Stat*, 6:218–230. pages 139, 164

Dryden, I. L. and Kent, J. T., editors (2015). *Geometry driven statistics*, Chichester. John Wiley and Sons. pages 167

Dryden, I. L., Kim, K.-R., Laughton, C. A., and Le, H. (2019a). Principal nested shape space analysis of molecular dynamics data. *Ann. Appl. Stat.*, 13(4):2213–2234. pages 164

Dryden, I. L., Kim, K.-R., and Le, H. (2019b). Bayesian linear size-and-shape regression with applications to face data. *Sankhya A*, 81(1):83–103. pages 167

Dryden, I. L., Koloydenko, A., and Zhou, D. (2009). Non-Euclidean statistics for covariance matrices, with applications to diffusion tensor imaging. *Ann. Appl. Stat.*, 3(3):1102–1123. pages 143, 146, 172, 173, 174

Dryden, I. L., Kume, A., Le, H., and Wood, A. T. A. (2008). A multi-dimensional scaling approach to shape analysis. *Biometrika*, 95(4):779–798. pages 134

Dryden, I. L., Kume, A., Le, H., and Wood, A. T. A. (2010). Statistical inference for functions of the covariance matrix in the stationary Gaussian time-orthogonal principal components model. *Ann. Inst. Statist. Math.*, 62(5):967–994. pages 164

Dryden, I. L., Kume, A., Paine, P. J., and Wood, A. T. A. (2020). Regression modeling for size-and-shape data based on a Gaussian model for landmarks. *J. Am. Stat. Assoc.*. pages 167

Dryden, I. L. and Mardia, K. V. (2016). *Statistical shape analysis with applications in R*. Wiley Series in Probability and Statistics. John Wiley & Sons, Ltd., Chichester, second edition. pages 12, 68, 141, 158, 166, 167, 202, 225, 328, 329

Dryden, I. L., Pennec, X., and Peyrat, J.-M. (2010). Power Euclidean metrics for covariance matrices with application to diffusion tensor imaging. *arXiv e-prints*, page arXiv:1009.3045. pages 146

Dryden, I. L. and Walker, G. (1999). Highly resistant regression and object matching. *Biometrics*, 55:820–825. pages 328, 329

Du, J., Dryden, I. L., and Huang, X. (2015). Size and shape analysis of error-prone shape data. *J. Am. Stat. Assoc.*, 110(509):368–379. pages 167

Dubey, P. and Müller, H.-G. (2020). Functional models for time-varying random objects. *J. R. Stat. Soc. Series B Stat. Methodol.*, 82(2):275–327. pages 213

Duda, R. O., Hart, P. E., and Stork, D. G. (2001). *Pattern classification*. Wiley-Interscience, New York, second edition. pages 70, 120, 216, 223, 224

Duin, R. P. W. and Pekalska, E. (2005). *The dissimilarity representation for pattern recognition, : Foundations and applications*, volume 64. World scientific. pages 134

Dupuis, P., Grenander, U., and Miller, M. I. (1998). Variational problems on flows of diffeomorphisms for image matching. *Quart. Appl. Math.*, 56(3):587–600. pages 152

Eckart, C. and Young, G. (1936). The approximation of one matrix by another of lower rank. *Psychometrika*, 1(3):211–218. pages 132

Efromovich, S. (1999). *Nonparametric curve estimation*. Springer Series in Statistics. Springer-Verlag, New York. Methods, theory, and applications. pages 294

Efron, B. and Tibshirani, R. J. (1993). *An introduction to the bootstrap*, volume 57 of *Monographs on Statistics and Applied Probability*. Chapman and Hall, New York. pages 294

Eick, S. G., Graves, T. L., Karr, A. F., Marron, J. S., and Mockus, A. (2001). Does code decay? Assessing the evidence from change management data. *IEEE Trans. Softw. Eng.*, 27(1):1–12. pages 311

Eilers, P. H. C. and Marx, B. D. (1996). Flexible smoothing with B-splines and penalties. *Statist. Sci.*, 11(2):89–121. With comments and a rejoinder by the authors. pages 40

Eisen, M. B., Spellman, P. T., Brown, P. O., and Botstein, D. (1998). Cluster analysis and display of genome-wide expression patterns. *Proc. Natl. Acad. Sci.*, 95(25):14863–14868. pages 97

El Karoui, N. (2010). The spectrum of kernel random matrices. *Ann. Statist.*, 38(1):1–50. pages 231, 232

Ellerby, Z., McCulloch, J., Wilson, M., and Wagner, C. (2020). Exploring how component factors and their uncertainty affect judgements of risk in cyber-security. In Nadjm-Tehrani, S., editor, *Critical Information Infrastructures Security*, pages 31–42, Cham. Springer International Publishing. pages 365

Eltzner, B., Huckemann, S., and Mardia, K. V. (2018). Torus principal component analysis with applications to RNA structure. *Ann. Appl. Stat.*, 12(2):1332–1359. pages 156

Eltzner, B. and Huckemann, S. F. (2019). A smeary central limit theorem for manifolds with application to high-dimensional spheres. *Ann. Statist.*, 47(6):3360–3381. pages 168

Eltzner, B., Jung, S., and Huckemann, S. (2015). Dimension reduction on polyspheres with application to skeletal representations. In *Geometric science of information*, volume 9389 of *Lecture Notes in Comput. Sci.*, pages 22–29. Springer, Cham. pages 156

Enders, C. K. (2010). *Applied missing data analysis*. Guilford Press. pages 78

Erästö, P. and Holmström, L. (2005). Bayesian multiscale smoothing for making inferences about features in scatterplots. *J. Comput. Graph. Statist.*, 14(3):569–589. pages 307

Eubank, R. L. (1999). *Nonparametric regression and spline smoothing*, volume 157 of *Statistics: Textbooks and Monographs*. Marcel Dekker, Inc., New York, second edition. pages 293

Fan, J. and Gijbels, I. (1996). *Local polynomial modeling and its applications*, volume 66 of *Monographs on Statistics and Applied Probability*. Chapman & Hall, London. pages 293, 300

Fan, J. and Lv, J. (2008). Sure independence screening for ultrahigh dimensional feature space. *J. R. Stat. Soc. Ser. B Stat. Methodol.*, 70(5):849–911. pages 290, 291

Fan, J. and Marron, J. S. (1994). Fast implementations of nonparametric curve estimators. *J. Comput. Graph. Stat.*, 3(1):35–56. pages 302

Farrell, R. H. (1972). On the best obtainable asymptotic rates of convergence in estimation of a density function at a point. *Ann. Math. Statist.*, 43:170–180. pages 300

Fasy, B. T., Lecci, F., Rinaldo, A., Wasserman, L., Balakrishnan, S., and Singh, A. (2014). Confidence sets for persistence diagrams. *Ann. Statist.*, 42(6):2301–2339. pages 311

Febrero-Bande, M. and Oviedo de la Fuente, M. (2012). Statistical computing in functional data analysis: The r package fda. usc. *J. Stat. Softw.*, 51(4):1–28. pages 54

Feng, Q., Foskey, M., Chen, W., and Shen, D. (2010). Segmenting ct prostate images using population and patient-specific statistics for radiotherapy. *Med. Phys.*, 37(8):4121–4132. pages 17

Feng, Q., Hannig, J., and Marron, J. (2016). A note on automatic data transformation. *Stat*, 5(1):82–87. pages 94

Feng, Q., Jiang, M., Hannig, J., and Marron, J. S. (2018). Angle-based joint and individual variation explained. *J. Multivariate Anal.*, 166:241–265. pages 4, 359

Feragen, A., Lauze, F., Lo, P., de Bruijne, M., and Nielsen, M. (2011). Geometries on spaces of treelike shapes. *Computer Vision–ACCV 2010*, pages 160–173. pages 197

Feragen, A. and Nye, T. (2020). Statistics on stratified spaces. In *Riemannian Geometric Statistics in Medical Image Analysis*, pages 299–342. Elsevier. pages 202

Feragen, A., Owen, M., Petersen, J., Wille, M. M., Thomsen, L. H., Dirksen, A., and de Bruijne, M. (2013). Tree-space statistics and approximations for large-scale analysis of anatomical trees. In *Inf. Process Med. Imaging*, pages 74–85. Springer. pages 197

Ferraty, F. and Vieu, P. (2006). *Nonparametric functional data analysis*. Springer Series in Statistics. Springer, New York. Theory and practice. pages 106, 361

Fillard, P., Arsigny, V., Pennec, X., and Ayache, N. (2007). Clinical DT-MRI estimation, smoothing, and fiber tracking with log-Euclidean metrics. *IEEE Trans. Med. Imaging*, 26(11):1472–1482. PMID: 18041263. pages 173

Fischler, M. A. and Bolles, R. C. (1981). Random sample consensus: A paradigm for model fitting with applications to image analysis and automated cartography. *Commun. ACM*, 24(6):381–395. pages 123

Fisher, N. I. (1993). *Statistical analysis of circular data*. Cambridge University Press, Cambridge. pages 69, 147

Fisher, N. I., Lewis, T., and Embleton, B. J. J. (1993). *Statistical analysis of spherical data*. Cambridge University Press, Cambridge. Revised reprint of the 1987 original. pages 69, 147

Fisher, N. I. and Marron, J. S. (2001). Mode testing via the excess mass estimate. *Biometrika*, 88(2):499–517. pages 294, 306

Fisher, R. A. (1936). The use of multiple measurements in taxonomic problems. *Ann. Eugen.*, 7(2):179–188. pages 215, 218

Fletcher, P. T. (2004). *Statistical variability in nonlinear spaces: Application to shape analysis and DT-MRI*. PhD thesis, The University of North Carolina at Chapel Hill. pages 149

Fletcher, P. T. and Joshi, S. (2004). Principal geodesic analysis on symmetric spaces: Statistics of diffusion tensors. In *Computer Vision and Mathematical Methods in Medical and Biomedical Image Analysis*, pages 87–98. Springer. pages 172

Fletcher, P. T. and Joshi, S. (2007). Riemannian geometry for the statistical analysis of diffusion tensor data. *Signal Process.*, 87(2):250–262. pages 145, 173

Fletcher, P. T., Lu, C., Pizer, S. M., and Joshi, S. (2004). Principal geodesic analysis for the study of nonlinear statistics of shape. *IEEE Trans. Med. Imaging*, 23(8):995–1005. pages 15, 152, 153

Fletcher, P. T., Venkatasubramanian, S., and Joshi, S. (2009). The geometric median on riemannian manifolds with application to robust atlas estimation. *NeuroImage*, 45(1):S143–S152. pages 128, 318

Fogel, P., Young, S. S., Hawkins, D. M., and Ledirac, N. (2006). Inferential, robust non-negative matrix factorization analysis of microarray data. *Bioinformatics*, 23(1):44–49. pages 122

Frazier, M. W. (2006). *An introduction to wavelets through linear algebra*. Springer Science & Business Media. pages 40

Fréchet, M. (1948). Les éléments aléatoires de nature quelconque dans un espace distancié. *Ann. Inst. H. Poincaré*, 10(3):215–310. pages 127

Fréchet, M. M. (1906). Sur quelques points du calcul fonctionnel. *Rendiconti del Circolo Matematico di Palermo (1884-1940)*, 22(1):1–72. pages 125

Freund, Y., Schapire, R., and Abe, N. (1999). A short introduction to boosting. *Journal-Japanese Society For Artificial Intelligence*, 14(771-780):1612. pages 241

Freund, Y. and Schapire, R. E. (1995). A desicion-theoretic generalization of on-line learning and an application to boosting. In *European conference on computational learning theory*, pages 23–37. Springer. pages 241

Friedman, J., Hastie, T., and Tibshirani, R. (2000). Additive logistic regression: A statistical view of boosting. *Ann. Statist.*, 28(2):337–407. With discussion and a rejoinder by the authors. pages 241

Friedman, J. H. (1996). Another approach to polychotomous classification. *Technical Report, Statistics Department, Stanford University*. pages 234, 247

Fukuyama, J., McMurdie, P. J., Dethlefsen, L., Relman, D. A., and Holmes, S. (2012). Comparisons of distance methods for combining covariates and abundances in microbiome studies. In *Biocomputing 2012*, pages 213–224. World Scientific. pages 181

Gabriel, K. R. (1971). The biplot graphic display of matrices with application to principal component analysis. *Biometrika*, 58:453–467. pages 346

Gao, C., Ma, Z., Ren, Z., and Zhou, H. H. (2015). Minimax estimation in sparse canonical correlation analysis. *Ann. Stat.*, 43(5):2168–2197. pages 351

Gaydos, T. L., Heckman, N. E., Kirkpatrick, M., Stinchcombe, J., Schmitt, J., Kingsolver, J., and Marron, J. S. (2013). Visualizing genetic constraints. *Ann. Appl. Stat.*, 7(2):860–882. pages 54, 121, 122

Gaynanova, I. and Li, G. (2019). Structural learning and integrative decomposition of multi-view data. *Biometrics*, 75(4):1121–1132. pages 359

Gelman, A. and Hill, J. (2007). *Data analysis using regression and multilevelhierarchical models*, volume 1. Cambridge University Press New York, NY, USA. pages 78

Geman, S. and Geman, D. (1984). Stochastic relaxation, Gibbs distributions, and the Bayesian restoration of images. *IEEE Trans. Pattern Anal. Mach. Intell.*, PAMI-6(6):721–741. pages 28

Gervini, D. and Gasser, T. (2004). Self-modeling warping functions. *J. R. Stat. Soc. Ser. B Stat. Methodol.*, 66(4):959–971. pages 175

Ghosh, A. K. and Biswas, M. (2016). Distribution-free high-dimensional two-sample tests based on discriminating hyperplanes. *TEST*, 25(3):525–547. pages 262, 290

Ginestet, C. E., Li, J., Balachandran, P., Rosenberg, S., and Kolaczyk, E. D. (2017). Hypothesis testing for network data in functional neuroimaging. *Ann. Appl. Stat.*, 11(2):725–750. pages 207, 208

Glavic, B. and Dittrich, K. (2007). Data provenance: A categorization of existing approaches. In Kemper, A., Schöning, H., Rose, T., Jarke, M., Seidl, T., Quix, C., and Brochhaus, C., editors, *Datenbanksysteme in Business, Technologie und Web (BTW 2007) - 12. Fachtagung des GI-Fachbereichs "Datenbanken und Informationssysteme" (DBIS)*, pages 227–241, Bonn. Gesellschaft für Informatik e. V. pages 71

Godtliebsen, F., Marron, J. S., and Chaudhuri, P. (2002). Significance in scale space for bivariate density estimation. *J. Comput. Graph. Statist.*, 11(1):1–21. pages 311

Godtliebsen, F., Marron, J. S., and Chaudhuri, P. (2004). Statistical significance of features in digital images. *Image Vis. Comput.*, 22(13):1093–1104. pages 311

Good, I. J. and Gaskins, R. A. (1980). Density estimation and bump-hunting by the penalized likelihood method exemplified by scattering and meteorite data. *J. Amer. Statist. Assoc.*, 75(369):42–73. With comments by Emanuel Parzen, William F. Eddy, David W. Scott, Michael E. Tarter, B. W. Silverman, Patrick C. Pointer and Richard Askey, and with a reply by the authors. pages 306

Goodall, C. R. (1991). Procrustes methods in the statistical analysis of shape (with discussion). *J. R. Stat. Soc. Series B.*, 53:285–339. pages 139, 141

Goodfellow, I., Bengio, Y., and Courville, A. (2016). *Deep learning*. MIT press. pages 241

Gower, J. C. (1966). Some distance properties of latent root and vector methods used in multivariate analysis. *Biometrika*, 53(3-4):325–338. pages 132, 318, 346

Gower, J. C. (1975). Generalized Procrustes analysis. *Psychometrika*, 40:33–50. pages 139, 141

Gower, J. C. and Hand, D. J. (1996). *Biplots*, volume 54 of *Monographs on Statistics and Applied Probability*. Chapman and Hall, Ltd., London. pages 346

Gradshteyn, I. S. and Ryzhik, I. M. (2015). *Table of integrals, series, and products*. Elsevier/Academic Press, Amsterdam, eighth edition. Translated from the Russian, Translation edited and with a preface by Daniel Zwillinger and Victor Moll, Revised from the seventh edition. pages 40

Green, P. J. and Silverman, B. W. (1994). *Nonparametric regression and generalized linear models: A roughness penalty approach*, volume 58 of *Monographs on Statistics and Applied Probability*. Chapman & Hall, London. pages 293

Gregory, J. S., Waarsing, J. H., Day, J., Pols, H. A., Reijman, M., Weinans, H., and Aspden, R. M. (2007). Early identification of radiographic osteoarthritis of the hip using an active shape model to quantify changes in bone morphometric features: can hip shape tell us anything about the progression of osteoarthritis? *Arthritis Rheum*, 56(11):3634–3643. pages 68

Grenander, U. and Miller, M. I. (2007). *Pattern theory: From representation to inference*. Oxford University Press, Oxford. pages 167

Groisser, D., Jung, S., and Schwartzman, A. (2017). Geometric foundations for scaling-rotation statistics on symmetric positive definite matrices: Minimal smooth scaling-rotation curves in low dimensions. *Electron. J. Stat.*, 11(1):1092–1159. pages 146, 174

Gu, C. (2013). *Smoothing spline ANOVA models*, volume 297 of *Springer Series in Statistics*. Springer, New York, second edition. pages 293

Guo, X. and Srivastava, A. (2020). Representations, metrics, and statistics for shape analysis of elastic graphs. In *Proceedings of the IEEE/CVF Conference on Computer Vision and Pattern Recognition Workshops*, pages 832–833. pages 206, 214

Györfi, L., Kohler, M., Krzyżak, A., and Walk, H. (2002). *A distribution-free theory of nonparametric regression*. Springer Series in Statistics. Springer-Verlag, New York. pages 293

Hadjipantelis, P. Z., Aston, J. A., and Evans, J. P. (2012). Characterizing fundamental frequency in mandarin: A functional principal component approach utilizing mixed effect models. *J. Acoust. Soc. Am.*, 131(6):4651–4664. pages 25

Hadjipantelis, P. Z., Aston, J. A. D., Müller, H. G., and Evans, J. P. (2015). Unifying amplitude and phase analysis: A compositional data approach to functional multivariate mixed-effects modeling of Mandarin Chinese. *J. Amer. Statist. Assoc.*, 110(510):545–559. pages 25

Haeckel, E. H. (1866). *Generelle Morphologie der Organismen allgemeine Grundzuge der organischen Formen-Wissenschaft, mechanisch begrundet durch die von Charles Darwin reformirte Descendenz-Theorie von Ernst Haeckel: Allgemeine Entwickelungsgeschichte der Organismen kritische Grundzuge der mechanischen Wissenschaft von den entstehenden Formen der Organismen, begrundet durch die Descendenz-Theorie*, volume 2. Verlag von Georg Reimer. pages 199

Hagan, M. T., Demuth, H. B., Beale, M. H., and De Jesus, O. (2014). *Neural network design, second edition*. Martin Hagan and Howard Demuth. pages 368

Haldane, J. (1948). Note on the median of a multivariate distribution. *Biometrika*, 35(3-4):414–417. pages 316

Hall, P. and Marron, J. S. (1987a). Extent to which least-squares cross-validation minimises integrated square error in nonparametric density estimation. *Probab. Theory Related Fields*, 74(4):567–581. pages 302

Hall, P. and Marron, J. S. (1987b). On the amount of noise inherent in bandwidth selection for a kernel density estimator. *Ann. Statist.*, 15(1):163–181. pages 302

Hall, P., Marron, J. S., and Neeman, A. (2005). Geometric representation of high dimension, low sample size data. *J. R. Stat. Soc. Ser. B Stat. Methodol.*, 67(3):427–444. pages 281, 283, 287

Hamelryck, T., Mardia, K. V., and Ferkinghoff-Borg, J., editors (2012). *Bayesian methods in structural bioinformatics*, New York. Springer. pages 167

Hamming, R. W. (1950). Error detecting and error correcting codes. *Bell Syst. Tech. J.*, 29(2):147–160. pages 122

Hampel, F. R., Ronchetti, E. M., Rousseeuw, P. J., and Stahel, W. A. (2011). *Robust statistics: The approach based on influence functions*, volume 114. John Wiley & Sons. pages 70, 128, 313, 315

Hanley, J. A. and McNeil, B. J. (1982). The meaning and use of the area under a receiver operating characteristic (ROC) curve. *Radiology*, 143(1):29–36. pages 93

Hannig, J. and Marron, J. S. (2006). Advanced distribution theory for SiZer. *J. Amer. Statist. Assoc.*, 101(474):484–499. pages 307

Hannig, J., Marron, J. S., and Riedi, R. H. (2001). Zooming statistics: Inference across scales. *J. Korean Statist. Soc.*, 30(2):327–345. 30th anniversary of the Korean Statistical Society (Seoul, 2001). pages 311

Harary, F. (2015). *A seminar on graph theory*. Courier Dover Publications. pages 197

Härdle, W. (1990). *Applied nonparametric regression*, volume 19 of *Econometric Society Monographs*. Cambridge University Press, Cambridge. pages 293

Härdle, W. and Marron, J. S. (1991). Bootstrap simultaneous error bars for nonparametric regression. *Ann. Statist.*, 19(2):778–796. pages 303

Härdle, W. and Schimek, M. G., editors (1996). *Statistical theory and computational aspects of smoothing: Proceedings of the COMPSTAT94 Satellite Meeting held in Semmering, Austria, 27–28 August*, Heidelberg. Physica-Verlag. pages 302

Harris, T. E. (1952). First passage and recurrence distributions. *Trans. Amer. Math. Soc.*, 73:471–486. pages 202

Hart, J. D. (1997). *Nonparametric smoothing and lack-of-fit tests*. Springer Series in Statistics. Springer-Verlag, New York. pages 294

Hartigan, J. A. (1975). *Clustering algorithms*. John Wiley & Sons, New York-London-Sydney. Wiley Series in Probability and Mathematical Statistics. pages 70, 243

Hastie, T. and Stuetzle, W. (1989). Principal curves. *J. Amer. Statist. Assoc.*, 84(406):502–516. pages 151

Hastie, T., Tibshirani, R., and Friedman, J. (2009). *The elements of statistical learning: Data mining, inference, and prediction.* Springer Series in Statistics. Springer, New York, second edition. pages 70, 215

Hastings, W. K. (1970). Monte Carlo sampling methods using Markov chains and their applications. *Biometrika*, 57(1):97–109. pages 364

Hellton, K. and Thoresen, M. (2014). Integrative clustering of high-dimensional data with joint and individual clusters, with an application to the Metabric study. *arXiv preprint arXiv:1410.8679.* pages 256

Hellton, K. H. and Thoresen, M. (2017). When and why are principal component scores a good tool for visualizing high-dimensional data? *Scand. Stat. Theory. Appl.*, 44(3):581–597. pages 290

Ho, Y. Y. (2011). *Protein profiling of acute myeloid leukemia-specific membrane proteins using label-free liquid chromatography-mass spectrometry.* Honours thesis in Chemistry, The University of Adelaide (available from peter.hoffmann@adelaide.edu.au). pages 19

Hoadley, K. A., Yau, C., Wolf, D. M., Cherniack, A. D., Tamborero, D., Ng, S., Leiserson, M. D., Niu, B., McLellan, M. D., Uzunangelov, V., et al. (2014). Multiplatform analysis of 12 cancer types reveals molecular classification within and across tissues of origin. *Cell*, 158(4):929–944. pages 60, 89

Holmström, L. and Erästö, P. (2002). Making inferences about past environmental change using smoothing in multiple time scales. *Comput. Statist. Data Anal.*, 41(2):289–309. pages 307

Hong, J., Vicory, J., Schulz, J., Styner, M., Marron, J., and Pizer, S. M. (2016). Non-Euclidean classification of medically imaged objects via s-reps. *Med. Image Anal.*, 31:37–45. pages 16

Hotelling, H. (1933). Analysis of a complex of statistical variables into principal components. *J. Educ. Psychol.*, 24(6):417. pages 331

Hotelling, H. (1936). Relations between two sets of variates. *Biometrika*, 28(3/4):321–377. pages 354

Hotz, T. (2013). Extrinsic vs intrinsic means on the circle. In *Geometric science of information*, volume 8085 of *Lecture Notes in Comput. Sci.*, pages 433–440. Springer, Heidelberg. pages 169

Hotz, T. and Huckemann, S. (2015). Intrinsic means on the circle: Uniqueness, locus, and asymptotics. *Ann. Inst. Statist. Math.*, 67(1):177–193. pages 169

Hotz, T., Huckemann, S., Le, H., Marron, J. S., Mattingly, J. C., Miller, E., Nolen, J., Owen, M., Patrangenaru, V., and Skwerer, S. (2013). Sticky central limit theorems on open books. *Ann. Appl. Probab.*, 23(6):2238–2258. pages 131, 169, 202

Hron, K., Menafoglio, A., Templ, M., Hruzova, K., and Filzmoser, P. (2016). Simplicial principal component analysis for density functions in Bayes spaces. *Comput. Statist. Data Anal.*, 94:330–350. pages 45

Hsing, T. and Eubank, R. (2015). *Theoretical foundations of functional data analysis, with an introduction to linear operators*. Wiley Series in Probability and Statistics. John Wiley & Sons, Ltd., Chichester. pages 106

Huang, H., Liu, Y., Du, Y., Perou, C. M., Hayes, D. N., Todd, M. J., and Marron, J. S. (2013). Multiclass distance-weighted discrimination. *J. Comput. Graph. Statist.*, 22(4):953–969. pages 239

Huang, H., Liu, Y., and Marron, J. S. (2012). Bidirectional discrimination with application to data visualization. *Biometrika*, 99(4):851–864. pages 231

Huang, H., Liu, Y., Yuan, M., and Marron, J. (2015). Statistical significance of clustering using soft thresholding. *J. Comput. Graph. Stat.*, 24(4):975–993. pages 263, 268, 270

Huber, P. J. and Ronchetti, E. M. (2009). *Robust statistics*. Wiley Series in Probability and Statistics. John Wiley & Sons, Inc., Hoboken, NJ, second edition. pages 70, 128, 313, 315, 316

Huckemann, S. and Hotz, T. (2014). On means and their asymptotics: Circles and shape spaces. *J. Math. Imaging Vision*, 50(1-2):98–106. pages 169

Huckemann, S. and Hotz, T. (2016). Nonparametric statistics on manifolds and beyond. In Denker, M. and Waymire, E. C., editors, *Rabi N. Bhattacharya*, pages 599–609. Springer. pages 131

Huckemann, S., Hotz, T., and Munk, A. (2010). Intrinsic shape analysis: geodesic PCA for Riemannian manifolds modulo isometric Lie group actions. *Statist. Sinica*, 20(1):1–58. pages 153

Huckemann, S., Kim, K. R., Munk, A., Rehfeldt, F., Sommerfeld, M., Weickert, J., and Wollnik, C. (2016). The circular SiZer, inferred persistence of shape parameters and application to early stem cell differentiation. *Bernoulli*, 22(4):2113–2142 pages 153, 311

Huckemann, S. F. and Eltzner, B. (2017). Essentials of backward nested descriptors inference. In *Functional Statistics and Related Fields*, pages 137–144. Springer. pages 131

Huckemann, S. F. and Eltzner, B. (2018). Backward nested descriptors asymptotics with inference on stem cell differentiation. *Ann. Stat.*, 46(5):1994–2019. pages 168, 172

Huettel, S. A., Song, A. W., and McCarthy, G. (2004). *Functional magnetic resonance imaging*, volume 1. Sinauer Associates, Sunderland, MA. pages 368

Hurley, J. R. and Cattell, R. B. (1962). The Procrustes program: Producing direct rotation to test a hypothesised factor structure. *Behav. Sci.*, 7:258–262. pages 139

Hyvärinen, A., Karhunen, J., and Oja, E. (2001). *Independent component analysis*. John Wiley & Sons Inc., New York. pages 120

Inselberg, A. (1985). The plane with parallel coordinates. *Visual. Comput.*, 1(2):69–91. pages 40

Inselberg, A. (2009). *Parallel coordinates: Visual multidimensional geometry and its applications*. Springer, New York. pages 40

Izem, R. and Kingsolver, J. G. (2005). Variation in continuous reaction norms: Quantifying directions of biological interest. *Am. Nat.*, 166(2):277–289. pages 121

Izem, R. and Marron, J. S. (2007). Analysis of nonlinear modes of variation for functional data. *Electron. J. Stat.*, 1:641–676. pages 121

Izenman, A. J. and Sommer, C. J. (1988). Philatelic mixtures and multimodal densities. *J. Am. Stat. Assoc.*, 83(404):941–953. pages 294

Jaccard, P. (1901). Étude comparative de la distribution florale dans une portion des alpes et des jura. *Bull Soc Vaudoise Sci Nat*, 37:547–579. pages 123

James, G., Witten, D., Hastie, T., and Tibshirani, R. (2013). *An introduction to statistical learning, with applications in R*, volume 103 of *Springer Texts in Statistics*. Springer, New York. pages 215

James, G. M. (2007). Curve alignment by moments. *Ann. Appl. Stat.*, 1(2):480–501. pages 175

Jammalamadaka, S. R. and SenGupta, A. (2001). *Topics in circular statistics*, volume 5 of *Series on Multivariate Analysis*. World Scientific Publishing Co., Inc., River Edge, NJ. pages 147

Jeong, J.-Y. (2009). *Estimation of probability distributions on multiple anatomical objects and evaluation of statistical shape models*. PhD thesis, The University of North Carolina at Chapel Hill. pages 153, 154

Jeong, J.-Y., Stough, J. V., Marron, J. S., and Pizer, S. M. (2008). Conditional-mean initialization using neighboring objects in deformable model segmentation. In *Medical Imaging*, pages 69144R–69144R. International Society for Optics and Photonics. pages 17, 153

Joachims, T. (2000). Estimating the generalization performance of an SVM efficiently. In Langley, P., editor, *ICML '00 Proc. 17th International Conference on Machine Learning*, pages 431–438, San Francisco. Morgan Kaufmann Publishers Inc. pages 234

Johnson, N. L. (1949). Systems of frequency curves generated by methods of translation. *Biometrika*, 36(1/2):149–176. pages 94

Johnson, W. E., Li, C., and Rabinovic, A. (2007). Adjusting batch effects in microarray expression data using empirical Bayes methods. *Biostatistics*, 8(1):118–127. pages 239

Johnstone, I. M. (2001). On the distribution of the largest eigenvalue in principal components analysis. *Ann. Statist.*, 29(2):295–327. pages 279

Jolliffe, I. T. (2002). *Principal component analysis*. Springer Series in Statistics. Springer-Verlag, New York, second edition. pages 5, 85, 331, 342, 349

Jones, M. C., Marron, J. S., and Park, B. U. (1991). A simple root n bandwidth selector. *Ann. Statist.*, 19(4):1919–1932. pages 302

Jones, M. C., Marron, J. S., and Sheather, S. J. (1996a). A brief survey of band-width selection for density estimation. *J. Amer. Statist. Assoc.*, 91(433):401–407. pages 302, 303

Jones, M. C., Marron, J. S., and Sheather, S. J. (1996b). Progress in data-based bandwidth selection for kernel density estimation. *Comput. Statist.*, 11(3):337–381. pages 302

Jung, S., Dryden, I. L., and Marron, J. S. (2012a). Analysis of principal nested spheres. *Biometrika*, 99(3):551–568. pages 23, 154, 155, 164

Jung, S., Foskey, M., and Marron, J. S. (2011). Principal arc analysis on direct product manifolds. *Ann. Appl. Stat.*, 5(1):578–603. pages 154

Jung, S., Lee, M. H., and Ahn, J. (2018). On the number of principal components in high dimensions. *Biometrika*, 105(2):389–402. pages 288

Jung, S. and Marron, J. S. (2009). PCA consistency in high dimension, low sample size context. *Ann. Statist.*, 37(6B):4104–4130. pages 288

Jung, S., Schwartzman, A., and Groisser, D. (2015). Scaling-rotation distance and interpolation of symmetric positive-definite matrices. *SIAM J. Matrix Anal. Appl.*, 36(3):1180–1201. pages 146, 174

Jung, S., Sen, A., and Marron, J. S. (2012b). Boundary behavior in high dimension, low sample size asymptotics of PCA. *J. Multivariate Anal.*, 109:190–203. pages 288

Kaiser, H. F. (1958). The varimax criterion for analytic rotation in factor analysis. *Psychometrika*, 23(3):187–200. pages 120

Kang, H. B., Reimherr, M., Shriver, M., and Claes, P. (2017). Manifold data analysis with applications to high-frequency 3d imaging. *arXiv preprint arXiv:1710.01619*. pages 150

Karcher, H. (2014). Riemannian center of mass and so called Karcher mean. *arXiv preprint arXiv:1407.2087*. pages 131, 186

Kaufman, L. and Rousseeuw, P. J. (2009). *Finding groups in data: An introduction to cluster analysis*, volume 344. John Wiley & Sons. pages 70, 243

Kelemen, A., Székely, G., and Gerig, G. (1999). Elastic model-based segmentation of 3-d neuroradiological data sets. *IEEE Trans. Med. Imaging.*, 18(10):828–839. pages 13

Kendall, D. G. (1970). A mathematical approach to seriation. *Philos. Trans. Royal Soc. A*, 269(1193):125–134. pages 133

Kendall, D. G. (1983). The shape of Poisson-Delaunay triangles. In Demetrescu, M. C. and Iosifescu, M., editors, *Studies in Probability and Related Topics*, pages 321–330. Nagard, Montreal. pages 158

Kendall, D. G. (1984). Shape manifolds, Procrustean metrics, and complex projective spaces. *Bull. London Math. Soc.*, 16(2):81–121. pages 12, 139, 157, 158

Kendall, D. G. (1989). A survey of the statistical theory of shape (with discussion). *Stat. Sci.*, 4:87–120. pages 158, 159

Kendall, D. G., Barden, D., Carne, T. K., and Le, H. (1999). *Shape and shape theory*. Wiley Series in Probability and Statistics. John Wiley & Sons Ltd., Chichester. pages 157, 158

Kendall, W. S. (1988). Symbolic computation and the diffusion of shapes of triads. *Adv. in Appl. Probab.*, 20(4):775–797. pages 282

Kent, J. T. (1994). The complex Bingham distribution and shape analysis. *J. R. Stat. Soc. Series B.*, 56:285–299. pages 158

Kent, J. T. and Mardia, K. V. (2001). Shape, Procrustes tangent projections, and bilateral symmetry. *Biometrika*, 88(2):469–485. pages 158, 160

Kettenring, J. R. (1971). Canonical analysis of several sets of variables. *Biometrika*, 58:433–451. pages 351, 356

Kim, B., Huckemann, S., Schulz, J., and Jung, S. (2019). Small-sphere distributions for directional data with application to medical imaging. *Scand. Stat. Theory. Appl.*, 46(4):1047–1071. pages 156

Kim, C. S. and Marron, J. S. (2006). SiZer for jump detection. *J. Nonparametr. Stat.*, 18(1):13–20. pages 311

Kim, K.-R., Dryden, I. L., Le, H., and Severn, K. E. (2021). Smoothing splines on Riemannian manifolds, with applications to 3D shape space. *J. R. Stat. Soc. Series. B. Stat. Methodol.*, 83(1):108–132. pages 167

Kimes, P. K., Cabanski, C. R., Wilkerson, M. D., Zhao, N., Johnson, A. R., Perou, C. M., Makowski, L., Maher, C. A., Liu, Y., Marron, J. S., et al. (2014). Sigfuge: Single gene clustering of rna-seq reveals differential isoform usage among cancer samples. *Nucleic Acids Res.*, 42(14):e113–e113. pages 59

Kingsolver, J. G., Heckman, N., Zhang, J., Carter, P. A., Knies, J. L., Stinchcombe, J. R., and Meyer, K. (2015). Genetic variation, simplicity, and evolutionary constraints for function-valued traits. *Am. Nat.*, 185(6):E166–E181. pages 121, 122

Klein, R. J., Zeiss, C., Chew, E. Y., Tsai, J.-Y., Sackler, R. S., Haynes, C., Henning, A. K., SanGiovanni, J. P., Mane, S. M., Mayne, S. T., et al. (2005). Complement factor H polymorphism in age-related macular degeneration. *Science*, 308(5720):385–389. pages 287, 325

Klemelä, J. (2009). *Smoothing of multivariate data: Density estimation and visualization*. Wiley Series in Probability and Statistics. John Wiley & Sons, Inc., Hoboken, NJ. pages 293

Koch, I. (2014). *Analysis of multivariate and high-dimensional data*. Cambridge Series in Statistical and Probabilistic Mathematics, [32]. Cambridge University Press, New York. pages 2

Koch, I., Hoffmann, P., and Marron, J. S. (2014). Proteomics profiles from mass spectrometry. *Electron. J. Stat.*, 8(2):1703–1713. pages 19, 20

Koenker, R. (2006). The median is the message: Toward the Fréchet median. *J. Soc. Fr. Stat.*, 147(2):61–64 (2007). pages 128

Kohavi, R. et al. (1995). A study of cross-validation and bootstrap for accuracy estimation and model selection. In *IJCAI'95: Proceedings of the 14th International Joint Conference on Artificial Intelligence*, volume 2:, pages 1137–1145. Montreal, Canada. pages 241

Kohn, R., Marron, J. S., and Yau, P. (2000). Wavelet estimation using Bayesian basis selection and basis averaging. *Statist. Sinica*, 10(1):109–128. pages 300

Kokoszka, P. and Reimherr, M. (2017). *Introduction to functional data analysis*. Texts in Statistical Science Series. CRC Press, Boca Raton, FL. pages 106

Kolaczyk, E. D. (2009). *Statistical analysis of network data*. Springer Series in Statistics. Springer, New York. Methods and models. pages 197, 207

Kolaczyk, E. D., Lin, L., Rosenberg, S., Walters, J., and Xu, J. (2020). Averages of unlabeled networks: Geometric characterization and asymptotic behavior. *Ann. Statist.*, 48(1):514–538. pages 214

Koltchinskii, V. and Lounici, K. (2016). Asymptotics and concentration bounds for bilinear forms of spectral projectors of sample covariance. *Ann. Inst. Henri Poincaré Probab. Stat.*, 52(4):1976–2013. pages 290

Koltchinskii, V. and Lounici, K. (2017a). New asymptotic results in principal component analysis. *Sankhya A*, 79(2):254–297. pages 290

Koltchinskii, V. and Lounici, K. (2017b). Normal approximation and concentration of spectral projectors of sample covariance. *Ann. Statist.*, 45(1):121–157. pages 290

Kooperberg, C. and Stone, C. J. (1991). A study of logspline density estimation. *Comput. Statist. Data Anal.*, 12(3):327–347. pages 138

Kruskal, J. B. (1964). Multidimensional scaling by optimizing goodness of fit to a nonmetric hypothesis. *Psychometrika*, 29:1–27. pages 134

Kurtek, S., Srivastava, A., Klassen, E., and Ding, Z. (2012). Statistical modeling of curves using shapes and related features. *J. Amer. Statist. Assoc.*, 107(499):1152–1165. pages 21

Kurtek, S., Srivastava, A., Klassen, E., and Laga, H. (2013). Landmark-guided elastic shape analysis of spherically-parameterized surfaces. *Comput. Graph. Forum*, 32(2pt4):429–438. pages 13

Kurtek, S., Xie, Q., and Srivastava, A. (2014). Analysis of juggling data: Alignment, extraction, and modeling of juggling cycles. *Electron. J. Stat.*, 8(2):1865–1873. pages 189

Lam, X. Y., Marron, J. S., Sun, D., and Toh, K.-C. (2018). Fast algorithms for large-scale generalized distance weighted discrimination. *J. Comput. Graph. Statist.*, 27(2):368–379. pages 237

Lambert, D. (1992). Zero-inflated Poisson regression, with an application to defects in manufacturing. *Technometrics*, 34(1):1–14. pages 83

Langlois, J. H. and Roggman, L. A. (1990). Attractive faces are only average. *Psychol. Sci.*, 1(2):115–121. pages 29

Le, H. (1991). On geodesics in Euclidean shape spaces. *J. London Math. Soc.*, 44:360–372. pages 159

Le, H. and Kendall, D. G. (1993). The Riemannian structure of Euclidean shape spaces: A novel environment for statistics. *Ann. Statist.*, 21(3):1225–1271. pages 158, 159

LeBlanc, M. and Tibshirani, R. (1994). Adaptive principal surfaces. *J. Am. Stat. Assoc.*, 89(425):53–64. pages 151

Ledoux, M. (2001). *The concentration of measure phenomenon*, volume 89 of *Mathematical Surveys and Monographs*. American Mathematical Society, Providence, RI. pages 290

Lee, D. D. and Seung, H. S. (1999). Learning the parts of objects by non-negative matrix factorization. *Nature*, 401(6755):788. pages 122, 171

Lee, J. M. (2018). *Introduction to Riemannian manifolds*, volume 176 of *Graduate Texts in Mathematics*. Springer, Cham. Second edition. pages 150

Lee, M. H. (2007). *Continuum direction vectors in high dimensional low sample size data*. PhD thesis, The University of North Carolina at Chapel Hill. pages 357

Lee, Y., Lin, Y., and Wahba, G. (2004). Multicategory support vector machines: Theory and application to the classification of microarray data and satellite radiance data. *J. Amer. Statist. Assoc.*, 99(465):67–81. pages 234

Leek, J. T., Johnson, W. E., Parker, H. S., Jaffe, A. E., and Storey, J. D. (2012). The SVA package for removing batch effects and other unwanted variation in high-throughput experiments. *Bioinformatics*, 28(6):882–883. pages 239

Lenglet, C., Rousson, M., and Deriche, R. (2006). DTI segmentation by statistical surface evolution. *IEEE Trans. Med. Imaging*, 25:685–700. pages 145

Levina, E. and Bickel, P. (2001). The earth mover's distance is the Mallows distance: Some insights from statistics. In *Proceedings Eighth IEEE Int. J. Comput. Vis. ICCV 2001*, volume 2, pages 251–256. IEEE. pages 137

Levy, O., Goldberg, Y., and Dagan, I. (2015). Improving distributional similarity with lessons learned from word embeddings. *Trans. Assoc. Comput. Linguist.*, 3:211–225. pages 369

Li, G. and Chen, Z. (1985). Projection-pursuit approach to robust dispersion matrices and principal components: Primary theory and Monte Carlo. *J. Am. Stat. Assoc.*, 80(391):759–766. pages 323

Li, K.-C. (1991). Sliced inverse regression for dimension reduction. *J. Amer. Statist. Assoc.*, 86(414):316–342. With discussion and a rejoinder by the author. pages 123

Li, R. and Marron, J. S. (2005). Local likelihood SiZer map. *Sankhyā*, 67(3):476–498. pages 311

Li, W. and Yao, J. (2018). On structure testing for component covariance matrices of a high dimensional mixture. *J. R. Stat. Soc. Series B Stat. Methodol.*, 80(2):293–318. pages 280

Lila, E., Arridge, S., and Aston, J. A. D. (2020). Representation and reconstruction of covariance operators in linear inverse problems. *Inverse Probl.*, 36(8):085002. pages 151

Lila, E. and Aston, J. A. D. (2020). Statistical analysis of functions on surfaces, with an application to medical imaging. *J. Am. Stat. Assoc.*, 115(531):1420–1434. pages 151

Lila, E., Aston, J. A. D., and Sangalli, L. M. (2016). Smooth principal component analysis over two-dimensional manifolds with an application to neuroimaging. *Ann. Appl. Stat.*, 10(4):1854–1879. pages 151

Lin, Y., Wahba, G., Zhang, H., and Lee, Y. (2002). Statistical properties and adaptive tuning of support vector machines. *Mach. Learn.*, 48(1-3):115–136. pages 234

Lindeberg, T. (2013). *Scale-space theory in computer vision*, volume 256. Springer Science & Business Media. pages 304

Liu, R. Y., Parelius, J. M., and Singh, K. (1999). Multivariate analysis by data depth: Descriptive statistics, graphics, and inference. *Ann. Statist.*, 27(3):783–858. With discussion and a rejoinder by Liu and Singh. pages 327

Liu, X. and Müller, H.-G. (2004). Functional convex averaging and synchronization for time-warped random curves. *J. Amer. Statist. Assoc.*, 99(467):687–699. pages 175

Liu, X., Parker, J., Fan, C., Perou, C. M., and Marron, J. (2009). Visualization of cross-platform microarray normalization. *Batch Effects and Noise in Microarray Experiments: Sources and Solutions. Wiley, New York*, pages 167–181. pages 62, 235, 237

Liu, Y., Hayes, D. N., Nobel, A., and Marron, J. S. (2008). Statistical significance of clustering for high-dimension, low–sample size data. *J. Am. Stat. Assoc.*, 103(483):1281–1293. pages 263, 267, 270

Locantore, N., Marron, J., Simpson, D., Tripoli, N., Zhang, J., Cohen, K., Boente, G., Fraiman, R., Brumback, B., Croux, C., et al. (1999). Robust principal component analysis for functional data. *Test*, 8(1):1–73. pages 319, 321, 322, 323, 324, 325

Lock, E. F. and Dunson, D. B. (2013). Bayesian consensus clustering. *Bioinformatics*, 29(20):2610–2616. pages 274

Lock, E. F., Hoadley, K. A., Marron, J. S., and Nobel, A. B. (2013). Joint and individual variation explained (JIVE) for integrated analysis of multiple data types. *Ann. Appl. Stat.*, 7(1):523–542. pages 359

López-Pintado, S. and Romo, J. (2006). Depth-based classification for functional data. In *data depth: Robust multivariate analysis, computational geometry, and applications*, volume 72 of *DIMACS Ser. Discrete Math. Theoret. Comput. Sci.*, pages 103–119. Amer. Math. Soc., Providence, RI. pages 327

Lu, C., Pizer, S. M., Joshi, S., and Jeong, J.-Y. (2007). Statistical multi-object shape models. *INT. J. COMPUT. VISION.*, 75(3):387–404. pages 17

Lu, X. (2013). *Object oriented data analysis of cell images and analysis of elastic functions.* PhD thesis, The University of North Carolina at Chapel Hill. pages 184, 193

Lu, X., Koch, I., and Marron, J. S. (2014a). Analysis of proteomics data: Impact of alignment on classification. *Electron. J. Stat.*, 8(2):1742–1747. pages 189

Lu, X. and Marron, J. S. (2014a). Analysis of juggling data: Object oriented data analysis of clustering in acceleration functions. *Electron. J. Stat.*, 8(2):1842–1847. pages 22, 23, 189, 195

Lu, X. and Marron, J. S. (2014b). Analysis of spike train data: Comparison between the real and the simulated data. *Electron. J. Stat.*, 8(2):1793–1796. pages 189

Lu, X., Marron, J. S., and Haaland, P. (2014b). Object-oriented data analysis of cell images. *J. Amer. Statist. Assoc.*, 109(506):548–559. pages 362

Lu, Y., Herbei, R., and Kurtek, S. (2017). Bayesian registration of functions with a Gaussian process prior. *J. Comput. Graph. Stat.*, 26(4):894–904. pages 193

Maaten, L. v. d. and Hinton, G. (2008). Visualizing data using t-SNE. *J. Mach. Learn. Res.*, 9(Nov):2579–2605. pages 229

MacQueen, J. (1967). Some methods for classification and analysis of multivariate observations. In *Proc. Fifth Berkeley Sympos. Math. Statist. and Probability (Berkeley, Calif., 1965/66)*, pages Vol. I: Statistics, pp. 281–297. Univ. California Press, Berkeley, Calif. pages 243

Maggiora, G. M. (2006). On outliers and activity cliffs why QSAR often disappoints. *J Chem Inf Model.*, 46:1535. pages 89

Mahalanobis, P. C. (1936). *On the generalized distance in statistics.* National Institute of Science of India. pages 219

Mahlberg, M., Stockwell, P., de Joode, J., Smith, C., and O'Donnell, M. B. (2016). CLiC Dickens: Novel uses of concordances for the integration of corpus stylistics and cognitive poetics. *Corpora*, 11(3):433–463. pages 211

Mallat, S. (2009). *A wavelet tour of signal processing: The sparse way, 3rd edition.* Elsevier/Academic Press, Amsterdam. pages 40

Mallat, S. G. (1989). A theory for multiresolution signal decomposition: The wavelet representation. *IEEE Trans. Pattern Anal. Mach. Intell.*, 11(7):674–693. pages 40

Mallows, C. L. (1972). A note on asymptotic joint normality. *Ann. Math. Statist.*, 43:508–515. pages 137

Mammen, E., Marron, J. S., Turlach, B. A., and Wand, M. P. (2001). A general projection framework for constrained smoothing. *Statist. Sci.*, 16(3):232–248. pages 333

Mardia, K. V. and Jupp, P. E. (2000). *Directional statistics*. Wiley Series in Probability and Statistics. John Wiley & Sons, Ltd., Chichester. Revised reprint of *Statistics of directional data* by Mardia. pages 69, 147, 168

Mardia, K. V., Kent, J. T., and Bibby, J. M. (1979). *Multivariate analysis*. Academic Press [Harcourt Brace Jovanovich, Publishers], London-New York-Toronto, Ont. Probability and Mathematical Statistics: A Series of Monographs and Textbooks. pages 2, 85, 120, 133, 224, 225, 231

Maronna, R. A., Martin, R. D., Yohai, V. J., and Salibián-Barrera, M. (2019). *Robust statistics*. Wiley Series in Probability and Statistics. John Wiley & Sons, Inc., Hoboken, NJ. Theory and methods (with R), Second edition. pages 325

Marron, J. S. (1983). Optimal rates on convergence to Bayes risk in nonparametric discrimination. *Ann. Statist.*, 11(4):1142–1155. pages 224

Marron, J. S. (1989). Automatic smoothing parameter selection: A survey. In *Semiparametric and nonparametric econometrics*, pages 65–86. Springer. pages 302

Marron, J. S. (1999). Spectral view of wavelets and nonlinear regression. In *Bayesian Inference in Wavelet-Based Models*, pages 19–32. Springer. pages 41

Marron, J. S. (2017a). Big data in context and robustness against heterogeneity. *Econ. Stat.*, 2:73–80. pages 3, 239, 257, 314

Marron, J. S. (2017b). Marron matlab software. http://marron.web.unc.edu/sample-page/marrons-matlab-software/. pages xi

Marron, J. S. (2020). Companion webpage to book on object oriented data analysis. https://github.com/jsmarron/ObjectOrientedDataAnalysis.git/. pages xi, 3

Marron, J. S., Adak, S., Johnstone, I. M., Neumann, M. H., and Patil, P. (1998). Exact risk analysis of wavelet regression. *J. Comput. Graph. Stat.*, 7(3):278–309. pages 41, 301

Marron, J. S. and Alonso, A. M. (2014). Overview of object oriented data analysis. *Biom. J.*, 56(5):732–753. pages 1, 3, 32, 362

Marron, J. S., Koch, I., and Hoffmann, P. (2014a). Rejoinder: Analysis of proteomics data. *Electron. J. Stat.*, 8(2):1756–1758. pages 19, 20

Marron, J. S. and Nolan, D. (1988). Canonical kernels for density estimation. *Statist. Probab. Lett.*, 7(3):195–199. pages 296

Marron, J. S., Ramsay, J. O., Sangalli, L. M., and Srivastava, A. (2014b). Statistics of time warpings and phase variations. *Electron. J. Stat.*, 8(2):1697–1702. pages 19, 22, 189

Marron, J. S., Ramsay, J. O., Sangalli, L. M., and Srivastava, A. (2015). Functional data analysis of amplitude and phase variation. *Statist. Sci.*, 30(4):468–484. pages 19, 21, 22, 177

Marron, J. S. and Schmitz, H.-P. (1992). Simultaneous density estimation of several income distributions. *Econom. Theory*, 8(4):476–488. pages 298, 304, 305

Marron, J. S., Todd, M. J., and Ahn, J. (2007). Distance-weighted discrimination. *J. Amer. Statist. Assoc.*, 102(480):1267–1271. pages 28, 235, 236, 237

Marron, J. S. and Tsybakov, A. B. (1995). Visual error criteria for qualitative smoothing. *J. Amer. Statist. Assoc.*, 90(430):499–507. pages 41, 138

Marron, J. S. and Wand, M. P. (1992). Exact mean integrated squared error. *Ann. Statist.*, 20(2):712–736. pages 301

Marron, J. S., Wendelberger, J. R., and Kober, E. M. (2004). Time series functional data analysis, los alamos national lab, no. Technical report, LA-UR-04-3911. pages 180

Marron, J. S. and Zhang, J.-T. (2005). SiZer for smoothing splines. *Comput. Statist.*, 20(3):481–502. pages 311

Martin, D. I. and Berry, M. W. (2007). Mathematical foundations behind latent semantic analysis. *Handbook of Latent Semantic Analysis*, pages 35–56. pages 368

Marčenko, V. A. and Pastur, L. A. (1967). The spectrum of random matrices. *Teor. Funkciĭ Funkcional. Anal. i Priložen. Vyp.*, 4:122–145. pages 276

McCulloch, W. S. and Pitts, W. (1943). A logical calculus of the ideas immanent in nervous activity. *Bull. Math. Biophys.*, 5(4):115–133. pages 368

McInnes, L., Healy, J., and Melville, J. (2018). Umap: Uniform manifold approximation and projection for dimension reduction. *arXiv preprint arXiv:1802.03426.* pages 230

McLachlan, G. (2004). *Discriminant analysis and statistical pattern recognition*, volume 544. John Wiley & Sons. pages 216

McShane, L. M., Radmacher, M. D., Freidlin, B., Yu, R., Li, M.-C., and Simon, R. (2002). Methods for assessing reproducibility of clustering patterns observed in analyses of microarray data. *Bioinformatics*, 18(11):1462–1469. pages 274

Meinshausen, N. and Bühlmann, P. (2015). Maximin effects in inhomogeneous large-scale data. *Ann. Statist.*, 43(4):1801–1830. pages 239, 314

Menafoglio, A., Grasso, M., Secchi, P., and Colosimo, B. M. (2018). Profile monitoring of probability density functions via simplicial functional PCA with application to image data. *Technometrics*, 60(4):497–510. pages 45

Menafoglio, A. and Secchi, P. (2017). Statistical analysis of complex and spatially dependent data: A review of object oriented spatial statistics. *Eur. J. Oper. Res.*, 258(2):401–410. pages 172, 367

Merck, D., Tracton, G., Saboo, R., Levy, J., Chaney, E., Pizer, S., and Joshi, S. (2008). Training models of anatomic shape variability. *J. Med. Phys.*, 35(8):3584–3596. pages 17

Miao, D. (2015). *Class-sensitive principal components analysis.* PhD thesis, The University of North Carolina at Chapel Hill. pages 221

Milasevic, P. and Ducharme, G. R. (1987). Uniqueness of the spatial median. *Ann. Statist.*, 15(3):1332–1333. pages 316

Miller, E., Owen, M., and Provan, J. S. (2015). Polyhedral computational geometry for averaging metric phylogenetic trees. *Adv. in Appl. Math.*, 68:51–91. pages 200

Minnotte, M. C. (2010). Mode testing via higher-order density estimation. *Comput. Statist.*, 25(3):391–407. pages 294

Minnotte, M. C. and Scott, D. W. (1993). The mode tree: A tool for visualization of nonparametric density features. *J. Comput. Graph. Stat.*, 2(1):51–68. pages 294

Moakher, M. (2005). A differential geometric approach to the geometric mean of symmetric positive-definite matrices. *SIAM J. Matrix Anal. Appl.*, 26(3):735–747 (electronic). pages 145

Monti, S., Tamayo, P., Mesirov, J., and Golub, T. (2003). Consensus clustering: A resampling-based method for class discovery and visualization of gene expression microarray data. *Mach. Learn.*, 52(1-2):91–118. pages 274

Moore, R. E., Kearfott, R. B., and Cloud, M. J. (2009). *Introduction to interval analysis*, volume 110. SIAM, Philadelphia. pages 366

Morton, J. T., Toran, L., Edlund, A., Metcalf, J. L., Lauber, C., and Knight, R. (2017). Uncovering the horseshoe effect in microbial analyses. *mSystems*, 2(1):e00166–16. pages 134, 180

Mosier, C. I. (1939). Determining a simple structure when loadings for certain tests are known. *Psychometrika*, 4:149–162. pages 139

Möttönen, J. and Oja, H. (1995). Multivariate spatial sign and rank methods. *J. Nonparametr. Statist.*, 5(2):201–213. pages 316, 319

Muirhead, R. J. (1982). *Aspects of multivariate statistical theory*. John Wiley & Sons, Inc., New York. Wiley Series in Probability and Mathematical Statistics. pages 2, 85

Müller, D. W. and Sawitzki, G. (1991). Excess mass estimates and tests for multimodality. *J. Amer. Statist. Assoc.*, 86(415):738–746. pages 306

Müller, H.-G. (1988). *Nonparametric regression analysis of longitudinal data*, volume 46 of *Lecture Notes in Statistics*. Springer-Verlag, Berlin. pages 293

Murteira, J. M. and Ramalho, J. J. (2016). Regression analysis of multivariate fractional data. *Econom. Rev.*, 35(4):515–552. pages 364

National Center for Biotechnology Information (2019). Gene data base. https://www.ncbi.nlm.nih.gov/gene. pages 63

Nelson, A. E., Fang, F., Arbeeva, L., Cleveland, R. J., Schwartz, T. A., Callahan, L. F., Marron, J. S., and Loeser, R. (2019). A machine learning approach to knee osteoarthritis phenotyping: Data from the FNIH Biomarkers Consortium. *Osteoarthr. Cartil.*, 27(7):994–1001. pages 230

Nelson, A. E., Liu, F., Lynch, J. A., Renner, J. B., Schwartz, T. A., Lane, N. E., and Jordan, J. M. (2014). Association of incident symptomatic hip osteoarthritis with differences in hip shape by active shape modeling: The johnston county osteoarthritis project. *Arthritis Care. Res.*, 66(1):74–81. pages 68

Nelson, A. E., Shi, Y., Tiller, R., Schwartz, T., Renner, J. B., Jordan, J. M., Aspden, R., Gregory, J. S., and Marron, J. S. (2017). Baseline knee shape discriminates cases of incident knee radiographic OA from controls: A case-control study using novel methodology from the Johnston County osteoarthritis project. *Osteoarthr. Cartil.*, 25:S70–S71. pages 68

Nielsen, A. A. (2002). Multiset canonical correlations analysis and multispectral, truly multitemporal remote sensing data. *IEEE Trans. Image Process.*, 11(3):293–305. pages 351

Nye, T. M. W. (2011). Principal components analysis in the space of phylogenetic trees. *Ann. Statist.*, 39(5):2716–2739. pages 201

Nye, T. M. W., Tang, X., Weyenberg, G., and Yoshida, R. (2017). Principal component analysis and the locus of the Fréchet mean in the space of phylogenetic trees. *Biometrika*, 104(4):901–922. pages 202

Oguz, I., Cates, J., Fletcher, T., Whitaker, R., Cool, D., Aylward, S., and Styner, M. (2008). Cortical correspondence using entropy-based particle systems and local features. In *2008 5th IEEE Int. Symp. Biomed. Imaging*, pages 1637–1640. IEEE. pages 199, 200

Oja, H. (2010). *Multivariate nonparametric methods with R: An approach based on spatial signs and ranks*, volume 199 of *Lecture Notes in Statistics*. Springer, New York. pages 316

Oliva, J. B., Dubey, A., Wilson, A. G., Póczos, B., Schneider, J., and Xing, E. P. (2016). Bayesian nonparametric kernel-learning. In *Artificial Intelligence and Statistics*, pages 1078–1086. pages 303

Owen, M. and Provan, J. S. (2011). A fast algorithm for computing geodesic distances in tree space. *IEEE/ACM Trans. Comput. Biol. Bioinform.*, 8(1):2–13. pages 200, 202

Owen, S. J. (1998). A survey of unstructured mesh generation technology. In *Proceedings, 7th International Meshing Roundtable*, pages 239–267. Sandia National Lab. pages 13

Panaretos, V. M., Pham, T., and Yao, Z. (2014). Principal flows. *J. Amer. Statist. Assoc.*, 109(505):424–436. pages 171

Panaretos, V. M. and Zemel, Y. (2019). Statistical aspects of Wasserstein distances. *Annu. Rev. Stat. Appl.*, 6:405–431. pages 41

Panaretos, V. M. and Zemel, Y. (2020). *An invitation to statistics in Wasserstein space*. Springer International. pages 41, 137

Papke, L. E. and Wooldridge, J. M. (1993). Econometric methods for fractional response variables with an application to 401 (k) plan participation rates. pages 364

Papke, L. E. and Wooldridge, J. M. (2008). Panel data methods for fractional response variables with an application to test pass rates. *J. Econom.*, 145(1):121–133. pages 364

Park, B. and Turlach, B. (1992). Practical performance of several data driven bandwidth selectors. Technical report, Université Catholique de Louvain, Center for Operations Research and Economics. LIDAM Discussion Papers CORE 1992005. pages 302

Park, C., Godtliebsen, F., Taqqu, M., Stoev, S., and Marron, J. S. (2007). Visualization and inference based on wavelet coefficients, SiZer and SiNos. *Comput. Statist. Data Anal.*, 51(12):5994–6012. pages 311

Park, C., Marron, J. S., and Rondonotti, V. (2004). Dependent SiZer: Goodness-of-fit tests for time series models. *J. Appl. Stat.*, 31(8):999–1017. pages 311

Parzen, E. (2004). Quantile probability and statistical data modeling. *Statist. Sci.*, 19(4):652–662. pages 44

Patrangenaru, V. and Ellingson, L. (2015). *Nonparametric statistics on manifolds and their applications to object data analysis*. CRC Press, Boca Raton, FL. pages 131, 167, 168, 169, 362

Paul, D. (2007). Asymptotics of sample eigenstructure for a large dimensional spiked covariance model. *Statist. Sinica*, 17(4):1617–1642. pages 288

Pelleg, D. and Moore, A. (1999). Accelerating exact k-means algorithms with geometric reasoning. In *Proceedings of the Fifth ACM SIGKDD International Conference on Knowledge Discovery and Data Mining*, KDD '99, pages 277–281, New York, NY, USA. Association for Computing Machinery. pages 245

Pennec, X. (2018). Barycentric subspace analysis on manifolds. *Ann. Stat.*, 46(6A):2711 – 2746. pages 172

Pennec, X., Fillard, P., and Ayache, N. (2006). A Riemannian framework for tensor computing. *Int. J. Comput. Vision*, 66(1):41–66. pages 145

Pennec, X., Sommer, S., and Fletcher, T., editors (2019). *Riemannian geometric statistics in medical image analysis*. Academic Press. pages 150, 167

Perou, C. M., Sorlie, T., Eisen, M. B., Van De Rijn, M., et al. (2000). Molecular portraits of human breast tumours. *Nature*, 406(6797):747. pages 61, 97, 243, 309

Pesarin, F. and Salmaso, L. (2010). Finite-sample consistency of combination-based permutation tests with application to repeated measures designs. *J. Nonparametr. Stat.*, 22(5-6):669–684. pages 290

Petersen, A. and Müller, H.-G. (2019). Fréchet regression for random objects with Euclidean predictors. *Ann. Stat.*, 47(2):691 – 719. pages 213

Pewsey, A. and García-Portugués, E. (2021). Recent advances in directional statistics. *TEST*, 30:1–58. pages 147

Pigoli, D., Aston, J. A. D., Dryden, I. L., and Secchi, P. (2014a). Distances and inference for covariance operators. *Biometrika*, 101(2):409–422. pages 172, 174

Pigoli, D., Aston, J. A. D., Dryden, I. L., and Secchi, P. (2014b). Permutation tests for comparison of covariance operators. In *Contributions in infinite-dimensional statistics and related topics*, pages 215–220. Esculapio, Bologna. pages 174

Pigoli, D., Hadjipantelis, P. Z., Coleman, J. S., and Aston, J. A. (2018). The statistical analysis of acoustic phonetic data: exploring differences between spoken romance languages. *J. R. Stat. Soc. Ser. C. Appl. Stat.*, 67(5):1103–1145. pages 25, 26

Pigoli, D., Menafoglio, A., and Secchi, P. (2016). Kriging prediction for manifold-valued random fields. *J. Multivar. Anal.*, 145:117–131. pages 174

Pinsker, M. S. (1980). Optimal filtering of square-integrable signals in Gaussian noise. *Problemy Peredachi Informatsii*, 16(2):52–68. pages 300

Pitt-Francis, J. and Whiteley, J. (2012). *Guide to scientific computing in C++*. Springer-Verlag, London. Undergraduate topics in Computer Science. pages 362, 363

Pizer, S. M., Broadhurst, R. E., Jeong, J.-Y., Han, Q., Saboo, R., Stough, J., Tracton, G., and Chaney, E. L. (2006). Intra-patient anatomic statistical models for adaptive radiotherapy. In *MICCAI Workshop From Statistical Atlases to Personalized Models: Understanding Complex Diseases in Populations and Individuals*, pages 43–46. pages 16

Pizer, S. M., Broadhurst, R. E., Levy, J., Liu, X., Jeong, J.-Y., Stough, J., Tracton, G., and Chaney, E. L. (2007). Segmentation by posterior optimization of m-reps: Strategy and results. *Unpublished manuscript*. pages 17

Pizer, S. M., Hong, J., Jung, S., Marron, J. S., Schulz, J., and Vicory, J. (2014). Relative statistical performance of s-reps with principal nested spheres vs. pdms. In *Proc. Shape 2014-Symp. of Stat. Shape Models and Appl*, pages 11–13. pages 16

Pizer, S. M., Hong, J., Vicory, J., Liu, Z., Marron, J. S., Choi, H.-y., Damon, J., Jung, S., Paniagua, B., Schulz, J., et al. (2020). Object shape representation via skeletal models (s-reps) and statistical analysis. In *Riemannian Geometric Statistics in Medical Image Analysis*, pages 233–271. Elsevier. pages 152, 155

Pizer, S. M., Jeong, J.-Y., Broadhurst, R. E., Ho, S., and Stough, J. (2005a). Deep structure of images in populations via geometric models in populations. In *Deep Structure, Singularities, and Computer Vision*, pages 49–59. Springer. pages 16, 153

Pizer, S. M., Jeong, J.-Y., Lu, C., Muller, K., and Joshi, S. (2005b). Estimating the statistics of multi-object anatomic geometry using inter-object relationships. In *Deep Structure, Singularities, and Computer Vision*, pages 60–71. Springer. pages 16

Pizer, S. M., Jung, S., Goswami, D., Vicory, J., Zhao, X., Chaudhuri, R., Damon, J. N., Huckemann, S., and Marron, J. (2013). Nested sphere statistics of skeletal models. In *Inandnovations for Shape Analysis*, pages 93–115. Springer. pages 13, 16, 155, 156

Pizer, S. M. and Marron, J. S. (2017). Object statistics on curved manifolds. In Zheng, G., Li, S., and Szekely, G., editors, *Statistical Shape and Deformation Analysis*, pages 137–164. Elsevier. pages 152, 153, 155

Portnoy, S. (1984). Asymptotic behavior of M-estimators of p regression parameters when p^2/n is large. I. Consistency. *Ann. Statist.*, 12(4):1298–1309. pages 275

Portnoy, S. (1985). Asymptotic behavior of M estimators of p regression parameters when p^2/n is large. II. Normal approximation. *Ann. Statist.*, 13(4):1403–1417. pages 275

Portnoy, S. (1988). Asymptotic behavior of likelihood methods for exponential families when the number of parameters tends to infinity. *Ann. Statist.*, 16(1):356–366. pages 275

Prakasa Rao, B. L. S. (1983). *Nonparametric functional estimation*. Probability and Mathematical Statistics. Academic Press, Inc. [Harcourt Brace Jovanovich, Publishers], New York. pages 294

Prothero, J. B., Hannig, J., and Marron, J. (2021). New perspectives on centering. *arXiv preprint arXiv:2103.12176*. pages 107, 334

R Core Team (2020). *R: A language and environment for statistical computing*. R J., Vienna, Austria. pages 361

Ramsay, J. O., Gribble, P., and Kurtek, S. (2014). Description and processing of functional data arising from juggling trajectories. *Electron. J. Stat.*, 8(2):1811–1816. pages 22

Ramsay, J. O. and Silverman, B. W. (2002). *Applied functional data analysis*. Springer Series in Statistics. Springer-Verlag, New York. Methods and case studies. pages 2, 48, 54, 106, 361

Ramsay, J. O. and Silverman, B. W. (2005). *Functional data analysis, Second Edition*. Springer, New York. pages 2, 48, 54, 106, 361

Rao, C. R. (1945). Information and the accuracy attainable in the estimation of statistical parameters. *Bull. Calcutta Math. Soc.*, 37:81–91. pages 185

Rao, C. R. (1958). Some statistical methods for comparison of growth curves. *Biometrics*, 14(1):1–17. pages 48

Rice, J. A. and Silverman, B. W. (1991). Estimating the mean and covariance structure nonparametrically when the data are curves. *J. Roy. Statist. Soc. Ser. B*, 53(1):233–243. pages 54

Richardson, S. and Green, P. J. (1997). On Bayesian analysis of mixtures with an unknown number of components. *J. Roy. Statist. Soc. Ser. B*, 59(4):731–792. pages 274

Ricketts, C. J., De Cubas, A. A., Fan, H., Smith, C. C., Lang, M., Reznik, E., Bowlby, R., Gibb, E. A., Akbani, R., Beroukhim, R., et al. (2018). The cancer genome atlas comprehensive molecular characterization of renal cell carcinoma. *Cell Rep.*, 23(1):313–326. pages 270, 310

Robins, V. and Turner, K. (2016). Principal component analysis of persistent homology rank functions with case studies of spatial point patterns, sphere packing, and colloids. *Phys. D*, 334:99–117. pages 204

Rodriguez, O., Aguero, C., and Arce, J. (2020). *RSDA: R to symbolic data Analysis*. R package version 3.0.4. pages 365

Rogowitz, B. E., Treinish, L. A., and Bryson, S. (1996). How not to lie with visualization. *Comput. Phys.*, 10(3):268–273. pages 97

Rondonotti, V., Marron, J. S., and Park, C. (2007). SiZer for time series: A new approach to the analysis of trends. *Electron. J. Stat.*, 1:268–289. pages 311

Rosipal, R. and Krämer, N. (2005). Overview and recent advances in partial least squares. In Saunders, C., Grobelnik, M., Gunn, S., and Shawe-Taylor, J., editors, *International Statistical and Optimization Perspectives Workshop "Subspace, latent structure and feature selection"*, pages 34–51, Berlin. Springer-Verlag. pages 353

Rousseeuw, P. J. (1984). Least median of squares regression. *J. Am. Stat. Assoc.*, 79:871–880. pages 328

Rousseeuw, P. J. and Leroy, A. M. (1987). *Robust regression and outlier detection*. Wiley Series in Probability and Mathematical Statistics: Applied Probability and Statistics. John Wiley & Sons, Inc., New York. pages 323, 328

Rousseeuw, P. J. and Yohai, V. J. (1984). Robust regression by means of S-estimators. In Franke, J., Härdle, W., and Martin, R. D., editors, *Int. J. Robust Nonlinear Control.*, pages 256–272, New York. Springer-Verlag. pages 328

Roweis, S. T. and Saul, L. K. (2000). Nonlinear dimensionality reduction by locally linear embedding. *Science*, 290(5500):2323–2326. pages 151

Royer, J.-Y. and Chang, T. (1991). Evidence for relative motions between the indian and australian plates during the last 20 my from plate tectonic reconstructions: Implications for the deformation of the indo-australian plate. *Journal of Geophysical Research: Solid Earth*, 96(B7):11779–11802. pages 151

Ruppert, D. (1987). What is kurtosis? An influence function approach. *Am. Stat.*, 41(1):1–5. pages 85

Ruppert, D., Sheather, S. J., and Wand, M. P. (1995). An effective bandwidth selector for local least squares regression. *J. Amer. Statist. Assoc.*, 90(432):1257–1270. pages 303

Ruppert, D., Wand, M. P., and Carroll, R. J. (2003). *Semiparametric regression*, volume 12 of *Cambridge Series in Statistical and Probabilistic Mathematics*. Cambridge University Press, Cambridge. pages 40

Sangalli, L. M. (2018). The role of statistics in the era of big data. *Stat. Probabil. Lett.*, 136:1–3. pages 1

Sangalli, L. M., Secchi, P., and Vantini, S. (2014a). AneuRisk65: A data set of three-dimensional cerebral vascular geometries. *Electron. J. Stat.*, 8(2):1879–1890. pages 22

Sangalli, L. M., Secchi, P., and Vantini, S. (2014b). Object oriented data analysis: A few methodological challenges. *Biom. J.*, 56(5):774–777. pages 41

Sarle, W. and Kuo, A.-H. (1993). The MODECLUS procedure. *SAS Technical Report P-256. SAS Institute, Cary, North Carolina.* pages 263

Scealy, J. L., de Caritat, P., Grunsky, E. C., Tsagris, M. T., and Welsh, A. H. (2015). Robust principal component analysis for power transformed compositional data. *J. Amer. Statist. Assoc.*, 110(509):136–148. pages 365

Scealy, J. L. and Welsh, A. H. (2011). Regression for compositional data by using distributions defined on the hypersphere. *J. R. Stat. Soc. Ser. B Stat. Methodol.*, 73(3):351–375. pages 365

Scealy, J. L. and Welsh, A. H. (2014). Colours and cocktails: Compositional data analysis 2013 Lancaster lecture. *Aust. N. Z. J. Stat.*, 56(2):145–169. pages 365

Schimek, M. G., editor (2013). *Smoothing and regression: Approaches, computation, and application*, New York. John Wiley & Sons. pages 294

Schölkopf, B., Smola, A., and Müller, K.-R. (1997). Kernel principal component analysis. In Gerstner, W., Germond, A., Hasler, M., and Nicoud, J. D., editors, *International conference on artificial neural networks ICANN'97*, pages 583–588, Berlin. Springer. pages 228

Schölkopf, B. and Smola, A. J. (2002). *Learning with kernels: Support vector machines, regularization, optimization, and beyond.* MIT press. pages 125, 231, 233

Schulz, J., Jung, S., Huckemann, S., Pierrynowski, M., Marron, J. S., and Pizer, S. M. (2015). Analysis of rotational deformations from directional data. *J. Comput. Graph. Statist.*, 24(2):539–560. pages 155

Schulz, J., Pizer, S. M., Marron, J. S., and Godtliebsen, F. (2016). Non-linear hypothesis testing of geometric object properties of shapes applied to hippocampi. *J. Math. Imaging. Vis.*, 54(1):15–34. pages 16

Schwartzman, A. (2006). *Random ellipsoids and false discovery rates: Statistics for diffusion tensor imaging data.* PhD thesis, Stanford University. pages 145

Schwiegerling, J., Greivenkamp, J. E., and Miller, J. M. (1995). Representation of videokeratoscopic height data with Zernike polynomials. *JOSA A*, 12(10):2105–2113. pages 321

Scott, D. W. (2015). *Multivariate density estimation.* Wiley Series in Probability and Statistics. John Wiley & Sons, Inc., Hoboken, NJ, second edition. pages 43, 293, 295

Secchi, P., Stamm, A., and Vantini, S. (2013). Inference for the mean of large p small n data: A finite-sample high-dimensional generalization of hotellings theorem. *Electron. J. Stat.*, 7:2005–2031. pages 174

Seifert, B., Brockmann, M., Engel, J., and Gasser, T. (1994). Fast algorithms for nonparametric curve estimation. *J. Comput. Graph. Statist.*, 3(2):192–213. pages 302

Sen, S. K., Foskey, M., Marron, J. S., and Styner, M. A. (2008). Support vector machine for data on manifolds: An application to image analysis. In *2008 5th Proc. IEEE. Int. Symp. Biomed. Imaging.*, pages 1195–1198. IEEE. pages 234

Severn, K. E., Dryden, I. L., and Preston, S. P. (2019). Manifold valued data analysis of samples of networks, with applications in corpus linguistics. *arXiv e-prints.* arXiv:1902.08290. pages 207, 211, 213

Severn, K. E., Dryden, I. L., and Preston, S. P. (2021). Non-parametric regression for networks. *Stat.* e373 STAT-20-0276. pages 213

Shawe-Taylor, J. and Cristianini, N. (2004). *Kernel methods for pattern analysis.* Cambridge University Press. pages 125, 231, 233

Sheather, S. J. and Jones, M. C. (1991). A reliable data-based bandwidth selection method for kernel density estimation. *J. Roy. Statist. Soc. Ser. B*, 53(3):683–690. pages 302, 303

Shen, D., Shen, H., Bhamidi, S., Muñoz Maldonado, Y., Kim, Y., and Marron, J. S. (2014). Functional data analysis of tree data objects. *J. Comput. Graph. Statist.*, 23(2):418–438. pages 202, 203, 311

Shen, D., Shen, H., and Marron, J. S. (2013). Consistency of sparse PCA in high dimension, low sample size contexts. *J. Multivariate Anal.*, 115:317–333. pages 289

Shen, D., Shen, H., and Marron, J. S. (2016a). A general framework for consistency of principal component analysis. *J. Mach. Learn. Res.*, 17(1):5218–5251. pages 275, 288, 291

Shen, D., Shen, H., Zhu, H., and Marron, J. S. (2016b). The statistics and mathematics of high dimension low sample size asymptotics. *Statist. Sinica*, 26(4):1747–1770. pages 275, 288, 289

Shiers, N., Aston, J. A. D., Smith, J. Q., and Coleman, J. S. (2017). Gaussian tree constraints applied to acoustic linguistic functional data. *J. Multivariate Anal.*, 154:199–215. pages 25

Shu, H., Wang, X., and Zhu, H. (2020). D-CCA: A decomposition-based canonical correlation analysis for high-dimensional data sets. *J. Amer. Statist. Assoc.*, 115(529):292–306. pages 356

Siddiqi, K. and Pizer, S. M. (2008). *Medial representations: Mathematics, algorithms, and applications*, volume 37. Springer Science & Business Media, New York. pages 13

Silverman, B. W. (1981). Using kernel density estimates to investigate multimodality. *J. Roy. Statist. Soc. Ser. B*, 43(1):97–99. pages 306

Silverman, B. W. (1982). Algorithm AS 176: Kernel density estimation using the fast fourier transform. *J. R. Stat. Soc. Ser. C*, 31(1):93–99. pages 302

Silverman, B. W. (1986). *Density estimation for statistics and data analysis.* Monographs on Statistics and Applied Probability. Chapman & Hall, London. pages 43, 293

Simonoff, J. S. (1996). *Smoothing methods in statistics*. Springer Series in Statistics. Springer-Verlag, New York. pages 294

Singh, N., Fletcher, P. T., Preston, J. S., Ha, L., King, R., Marron, J. S., Wiener, M., and Joshi, S. (2010). Multivariate statistical analysis of deformation momenta relating anatomical shape to neuropsychological measures. In *International Conference on Med. Image. Comput. Comput. Assist. Interv.*, pages 529–537. Springer. pages 352

Skwerer, S., Bullitt, E., Huckemann, S., Miller, E., Oguz, I., Owen, M., Patrangenaru, V., Provan, S., and Marron, J. S. (2014). Tree-oriented analysis of brain artery structure. *J. Math. Imaging Vision*, 50(1-2):126–143. pages 199, 200, 201

Skwerer, S., Provan, S., and Marron, J. S. (2018). Relative optimality conditions and algorithms for treespace Fréchet means. *SIAM J. Optim.*, 28(2):959–988. pages 200

Small, C. G. (1990). A survey of multidimensional medians. *International Statistical Review/Revue Internationale de Statistique*, 58:263–277. pages 327

Small, C. G. (1996). *The statistical theory of shape*. Springer Series in Statistics. Springer-Verlag, New York. pages 158

Sommer, S. (2020). Probabilistic approaches to geometric statistics: Stochastic processes, transition distributions, and fiber bundle geometry. In Pennec, X., Sommer, S., and Fletcher, T., editors, *Riemannian Geometric Statistics in Medical Image Analysis*, pages 377–416. Academic Press. pages 168

Sommerfeld, M., Heo, G., Kim, P., Rush, S. T., and Marron, J. S. (2017). Bump hunting by topological data analysis. *Stat*, 6:462–471. pages 294, 311

Spellman, P. T., Sherlock, G., Zhang, M. Q., Iyer, V. R., Anders, K., Eisen, M. B., Brown, P. O., Botstein, D., and Futcher, B. (1998). Comprehensive identification of cell cycle–regulated genes of the yeast saccharomyces cerevisiae by microarray hybridization. *Mol. Biol. Cell.*, 9(12):3273–3297. pages 121

Srebro, N., Rennie, J., and Jaakkola, T. S. (2005). Maximum-margin matrix factorization. In *Adv. Neural Inf. Process. Syst.*, pages 1329–1336. pages 122

Srivastava, A. and Klassen, E. P. (2016). *Functional and shape data analysis*. Springer, New York. pages 152, 167, 187

Srivastava, A., Wu, W., Kurtek, S., Klassen, E., and Marron, J. (2011). Registration of functional data using Fisher-Rao metric. *arXiv preprint arXiv:1103.3817*. pages 21, 175, 182, 183, 186, 187

Srivastava, M. S. and Du, M. (2008). A test for the mean vector with fewer observations than the dimension. *J. Multivariate Anal.*, 99(3):386–402. pages 262

Staicu, A.-M. and Lu, X. (2014). Analysis of AneuRisk65 data: Classification and curve registration. *Electron. J. Stat.*, 8(2):1914–1919. pages 189, 195

Staudte, R. G. and Sheather, S. J. (1990). *Robust estimation and testing*. Wiley Series in Probability and Mathematical Statistics: Applied Probability and Statistics. John Wiley & Sons, Inc., New York. pages 70, 128, 313

Steinhaus, H. (1956). Sur la division des corp materiels en parties. *Bull. Acad. Polon. Sci*, 1(804):801. pages 243

Stewart, C. and Field, C. (2011). Managing the essential zeros in quantitative fatty acid signature analysis. *J. Agric. Biol. Environ. Stat.*, 16(1):45–69. pages 365

Stone, C. J. (1982). Optimal global rates of convergence for nonparametric regression. *Ann. Statist.*, 10(4):1040–1053. pages 300

Stone, C. J., Hansen, M. H., Kooperberg, C., and Truong, Y. K. (1997). Polynomial splines and their tensor products in extended linear modeling. *Ann. Statist.*, 25(4):1371–1470. With discussion and a rejoinder by the authors and Jianhua Z. Huang. pages 40, 293

Stone, M. (1974). Cross-validatory choice and assessment of statistical predictions. *J. Roy. Statist. Soc. Ser. B*, 36:111–147. pages 241

Stone, M. and Brooks, R. J. (1990). Continuum regression: Cross-validated sequentially constructed prediction embracing ordinary least squares, partial least squares and principal components regression. *J. Roy. Statist. Soc. Ser. B*, 52(2):237–269. With discussion and a reply by the authors. pages 357

Stough, J. V., Broadhurst, R. E., Pizer, S. M., and Chaney, E. L. (2007). Regional appearance in deformable model segmentation. In *Biennial International Conference on Information Processing in Medical Imaging*, pages 532–543. Springer. pages 17

Sturges, H. A. (1926). The choice of a class interval. *Journal of the American Statistical Association*, 21(153):65–66. pages 295

Sturm, K.-T. (2003). Probability measures on metric spaces of nonpositive curvature. In *Heat kernels and analysis on manifolds, graphs, and metric spaces (Paris, 2002)*, volume 338 of *Contemp. Math.*, pages 357–390. Amer. Math. Soc., Providence, RI. pages 200

Suzuki, R. and Shimodaira, H. (2006). Pvclust: An R package for assessing the uncertainty in hierarchical clustering. *Bioinformatics*, 22(12):1540–1542. pages 274

Szegő, G. (1975). *Orthogonal polynomials*. American Mathematical Society, Providence, R.I., fourth edition. American Mathematical Society, Colloquium Publications, Vol. XXIII. pages 40

Talagrand, M. (1991). A new isoperimetric inequality and the concentration of measure phenomenon. In *Geometric aspects of functional analysis (1989–90)*, volume 1469 of *Lecture Notes in Math.*, pages 94–124. Springer, Berlin. pages 290

Talagrand, M. (1995). Concentration of measure and isoperimetric inequalities in product spaces. *Inst. Hautes Études Sci. Publ. Math.*, 81:73–205. pages 290

Tapia, R. A. and Thompson, J. R. (1978). *Nonparametric probability density estimation*, volume 1 of *Johns Hopkins Series in the Mathematical Sciences*. Johns Hopkins University Press, Baltimore, MD. pages 293

Tavakoli, S., Pigoli, D., Aston, J. A. D., and Coleman, J. S. (2019). A spatial modeling approach for linguistic object data: Analyzing dialect sound variations across Great Britain. *J. Am. Stat. Assoc.*, 114(527):1081–1096. pages 25

Telschow, F. J., Huckemann, S. F., and Pierrynowski, M. (2014). Asymptotics for object descriptors. *Biom. J.*, 56(5):781–785. pages 33

Tench, C. R., Morgan, P. S., Wilson, M., and Blumhardt, L. D. (2002). White matter mapping using diffusion tensor MRI. *Magn. Reson. Med.*, 47(5):967–972. pages 174

Tenenbaum, J. B., De Silva, V., and Langford, J. C. (2000). A global geometric framework for nonlinear dimensionality reduction. *Science*, 290(5500):2319–2323. pages 151

Tenenhaus, A. and Tenenhaus, M. (2011). Regularized generalized canonical correlation analysis. *Psychometrika*, 76(2):257. pages 351

Terras, A. (1985). *Harmonic analysis on symmetric spaces and applications. I.* Springer-Verlag, New York. pages 168

Terras, A. (2013). *Harmonic analysis on symmetric spaces—Euclidean space, the sphere, and the Poincaré upper half-plane*. Springer, New York, second edition. pages 168

Terras, A. (2016). *Harmonic analysis on symmetric spaces—higher rank spaces, positive definite matrix space, and generalizations*. Springer, New York, second edition. pages 168

Thompson, J. R. and Tapia, R. A. (1990). *Nonparametric function estimation, modeling, and simulation*. Society for Industrial and Applied Mathematics (SIAM), Philadelphia, PA. pages 294

Tibshirani, R. (1996). Regression shrinkage and selection via the lasso. *J. Roy. Statist. Soc. Ser. B*, 58(1):267–288. pages 63

Tibshirani, R., Hastie, T., Narasimhan, B., and Chu, G. (2002). Diagnosis of multiple cancer types by shrunken centroids of gene expression. *Proc. Natl. Acad. Sci.*, 99(10):6567–6572. pages 93

Tibshirani, R. and Walther, G. (2005). Cluster validation by prediction strength. *J. Comput. Graph. Statist.*, 14(3):511–528. pages 274

Tibshirani, R., Walther, G., and Hastie, T. (2001). Estimating the number of clusters in a data set via the gap statistic. *J. R. Stat. Soc. Ser. B Stat. Methodol.*, 63(2):411–423. pages 263, 274

Toh, K.-C., Todd, M. J., and Tutuncu, R. H. (2009). Sdpt3 version 4.0. https://blog.nus.edu.sg/mattohkc/softwares/sdpt3/. pages 237

Torgerson, W. S. (1952). Multidimensional scaling. I. Theory and method. *Psychometrika*, 17:401–419. pages 132, 133

Torgerson, W. S. (1958). *Theory and methods of scaling*. Wiley, New York. pages 132, 133

Tracy, C. A. and Widom, H. (1994). Level-spacing distributions and the Airy kernel. *Commun. Math. Phys.*, 159(1):151–174. pages 279

Trosset, M. W. and Priebe, C. E. (2008). The out-of-sample problem for classical multidimensional scaling. *Comput. Statist. Data Anal.*, 52(10):4635–4642. pages 134, 201

Tsagris, M. T., Preston, S., and Wood, A. T. (2011). A data-based power transformation for compositional data. *arXiv preprint arXiv:1106.1451*. pages 365

Tucker, J. D., Wu, W., and Srivastava, A. (2013). Generative models for functional data using phase and amplitude separation. *Comput. Statist. Data Anal.*, 61:50–66. pages 187

Tucker, J. D., Wu, W., and Srivastava, A. (2014). Analysis of proteomics data: Phase amplitude separation using an extended Fisher-Rao metric. *Electron. J. Stat.*, 8(2):1724–1733. pages 189

Tufte, E. (1983). *The visual display of quantitative information*. Encyclopedia of Mathematics and its Applications. Graphics Press. pages 47

Tukey, J. and Tukey, P. (1990). Strips displaying empirical distributions: I. textured dot strips. Technical report, Bellcore Technical Memorandum, Morristown, NJ. pages 50

Tukey, J. W. (1977). *Exploratory data analysis*. Addison-Wesley, Reading, Mass. pages 3

Tukey, J. W. (1990). Data-based graphics: Visual display in the decades to come. *Statist. Sci.*, 5(3):327–339. pages 47

Turaga, P. K. and Srivastava, A. (2016). *Riemannian computing in computer vision*. Springer, New York. pages 167

Turner, K., Mukherjee, S., and Boyer, D. M. (2014). Persistent homology transform for modeling shapes and surfaces. *Inf. Inference.*, 3(4):310–344. pages 206

van Meegen, C., Schnackenberg, S., and Ligges, U. (2020). Unequal priors in linear discriminant analysis. *J. Classif.*, 37:598–615. pages 224

Vapnik, V. (1982). *Estimation of dependences based on empirical data*. Springer Series in Statistics. Springer-Verlag, New York-Berlin. Translated from the Russian by Samuel Kotz. pages 232

Vapnik, V. N. (1995). *The nature of statistical learning theory*. Springer-Verlag, New York. pages 232

Vardi, Y. and Zhang, C.-H. (2000). The multivariate L_1-median and associated data depth. *Proc. Natl. Acad. Sci. USA*, 97(4):1423–1426. pages 327

Venables, W. N. and Ripley, B. D. (1994). *Modern applied statistics with S-Plus.* Statistics and Computing. Springer-Verlag, New York. pages 361

Venables, W. N. and Ripley, B. D. (2013). *Modern applied statistics with S-PLUS.* Springer, New York. pages 361

Verde, R., Irpino, A., and Balzanella, A. (2016). Dimension reduction techniques for distributional symbolic data. *IEEE Trans. Cybern.*, 46(2):344–355. pages 365

Vert, J.-P. (2002). A tree kernel to analyse phylogenetic profiles. *Bioinformatics*, 18(suppl 1):S276–S284. pages 197

Vidal, R., Ma, Y., and Sastry, S. S. (2016). *Generalized principal component analysis*, volume 40 of *J. Interdiscip. Math.* Springer, New York. pages 256

Wagner, H. and Kneip, A. (2019). Nonparametric registration to low-dimensional function spaces. *Comput. Statist. Data Anal.*, 138:49–63. pages 192

Wahba, G. (1990). *Spline models for observational data*, volume 59 of *CBMS-NSF Regional Conference Series in Applied Mathematics.* Society for Industrial and Applied Mathematics (SIAM), Philadelphia, PA. pages 293

Wahba, G., Lin, Y., Lee, Y., and Zhang, H. (2003). Optimal properties and adaptive tuning of standard and nonstandard support vector machines. In *Nonlinear estimation and classification (Berkeley, CA, 2001)*, volume 171 of *Lect. Notes Stat.*, pages 129–147. Springer, New York. pages 234

Wahba, G., Lin, Y., and Zhang, H. (2000). Generalized approximate cross validation for support vector machines, or, another way to look at margin-like quantities. In Smola, A. J., Bartlett, P. L., Schölkopf, B., and Schurmans, D., editors, *Advances in Large Margin Classifiers.* MIT Press. pages 234

Wainwright, M. J. (2019). *High-dimensional statistics: A non-asymptotic viewpoint*, volume 48 of *Cambridge Series in Statistical and Probabilistic Mathematics.* Cambridge University Press, Cambridge. pages 275, 290

Walther, G. (2002). Detecting the presence of mixing with multiscale maximum likelihood. *J. Amer. Statist. Assoc.*, 97(458):508–513. pages 294

Wand, M. P. and Jones, M. C. (1993). Comparison of smoothing parameterizations in bivariate kernel density estimation. *J. Amer. Statist. Assoc.*, 88(422):520–528. pages 301

Wand, M. P. and Jones, M. C. (1995). *Kernel smoothing*, volume 60 of *Monographs on Statistics and Applied Probability.* Chapman and Hall, Ltd., London. pages 7, 294

Wang, B. and Zou, H. (2016). Sparse distance weighted discrimination. *J. Comput. Graph. Stat.*, 25(3):826–838. pages 239

Wang, B. and Zou, H. (2018). Another look at distance-weighted discrimination. *J. R. Stat. Soc. Ser. B. Stat. Methodol.*, 80(1):177–198. pages 239

Wang, H. and Marron, J. S. (2007). Object oriented data analysis: Sets of trees. *Ann. Statist.*, 35(5):1849–1873. pages 1, 198, 361

Wang, X. and Marron, J. S. (2008). A scale-based approach to finding effective dimensionality in manifold learning. *Electron. J. Stat.*, 2:127–148. pages 151

Wang, Y. (2011). *Smoothing splines*, volume 121 of *Monographs on Statistics and Applied Probability*. CRC Press, Boca Raton, FL. Methods and applications. pages 293

Wang, Y., Marron, J. S., Aydin, B., Ladha, A., Bullitt, E., and Wang, H. (2012). A nonparametric regression model with tree-structured response. *J. Amer. Statist. Assoc.*, 107(500):1272–1285. pages 199

Wang, Z., Gerstein, M., and Snyder, M. (2009). RNA-Seq: A revolutionary tool for transcriptomics. *Nat. Rev. Genet.*, 10(1):57–63. pages 56

Wang, Z., Vemuri, B., Chen, Y., and Mareci, T. (2004). A constrained variational principle for direct estimation and smoothing of the diffusion tensor field from complex DWI. *IEEE Trans. Med. Imaging*, 23(8):930–939. pages 145

Ward, Jr., J. H. (1963). Hierarchical grouping to optimize an objective function. *J. Amer. Statist. Assoc.*, 58:236–244. pages 249

Wasserman, L. (2018). Topological data analysis. *Annu. Rev. Stat. Appl.*, 5:501–535. pages 204

Wegelin, J. A. (2000). A survey of partial least squares (PLS) methods, with emphasis on the two-block case. *University of Washington, Department of Statistics, Tech. Report 371*. pages 351, 352, 353

Wei, S., Lee, C., Wichers, L., and Marron, J. S. (2016). Direction-projection-permutation for high-dimensional hypothesis tests. *J. Comput. Graph. Statist.*, 25(2):549–569. pages 67, 257, 260

Weihs, C., Jannach, D., Vatolkin, I., and Rudolph, G. (2016). Music data analysis: Foundations and applications. pages 28

Weinstein, J. N., Collisson, E. A., Mills, G. B., Shaw, K. R. M., Ozenberger, B. A., Ellrott, K., Shmulevich, I., Sander, C., Stuart, J. M., Network, C. G. A. R., et al. (2013). The cancer genome atlas pan-cancer analysis project. *Nat. Genet.*, 45(10):1113–1120. pages 56, 239

West, D. B. (1996). *Introduction to graph theory*. Prentice Hall, Inc., Upper Saddle River, NJ. pages 197

White, H. (2014). *Asymptotic theory for econometricians*. Academic press. pages 287

Wigner, E. P. (1955). Characteristic vectors of bordered matrices with infinite dimensions. *Ann. of Math. (2)*, 62:548–564. pages 276

Wilkinson, L., Anand, A., and Grossman, R. L. (2005). Graph-theoretic scagnostics. In *INFOVIS*, volume 5, page 21. pages 71

Wilkinson, L. and Friendly, M. (2009). The history of the cluster heat map. *Amer. Statist.*, 63(2):179–184. pages 97

Wilkinson, L. and Wills, G. (2008). Scagnostics distributions. *J. Comput. Graph. Stat.*, 17(2):473–491. pages 71

Willis, A. (2018). Confidence sets for phylogenetic trees. *J. Am. Stat. Assoc.*, pages 1–10. pages 202

Willis, A. and Bell, R. (2018). Uncertainty in phylogenetic tree estimates. *J. Comput. Graph. Statist.*, 27(3):542–552. pages 202

Wilmoth, J. R. and Shkolnikov, V. (2008). Human mortality database. *University of California, Berkeley (USA), and Max Planck Institute for Demographic Research (Germany).* pages 3, 9

Wold, H. (1975). Soft modeling by latent variables: The non-linear iterative partial least squares (NIPALS) approach. *J. Appl. Probab.*, 12 (S1):117–142. pages 351, 352

Wold, H. (1985). Partial least squares. In Kotz, S. and Johnson, N. L., editors, *Encyclopedia of statistical sciences (vol. 6).* Wiley, New York. pages 351, 352

Wright, F. A., Strug, L. J., Doshi, V. K., Commander, C. W., Blackman, S. M., Sun, L., Berthiaume, Y., Cutler, D., Cojocaru, A., Collaco, J. M., et al. (2011). Genome-wide association and linkage identify modifier loci of lung disease severity in cystic fibrosis at 11p13 and 20q13. 2. *Nat. Genet.*, 43(6):539. pages 326

Wu, W., Hatsopoulos, N. G., and Srivastava, A. (2014). Introduction to neural spike train data for phase-amplitude analysis. *Electron. J. Stat.*, 8(2):1759–1768. pages 22, 175

Wu, W. and Srivastava, A. (2014). Analysis of spike train data: Alignment and comparisons using the extended Fisher-Rao metric [mr3273592]. *Electron. J. Stat.*, 8(2):1776–1785. pages 189

Xie, Q., Kurtek, S., and Srivastava, A. (2014). Analysis of AneuRisk65 data: Elastic shape registration of curves [mr3273608]. *Electron. J. Stat.*, 8(2):1920–1929. pages 189

Xiong, J. (2015). *Radial distance weighted discrimination.* PhD thesis, The University of North Carolina at Chapel Hill. pages 239

Xiong, J., Dittmer, D. P., and Marron, J. S. (2015). "Virus hunting" using radial distance weighted discrimination. *Ann. Appl. Stat.*, 9(4):2090–2109. pages 90, 239, 240, 365

Xu, C., Tao, D., and Xu, C. (2013). A survey on multi-view learning. *arXiv preprint arXiv:1304.5634.* pages 353

Yamanishi, Y., Bach, F., and Vert, J.-P. (2007). Glycan classification with tree kernels. *Bioinformatics*, 23(10):1211–1216. pages 197

Yang, X., Hannig, J., and Marron, J. S. (2021). Visual high dimensional hypothesis testing. *arXiv preprint arXiv:2101.00362.* pages 262

Yao, F., Müller, H.-G., and Wang, J.-L. (2005). Functional data analysis for sparse longitudinal data. *J. Amer. Statist. Assoc.*, 100(470):577–590. pages 54, 106

Yao, J., Zheng, S., and Bai, Z. (2015). *Large sample covariance matrices and high-dimensional data analysis*, volume 39 of *Cambridge Series in Statistical and Probabilistic Mathematics*. Cambridge University Press, New York. pages 277

Yata, K. and Aoshima, M. (2009). Pca consistency for non-gaussian data in high dimension, low sample size context. *Commun. Stat. Theory Methods*, 38(16-17):2634–2652. pages 289

Yata, K. and Aoshima, M. (2010a). Effective pca for high-dimension, low-sample-size data with singular value decomposition of cross data matrix. *J. Multivar. Anal.*, 101(9):2060–2077. pages 289, 290

Yata, K. and Aoshima, M. (2010b). Intrinsic dimensionality estimation of high-dimension, low sample size data with d-asymptotics. *Commun. Stat. Theory Methods*, 39(8-9):1511–1521. pages 289

Yata, K. and Aoshima, M. (2012). Effective pca for high-dimension, low-sample-size data with noise reduction via geometric representations. *J. Multivar. Anal.*, 105(1):193–215. pages 287, 289, 290

Yata, K. and Aoshima, M. (2013). Pca consistency for the power spiked model in high-dimensional settings. *J. Multivar. Anal.*, 122:334–354. pages 289, 290, 291

Younes, L. (2010). *Shapes and diffeomorphisms*, volume 171 of *Appl. Math. Sci.*. Springer-Verlag, Berlin. pages 167

Young, G. and Householder, A. S. (1938). Discussion of a set of points in terms of their mutual distances. *Psychometrika*, 3(1):19–22. pages 132

Yu, Q. (2017). *Curve registration and human connectome data*. PhD thesis, The University of North Carolina at Chapel Hill. pages 190

Yu, Q., Lu, X., and Marron, J. S. (2017a). Principal nested spheres for time-warped functional data analysis. *J. Comput. Graph. Statist.*, 26(1):144–151. pages 23, 156, 184, 193, 195

Yu, Q., Risk, B. B., Zhang, K., and Marron, J. (2017b). Jive integration of imaging and behavioral data. *NeuroImage*, 152:38–49. pages 359, 368

Yuan, Y., Zhu, H., Lin, W., and Marron, J. S. (2012). Local polynomial regression for symmetric positive definite matrices. *J. R. Stat. Soc. Ser. B. Stat. Methodol.*, 74(4):697–719. pages 174

Yuan, Y., Zhu, H., Styner, M., Gilmore, J. H., and Marron, J. S. (2013). Varying coefficient model for modeling diffusion tensors along white matter tracts. *Ann. Appl. Stat.*, 7(1):102–125. pages 174

Zadeh, L. A. (1965). Fuzzy sets. *Information and control*, 8(3):338–353. pages 216

Zemel, Y. and Panaretos, V. M. (2019). Fréchet means and Procrustes analysis in Wasserstein space. *Bernoulli*, 25(2):932–976. pages 41

Zhai, H. (2016). *Principal component analysis in phylogenetic tree space*. PhD thesis, The University of North Carolina at Chapel Hill. pages 134, 199, 200, 201, 202

Zhang, J., Heckman, N., Cubranic, D., Kingsolver, J. G., Gaydos, T., and Marron, J. (2014). Prinsimp. *R Journal*, 6(2). pages 122

Zhang, J.-T. (2014). *Analysis of variance for functional data*, volume 127 of *Monographs on Statistics and Applied Probability*. CRC Press, Boca Raton, FL. pages 106

Zhang, L., Lu, S., and Marron, J. S. (2015). Nested nonnegative cone analysis. *Comput. Statist. Data Anal.*, 88:100–110. pages 122, 171

Zhang, L., Marron, J. S., Shen, H., and Zhu, Z. (2007a). Singular value decomposition and its visualization. *J. Comput. Graph. Statist.*, 16(4):833–854. pages 339

Zhang, Z., Li, T., Ding, C., and Zhang, X. (2007b). Binary matrix factorization with applications. In *Data Mining, 2007. ICDM 2007. Seventh IEEE International Conference on*, pages 391–400. IEEE. pages 84, 122

Zhao, X., Marron, J. S., and Wells, M. T. (2004). The functional data analysis view of longitudinal data. *Statist. Sinica*, 14(3):789–808. pages 121

Zheng, G., Li, S., and Szekely, G. (2017). *Statistical shape and deformation analysis: Methods, implementation and applications*. Elsevier Science. pages 167

Zhou, Y.-H., Marron, J., and Wright, F. A. (2018a). Eigenvalue significance testing for genetic association. *Biometrics*, 74(2):439–447. pages 279, 287

Zhou, Y.-H. and Marron, J. S. (2015). High dimension low sample size asymptotics of robust PCA. *Electron. J. Stat.*, 9(1):204–218. pages 321

Zhou, Y.-H. and Marron, J. S. (2016). Visualization of robust L1PCA. *Stat*, 5:173–184. pages 326, 327, 328, 329

Zhou, Y.-H., Marron, J. S., and Wright, F. A. (2018b). Computation of ancestry scores with mixed families and unrelated individuals. *Biometrics*, 74(1):155–164. pages 279, 287

Zomorodian, A. (2012). Topological data analysis. In *Advances in applied and computational topology*, volume 70 of *Proc. Sympos. Appl. Math.*, pages 1–39. Amer. Math. Soc., Providence, RI. pages 204

Index

L^1 M-estimate, 131, 316
2-d Toy data, 33, 38, 107, 331

alternating direction method of multipliers, 237
amplitude, 183–186
  data objects, 21–27, 151, 188
  variation, 2, 31, 39, 94, 175–362
ANOVA, 153, 170
Area Under the Curve, 93, 241
asymptotic analysis, 202, 275, 302, 354
asymptotic domains, 275
auto-encoding, 229

backwards mean, 155–164
backwards methods, 131, 155, 172, 327
Backwards PCA, 156
backwards PCA, 169
balanced permutations, 262
bandwidth, 228, 296, 309
  common, 298
  selection, 217, 296–304
Barycentric Subspace Analysis, 172
Bayes risk, 216
Besov space, 293
best rank k approximation, 344
bi-orthogonal, 352–356
Bidirectional Discrimination, 231
Bimodal Phase Shift data, 22, 39, 94, 104, 181, 187
bin edge effect, 295
binary classification, 215–217, 247
binary data, 83, 261
Binary Matrix Factorization, 84, 122
Bladder-Prostate-Rectum data, 10–16, 39, 152, 154
Blind Source Separation, 120
Bonferroni adjustment, 68
Boolean operations, 84, 122

boosting, 241
Box-Cox transformation, 94
Brain Artery data, 24, 198–207
Bralower Fossils data, 299–308
branch length representation, 203
breakdown, 328
British Family Incomes data, 304–306
brushing, 23, 57–59, 105, 116, 271

Canonical Correlation Analysis, 70, 123, 225, 350, 354–358
canonical kernel, 296
Canonical Variate Analysis, 120, 224–225, 247, 356
CART, 241
centering, 107–116, 148, 267, 279, 332–348, 368
  double, 112–116, 338
Central Limit Theorem, 167, 232, 286, 290
Circle of Willis, 197–199
classification, 69, 215–242, 368
  binary, 247
  directions, 120
  in drug discovery, 81
  kernel, 303
  male vs. female, 28
  relation to CCA, 356
  shapes, 15
classification direction vector, 216
classification rule, 215–241
classifier
  centroid, 217
  Gaussian Likelihood Ratio, 224
  margin, 232
  Mean Difference, 217, 257
  Naive Bayes, 217
  nearest neighbor, 217
  support vectors, 232
cluster, 58
cluster analysis, 125, 211, 215, 243, 366

For Product Safety Concerns and Information please contact our
EU representative GPSR@taylorandfrancis.com Taylor & Francis
Verlag GmbH, Kaufingerstraße 24, 80331 München, Germany